Chemical Analysis: Modern Methods and Techniques

Chemical Analysis: Modern Methods and Techniques

Edited by
Alana Wood

www.willfordpress.com

Published by Willford Press,
118-35 Queens Blvd., Suite 400,
Forest Hills, NY 11375, USA

ISBN: 978-1-64728-519-7

Cataloging-in-Publication Data

Chemical analysis : modern methods and techniques / edited by Alana Wood.
    p. cm.
Includes bibliographical references and index.
ISBN 978-1-64728-519-7
1. Analytical chemistry. 2. Chemistry. I. Wood, Alana.
QD75.22 .C445 2023
543--dc24

For information on all Willford Press publications
visit our website at www.willfordpress.com

Contents

Preface

The world is advancing at a fast pace like never before. Therefore, the need is to keep up with the latest developments. This book was an idea that came to fruition when the specialists in the area realized the need to coordinate together and document essential themes in the subject. That's when I was requested to be the editor. Editing this book has been an honour as it brings together diverse authors researching on different streams of the field. The book collates essential materials contributed by veterans in the area which can be utilized by students and researchers alike.

Chemical analysis is a process that involves determining the physical properties or chemical composition of samples of matter. In the field of analytical chemistry, various techniques are used for detecting, identifying, characterizing and quantifying chemical compounds. Mass spectrometry is an important technique of chemical analysis that is used extensively in proteomics for detecting and identifying molecules according to their masses. High-performance liquid chromatography (HPLC) is another analytical technique used to separate, identify and quantify the different components of a mixture according to different characteristics such as size of the components and their affinity for a ligand. Calorimetry is yet another chemical analysis technique used in drug design, quality control and metabolic rate examination. This book provides significant information of analytical chemistry to help develop a good understanding of the different chemical analysis procedures. A number of latest researches have been included to keep the readers updated with the global concepts in this area of study. This book is a vital tool for all researching or studying chemical analysis procedures as it gives incredible insights into emerging methods and techniques.

Each chapter is a sole-standing publication that reflects each author's interpretation. Thus, the book displays a multi-facetted picture of our current understanding of application, resources and aspects of the field. I would like to thank the contributors of this book and my family for their endless support..

Editor

Non-Chromatographic Speciation of As by HG Technique—Analysis of Samples with Different Matrices

Maja Welna *, Anna Szymczycha-Madeja⬦ and Pawel Pohl⬦

Department of Analytical Chemistry and Chemical Metallurgy, Faculty of Chemistry,
Wroclaw University of Science and Technology, Wybrzeze Wyspianskiego 27, 50-370 Wroclaw, Poland;
anna.szymczycha-madeja@pwr.edu.pl (A.S.-M.); pawel.pohl@pwr.edu.pl (P.P.)
* Correspondence: maja.welna@pwr.edu.pl

Academic Editor: Constantinos K. Zacharis

Abstract: The applicability of the hydride generation (HG) sample introduction technique combined with different spectrochemical detection methods for non-chromatographic speciation of toxic As species, i.e., As(III), As(V), dimethylarsinate (DMA) and monomethylarsonate (MMA), in waters and other environmental, food and biological matrices is presented as a promising tool to speciate As by obviating chromatographic separation. Different non-chromatographic procedures along with speciation protocols reported in the literature over the past 20 year are summarized. Basic rules ensuring species selective generation of the corresponding hydrides are presented in detail. Common strategies and alternative approaches are highlighted. Aspects of proper sample preparation before analysis and the selection of adequate strategies for speciation purposes are emphasized.

Keywords: arsenic; arsenic species; hydride generation; atomic spectrometry; non-chromatographic speciation; selective reduction; speciation protocols; environmental; food and biological matrices; sample preparation

1. Introduction

Because the toxicity and physiological behavior of various As compounds differ greatly, knowledge regarding As species is crucial to understand their potential harmful effects to human beings. Arsenic has a variety of inorganic and organic forms ranging from highly hazardous inorganic arsenicals (i-As), i.e., arsenite (As(III)) and arsenate (As(V)), to relatively less toxic methyl-substituted organic arsenicals (o-As), i.e., monomethylarsonate (MMA) and dimethylarsinate (DMA). Other o-As compounds, e.g., arsenobetaine (AsB), arsenocholine (AsC) and As-sugars, typically present in marine organisms, are generally considered to be non-toxic [1]. Accordingly, today, determination of the total As content is insufficient, and speciation information is essential to reflect the risk associated with exposure to this element.

As shown in Table 1, contamination with As can be found in various environment compartments including atmosphere (as gaseous compounds and in particulate matter), water, rocks, soil and plants [2–59]. It can be emitted naturally from volcanic activity or from anthropogenic sources, such as mines, coal-fired plants for energy production, pesticides, phosphate fertilizer factories, irrigation and oxidation of volatile arsines in air, dust from burned fossil fuels as well as the disposal of industrial, municipal and animal waste [32,34]. The widespread use of phosphate rocks in the production of phosphate fertilizers is a significant source of As contamination and cause of exposure to this element [33,36]. The tendency of As to accumulate in plant materials, e.g., cereals, tea plants, vegetables, fruits or herbs, causes it to subsequently appear in food products, beverages or (natural) pharmaceuticals (see Table 1). Among them, rice (together with water) is considered to be the highest contributor to i-As intake among all products of vegetable origin. Although AsB and As-sugars are major components (>90%) in seafood, e.g., fish (AsB) [52,53] and seaweed, e.g., algae (As-sugars) [44,53–55], the level of minor i-As, which is responsible for As toxicity in these samples, should also be monitored. Therefore, analyses of the contents of toxic As species in various environmental and food samples would provide information about their quality and safety. On the other hand, analyses of biological/clinical samples like human urine, tissue, cells or blood [4,31,60–63] would be helpful in recognizing As species in body fluids and tissue to evaluate the pattern of i-As metabolism and assess the risk of adverse health effects associated with i-As exposure. It is worth mentioning that As was also applied historically as a medicinal treatment [56]; nowadays, it is used, e.g., in glass production as a fining agent [35].

Table 1. The concentrations of As species in different samples obtained by non-chromatographic methods based on the hydride generation technique [a].

Matrix	As(III), μg L⁻¹	As(V), μg L⁻¹	i-As, μg L⁻¹	DMA, μg L⁻¹	MMA, μg L⁻¹	o-As [b], μg L⁻¹
Liquid Matrices						
Waters						
drinking	0.17–4.7 [2-4]	0.05–63.9 [2-5]				
ground	0.080–395 [4,6–11]	0.01–312 [4,6–11]				
underground	1.25–1016 [3,15,16]	0.97–554.3 [3,15,16]	0.31–308 [12–14]			
tap	0.015–12.7 [4,9,16–19]	0.050–35 [4,9,15–21]		1.4–2.8 [3]	0.8–1.2 [3]	0.30–0.80 [19]
river	0.235–1.4 [4,10,20,22]	0.186–2.22 [4,10,20–22]	1.79–2.05 [14]			
lake	0.11–0.95 [10,20,23,24]	0.03–1.30 [10,20–24]	3.80 [14]	0.007 [24]		
sea	0.03–2.2 [3,10,11,17,24,25]	0.17–19.8 [3,10,11,17,24,25]		0.15 [24]	0.15–0.19 [3]	
waste		0.052–0.957 [4,21]				
rain	0.345 [4]	1.52 [26]	5.16 [26]			
snow		2.04 [26]	3.60 [26]			
Beverages						
wine	1.3–21.3 [27,28]					
tea	0.3–14.4 [29]	56–59.6 [29]				11.4–23.3 [29]
fruit juices	0.3–3.9 [30]	0.12–6.6 [30]				
Biological Samples (human fluids/tissue)						
serum	0.604–0.838 [4]	1.087–3.010 [4]				
urine	0.548–3.142 [4]	0.410–1.334 [4]				
blood/blood plasma			15.8–19.2 [31]	13.4–34.8 [31]	13.5–30.6 [31]	

Matrix	As(III), ng g⁻¹	As(V), ng g⁻¹	i-As, ng g⁻¹	DMA, ng g⁻¹	MMA, ng g⁻¹	o-As [b], ng g⁻¹
Solid Matrices						
Environmental						
soil	5.2–8.1 [c] [32]	16.0–20.4 [c] [32]				
phosphate rocks	2.1–3.9 μg g⁻¹ [33]		5.2–20.0 μg g⁻¹ [33]			
airborne particulate matter	2.7–10.5 ng m⁻³ [34]		3.8–20 ng m⁻³ [34]			
Industrial						
glass	13.6–395 μg g⁻¹ [35]	10.6–1205 μg g⁻¹ [35]				

Table 1. *Cont.*

Matrix	As(III), μg L⁻¹	As(V), μg L⁻¹	i-As, μg L⁻¹	DMA, μg L⁻¹	MMA, μg L⁻¹	o-As [b], μg L⁻¹
Agricultural Agents						
phosphate fertilizers	2.6–7.5 μg g⁻¹ [33,36]		11.79–69.02 μg g⁻¹ [33,36]			
pesticide	0.90 μg g⁻¹ [37]	0.81 μg g⁻¹ [37]				
herbicyde		1.47 μg g⁻¹ [37]				
Food						
rice	22–248 [22,38–40]	5–76 [22,38–40]	30–600 [41–46]	4.2–67.3 [22,39,40]	2.2–38.1 [22,39]	12.2–112.2 [45]
rice products	12.3–52.7 [47]	1.4–29.6 [47]	20–570 [41–48]	15–89 [48]	2.9–5.3 [48]	
wheat semolina	55 [39]	25 [39]		1.9 [39]	1.5 [39]	
milk	1.5–5.9 [49]	2.1–8.1 [49]				
mushrooms	81–624 [50]	59–380 [50]				
chard	89.2–90.6 [51]	14.2–15.3 [51]		4.1–4.3 [51]	3.5–3.7 [51]	
aubergine	20.6–20.9 [51]	61.0–61.9 [51]		1.1–1.2 [51]	1.2 [51]	
Marine Organism						
fish	80.3–230 [52]	108–310 [52]		510–1310 [52]	490–780 [52]	9.4–16 [a,c] μg g⁻¹ [52]
seaweed			0.05–57.5, μg/g [44,53–55]			
Pharmaceuticals						
TCM (herbs)	22.8 [c] [56]	145.1 [c] [56]				116.4 [b,d] [56]
herbaceous plant	0.030–8.32 μg g⁻¹ [57,58]	0.050–4.59 μg g⁻¹ [57]	1.08–6.91 μg g⁻¹ [58]			0.040, μg g⁻¹ [57]
dietary supplements	25–93 [59]	58–201 [59]				

i-As: the inorganic tri- and pentavalent As species (As(III) and As(V)). TCM: traditional Chinese medicines. [a] Concentration ranges based on reported results in works cited in this review. [b] Total organic As species concentrations (DMA+MMA) calculated as the difference between total As (determined after digestion) and i-As concentrations. [c] Water soluble fraction. [d] As non-toxic, i.e., unreactive As forms toward HG (mainly AsB) calculated as the difference between total As (determined after digestion) and the hydride-active toxic As species, i.e., the sum of As(III)+As(V)+DMA+MMA.

Speciation analysis involves two steps, i.e., separation of different forms and their subsequent quantification. The most common and recommended approach providing complete information on the distribution of species and their structures employs chromatographic separation of As species by high performance liquid chromatography (HPLC) interfaced with inductively coupled plasma mass spectrometry (ICP-MS) due to extremely low detection limits (LODs) of As species and their high selectivity [64].

Concurrently to HPLC-based As speciation schemes, there is considerable interest in developing simpler, more robust and reliable non-chromatographic methodologies for the determination of toxic As forms in different matrices at trace levels. Among them, hydride generation (HG) is a promising alternative to speciate As by obviating the need for chromatography separation. The generation of volatile hydrides in reactions with sodium/potassium tetrahydroborate ($NaBH_4/KBH_4$) in acidic media (usually HCl) is a well-known derivatization sample introduction technique applied in parallel with atomic and/or mass detection methods for the determination of trace levels of As. It can be easily coupled with various detectors like atomic absorption spectrometry (AAS), atomic fluorescence spectrometry (AFS), inductively coupled plasma optical emission spectrometry (ICP-OES) and ICP-MS, providing selectivity, sensitivity, detectability, separation of analytes from a sample matrix and efficiency that exceeds that offered by conventional pneumatic sample nebulization (PN) [65].

The HG technique, originally developed to separate As from the sample matrix, was found to be a viable method by which to discriminate among major toxic As compounds, i.e., i-As, DMA and MMA. Satisfactorily, all these forms are hydride-active and react with $NaBH_4$, forming the corresponding hydrides (arsines), i.e., AsH_3, $(CH_3)_2AsH$ and CH_3AsH_2 for As(III,V), DMA and MMA [66,67]. Other arseno-organic compounds like AsB and AsC or As-sugars are not reducible and are decomposed to i-As, usually by microwave [51,52] or UV-irradiation, prior to HG [19,68]. However, recent developments have shown that significant HG activity of As-sugars can be achieved under properly chosen conditions in a batch reaction mode [69].

The effectiveness of the HG processes of these individual As forms differs largely and depends strongly on experimental conditions used, i.e., basically, the type and concentration of acid, as well as the pH of the reacting medium. Generally, As(III) can be reduced in a wide HCl concentration range (pH 0–9), while for As(V), strong acidity is required (pH <2) [5,6,70]. In contrast, at low concentrations of HCl, both DMA and MMA hydrides can be effectively generated [71]. Trivalent arsenic can be also converted to arsine in the presence of weak organic acids (citric, acetic, tartaric) [3,40,71,72]. Moreover, this condition could also be applied to HG for DMA and MMA, but not for As(V) [3,71]. Nevertheless, there are considerable differences between the sensitivities obtained for the i-As and o-As forms, and the HG efficiency for V-valent As species is lower than that for As(III) [3,27,38,53,71]. Therefore, finding a compromise reaction medium under which the same response can be obtained for all four As species is problematic. Thus, a pre-reduction step is usually carried out with suitable pre-reductants (mostly KI-ascorbic acid, thiourea and L-cysteine) to ensure that any inorganic and methyl-substituted As(V) forms are present as As(III) before their reaction with $NaBH_4$ [3,40,71]. However, similar responses of As(III) and As(V) species under appropriately selected reaction conditions can be achieved, and hence, pre-treatment before HG can be avoided [20,35,72].

On the other hand, the ability of the HG technique to differentiate As forms by their oxidation state (III/V) or nature (inorganic/organic) using simple procedures certainly broadens the application range of HG regarding As speciation, without the use of chromatographic separation (see Table 1). Non-chromatographic approaches to differentiating the four As species are less time-consuming, simpler and more suitable and affordable compared to HPLC [64,73,74]. Generally, selective hydride formation can be achieved by adjusting the reducing conditions in terms of the reaction medium, the $NaBH_4$/HCl concentration and the sample pre-treatment with suitable additives. As a result, based on different responses of As species resulting from HG, non-chromatographic protocols for As speciation are proposed with variants for determinations of two- (e.g., [2,12,50,72]), three- (e.g., [7,75,76]) and four- (e.g., [3,28,39,40]) species of As in one sample.

The speciation of As by HG in sample solutions containing various inorganic and organic species is challenging, because several factors affecting the efficiency have to be carefully controlled to improve the accuracy. In reference to this, the present review covers the speciation of As in various matrices including environmental, industrial, food, biological and clinical samples using the non-chromatographic approach, based on the HG technique, and combined with different spectrochemical detection methods. It illustrates the main aspects of proper sample preparation before analysis and the choice of an adequate strategy for speciation (see Figure 1). Analytical methodologies for As speciation by HG are categorized according to their analytical performance, advantages and problems. Different non-chromatographic procedures, along with the speciation protocols reported in the literature over the past 20 year, are summarized. Additionally, the role of specific extraction–complexation, retention and co-precipitation enrichment methods before HG, in addition to the separation of evolved hydrides in cold traps or pervaporation modules to improve the detectability of As species and the selectivity of measurements, are discussed. Finally, alternative methods to conventional wet chemical HG providing selective generation of As hydrides are presented.

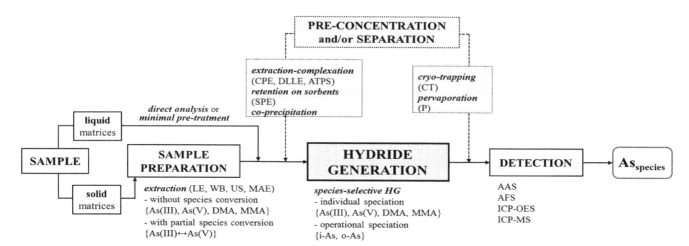

Figure 1. Analytical methodology for non-chromatographic speciation of As by HG technique. AAS: atomic absorption spectrometry. AFS: atomic fluorescence spectrometry. ATPS: liquid-liquid extraction of aqueous two-phase systems. CPE: cloud point extraction. CT: cryogenic cold trapping separation technique. DLLME: dispersive liquid-liquid microextraction. ICP-OES: inductively-coupled plasma optical emission spectrometry. ICP-MS: inductively-coupled plasma mass spectrometry. i-As: the inorganic tri- and penta- valent As species (As(III) and As(V)). LE: solvent extraction. MAE: microwave-assisted extraction. o-As: the organic, i.e., methylated pentavalent As species (DMA and MMA). P: pervaporation-based membrane separation technique. SPE: solid phase extraction. US: ultrasonication in an ultrasound water bath. WB: water bath.

2. Instrumental Techniques Used for the Determination of As Species and Ways of Verifying the Reliability of the Results

Hydride generation as a sample introduction technique was combined with different sensitive spectrometric detectors to measure the concentration of As species. A great majority of papers cited in the present review are devoted to its application in combination with AAS [3,6,7,12,15,16,18–20, 24,25,27,33–36,38,41,46–48,50,59–62,68,70,72,75–79] or AFS [2,5,9,10,13,14,17,21,22,26,28–30,32,39,40,42, 49,51–53,55–58,63,80–82], due to their lower costs in comparison with other techniques, particularly the most expensive ICP-MS, and higher availability in most laboratories. Occasionally, to improve sensitivity and extend the calibration range, ICP-OES is recommended [8,37,45,71,83,84]. Despite the intrinsic advantages of ICP-MS like extremely high sensitivity and a wide dynamic range, i.e., adequate for determination of (ultra)trace quantities of As, this method is rarely applied [11,31,44,53,54].

Generally, the quantification of As was carried out using external calibration with simple standards (e.g., [6,12,15,20,27–29,32–36,49,80,83]) or procedural blank-based standard (matrix-matched) solutions (e.g., [2,28,38,44–47,53,54,59,77]) prepared in the same way as samples (including pre-reduction) to keep the same acidification and the effect of additional reagents used in HG for As species (matrix-matched standards). To account for potential matrix-interfering effects from associated sample constituents, calibration by standard addition was also performed [6,8,17,27,33,34,36,59,72]. The lack of statistically significant differences between the slopes of these calibration curves indicated that there were no interferences, so simple external calibration was acceptable for analysis [6,27,33,34,36,59]. Frequently, to avoid or minimize intensive foam and bubble formation in the gas–liquid separator during HG, usually affecting the As response, silicone-based antifoam emulsions or alcohol-based agents (e.g., n-octanol) were added to the reductant or sample solutions [18,23,27,28,30,40,44–46,49,50,53,54,59].

It should be noted that detailed verification of the analytical performance of each method was made by evaluating several quality criteria, including accuracy, (procedural) blanks and analyte-specific figures of merit such as precision (as relative standard deviation, %RSD), LODs and linearity ranges of calibration curves.

The most reliable approach to demonstrate the accuracy of the whole method of As speciation was based on the analysis of certified reference materials (CRMs). Unfortunately, except for few such CRMs, e.g., NIST 2669 (Human Urine) [4], ERM-BC211 (Rice Flour) [38,46–48] and NIST 1568b (Rice Flour) [43], which were developed for As speciation, commercially available CRMs of different matrices with provided certified values of various As species are scare. Nevertheless, other CRMs could be used for As speciation. In this case, the content of As species was analyzed, and then the sum of their concentrations was compared with the certified value of the total As content [10,38,39,44,50–52,81]. Similarly, the content of As species in a real sample could also be analyzed, and then the sum of their concentrations could be compared with the total As content obtained after sample decomposition by wet digestion [2,7,40,44,48,49,51]. Due to a lack of proper CRMs, the reliability of the proposed methods was demonstrated by a spike-and-recovery test, based on the addition of known amounts of As species and the application of the whole procedure. Another way to check the accuracy of a newly developed methods is to compare their results with those obtained using well-established methods, e.g., HG-AAS with HPLC-HG-AFS [72], HPLC-ICP-MS [46,79], LC-ICP-MS [7], ETAAS [27] or HG-AFS [59]; HG-AFS with ICP-MS [2,49,50], IC-HG-AFS [28] or HPLC-HG-AFS [5]; HG-ICP-MS with HPLC-(HG)-ICP-MS [44,53,54] or HPLC-HG-AFS [53]. Two different sample preparation procedures, i.e., the reference and the one being examined, followed by measurements using the same detection technique, can also be applied for this purpose, i.e., a slurry sampling versus complete acid digestion in a digest block with a cold finger followed by HG-AAS [33,34], or in a microwave oven (microwave-assisted digestion) followed by ICP-MS [49].

3. Samples and Their Preparation

The usefulness of the HG technique for non-chromatographic speciation analysis of As is reflected in a wide spectrum of analyzed samples. This included various environmental, food and biological/clinical materials as follows:

- natural/environmental waters, i.e., mainly drinking [2–5,14,18,71,83], tap [4,9,15–21,84,85], ground [4,6–14,83], underground [3,15,16], sea [3,10,11,17,24,25,68,80], lake [10,14,20–24,80], river [10,14,20–22,75,82,86], waste [4,21] and snow and rain [26]
- sediments [72,75–77]
- soil [32,76,77]
- ash [72]
- phosphate rocks [33]
- airborne particular matter [34]
- plants [75]
- agricultural agents, i.e., phosphate fertilizers [33,36], herbicides [37] and pesticides [37]
- glass [35]

- beverages, i.e., alcoholic (wine [27,28]) and alcohol-free (fruit juices [30] and tea [29]),
- cereals, i.e., rice and/or rice products [22,38,41–48] and semolina [39],
- milk [49]
- mushrooms [50]
- vegetables, i.e., eggplant [51], chard [51]
- marine organisms, i.e., seafood [44,52,53,79,81], seaweed/algae [44,53–55] and plankton [70]
- pharmaceuticals, i.e., Chinese medicines [56–58,78], dietary supplements [59]
- biological fluids/tissue, i.e., human urine [4,62], serum [4], cells [60,61,63] and blood and blood plasma [31]

Unfortunately, to determine the concentrations of particular As species by atomic and/or mass detection methods along with the HG technique, samples need to be in a liquid form. When focusing on speciation analyses, special attention should be paid to the sample preparation step in order to maintain the original characteristics of the species and avoid any changes in a species distribution.

Sample preparation is unnecessary when analyzing simple liquid samples like water. Usually, before analysis, natural/environmental water samples were only filtered [9,13] through 0.22 [26] or 0.45 μm membrane filters [4,7,10,11,14,16–21,23–25], or centrifuged [84] to remove suspended solids or insoluble materials. They could be also acidified to 0.01 [CP2] or 0.1 mol L^{-1} HCl [3] or 0.1–1% HNO$_3$ [11,16] before [16,23] or after [11] filtering. In some cases, waters were stabilized with concentrated HNO$_3$ [6] or HCl [12] immediately after collection. However, the addition of an oxidizing acid would likely lead to erroneous results due to transformations between As(III) and As(V) forms [6], or an interfering effect in the determination of DMA [68]. Occasionally, EDTA [15] or NaF [80] were added to mask interferences in As determination; if precipitation took place afterward, samples were filtered [21]. Samples could also be preserved during storage with a high concentration of an Fe(II) salt, added to slow down the oxidation process of As(III) to As(V) in the presence of microbial activity or oxidizing substances like Fe(III) or Mn(IV) [12,83]. Interestingly, by accomplishing HG with a pervaporation technique used prior to the detection of As species, aqueous "dirty" samples (with a suspended particulate matter) could be analyzed as received, i.e., without any previous filtration [82,86]. Similarly, when detection was performed with the HG technique (not PN), simple cleaning comprising filtration with a 0.45 μm filter was sufficient to yield accurate results for fruit juices [30]. For more complex matrices such as alcoholic beverages, a minimal pre-treatment was required to overcome potential interference from matrix components. However, 20% evaporation of the sample volume [27] or only 5- to 10-fold dilution [28] were enough to completely remove ethanol and get rid of interference during analyses of wines by HG-AAS/AFS. Similarly, liquid rice-based products (wine, beer, vinegar) could simply be diluted (2- to 5-fold) with 0.28 mol L^{-1} HNO$_3$ prior to determination of As(III,V) by HG-AAS [47]. A direct dilution pre-treatment with water was found to be adequate to prepare biological/clinical samples such as serum, urine and blood prior to spectrometric measurements combined with HG [4,31,60–63].

In contrast to liquid samples, As speciation in solid matrices (e.g., food and environmental samples) is a challenge, since all As species must be firstly isolated from the sample matrix before further separation and detection. In contrast, to determine the total As content (usually preceded by wet digestion with aggressive reagents, i.e., commonly concentrated acids), mild but very efficient extraction is required for speciation purposes, ensuring a complete release of As compounds to be determined, albeit with no change in the original identity and concentrations of the individual As species, i.e., oxidation of As(III) to As(V) form and degradation of o-As to i-As forms. Frequently, this was achieved by solvent extraction (LE) with water [70], diluted solutions of HNO$_3$ [36,38,40,43,46,52], H$_3$PO$_4$ [39,50,51] and HCl [33,34,75] or a water:methanol mixture [81]. On the other hand, solvents with partial conversion between As species could also be used. They led to the solubilization of As species, followed by the oxidation of As(III) to As(V), but without decomposition of organic As species to As(V). Typically, extraction and oxidation were carried out in the presence of H$_2$O$_2$, which was most often used to facilitate the extraction of total i-As species from samples in the form of As(V) [44,53,54]. Agua regia was also found to be effective in this regard [45]. In general, extractions were carried out at elevated temperatures (<100 °C); hence, heating was applied from hot-plates [38,47], shaking water baths (WBs) [41,43] or microwave (MW)-assisted radiation. Nevertheless, MW-assisted extraction (MAE) using conventional MW systems was preferred [38,44,46,53,54,81] to accelerate the release of the analyte into solutions. For As speciation, several procedures were developed based on the application of classical LE. To selectively determine As(III,V) and i-As in rice and rice products by HG-AAS or HG-AFS, extraction with 0.02–0.28 mol L^{-1} HNO_3 by a WB (90 °C, 60 min) [43], under heating in a WB on a hot plate at 95 °C for 90 min [38,47] or MAE (95 °C for 30 min) [46] was found to be advantageous. Furthermore, in one work [38], the same results as for conventional heating with 0.28 mol L^{-1} HNO_3 were obtained using MAE (T_{max} 95 °C, 40 min) with 0.14 mol L^{-1} HNO_3. Otherwise, MAE (50 °C, 5 min) in a water:methanol (1:4, v:v) mixture was the procedure of choice to quantitatively extract all four As species from fish prior to HG-AAS measurements [81]. In cases of selective determination of i-As as As(V) in rice and samples of marine origin (seafood, seaweed), HG-ICP-MS, HG-AAS or HG-AAS, MAE (85–95 °C for 20–40 min) [42,44,53,54,79] or WB extraction (90 °C, 60 min) [41] with diluted HNO_3-H_2O_2 solutions (0.06–0.1 mol L^{-1}, 1–3% or 1–2% HNO_3 combined with 3% H_2O_2) were selected. In the case of the extraction procedure with water as the solvent, it was demonstrated that As(III) and As(V) were stable during MAE (800 W, 10 min) of plankton samples, and microwave heating was not able to reduce pentavalent inorganic As [70].

More recently, solvent extraction procedures supported by ultrasonic (US) agitation at room temperature (RT), realized using ultrasound WBs [36,39,40,45,50–52,75], have been recommended for the speciation of As. However, sample sonication at elevated temperature (80 °C) has also been practiced [40]. Accordingly, regarding individual speciation, a simple procedure for the determination of As(III) and i-As in phosphate fertilizers by HG-AAS was described by Rezende et al. [36], in which samples were sonicated for 35 min with 0.35% Triton X-114 and 6.5 mol L^{-1} HNO_3. In another work [75], for As speciation in environmental samples (sediment, plant) by HG-AFS, US using 6.0 mol L^{-1} HCl assured the quantitative extraction of As(III), As(V) and DMA; only 10 min was required for sample preparation. Fast US-assisted extraction (10 min) with 1 mol L^{-1} H_3PO_4-0.1% Triton XT-100 and of 0.1% EDTA (as a surface cleaning reagent) was also proposed by Gonzalvez et al. [50] to determine As(III) and As(V) by HG-AFS in cultivated and wild mushroom samples of different origins. Similarly, four As species (As(III,V), DMA, MMA) were selectively extracted during 10–20 min sonication of fish [52], cereals [39] and vegetables [51] samples in a mixture of 3 mol L^{-1} HNO_3 [52] or 1 mol L^{-1} H_3PO_4 [39,51] and 0.1% Triton-X114, combined with 0.1% EDTA and analyzed by HG-AFS. Finally, Wang et al. [40] extracted these four target As species from rice by sonication at 80 °C for 10 min with 1% HNO_3; the whole procedure was repeated three times. US extraction was also advantageous for operational As speciation. Sonication of samples with *aqua regia* for 15 min was used to release four As species from the rice sample matrix, leading to oxidation of As(III) into As(V), but without demethylation of DMA and MMA [45]. i-As species were then determined by HG-ICP-OES.

A promising excellent alternative to traditional extraction is slurry sampling (SS) based on direct analysis of solid particles of sample dispersed in a liquid phase (mainly diluted acids). Furthermore, it was demonstrated that this sample preparation made it possible to speciate i-As, whereas individual species of As did not undergo any changes in oxidation state. In view of this, recently, some analytical approaches to As speciation in various solid matrices, i.e., airborne particular matter [34], phosphate fertilizers and rocks [33], milk [49] and dietary supplements [59], have been developed, adopting the SS and HG techniques and atomic spectrometry detection. By contrast to traditional LE, Macedo et al. [33,34] employed SS with HG-AAS to determine the total i-As and As(III) in various environmental materials including phosphate fertilizers and rocks [33], and airborne particulate matter samples (filters) [34]. Under optimal conditions, sample portions were mixed with 4.0–6.0 mol L^{-1} HCl and sonicated at RT for 30 min. The resulting sample slurries were diluted with deionized water and further analyzed. Additionally, the bioavailability of As from a breathable particulate matter, which had entered lung fluids, was investigated [34]. In this scenario, deionized water instead of HCl was used as the extracting medium, following the same extraction and quantification methods for the As content. Cava-Montesinos et al. [49] developed a sensitive procedure for the determination of As(III) and As(V) in milk samples by HG-AFS. It was based on the leaching of As species from milk through sonication with *aqua regia* at RT for 10 min, followed by dilution with HCl. Importantly, it was demonstrated that neither As(III) nor As(V) were modified by the proposed sample treatment. This was in contrast to [39,45], where it was reported that the integrity of As(III) and As(V) in *aqua regia* was not preserved, leading to the oxidation of As(III) to As(V) during As species extraction from cereals such as rice [39,45] and wheat semolina [39]. The speciation of inorganic As in various dietary supplements (tablets, capsules) by using the SS and HG-AAS was described by Sun et al. [59], wherein As(III) and As(V) were isolated from the sample matrix during heating with 50% HCl in a WB for boiling for 5–10 min. Since adequate sample slurry preparation is of major importance in this sample technique, critical studies, aiming to achieve stability and uniformity of slurries, were undertaken. The best results were obtained for a sample particle size of 54 μm and a 0.3% agar reagent added to increase the viscosity of the medium. To achieve adequate homogeneity, ultrasonic agitation in an ultrasonic bath at 350 W for 30 min was applied.

Specific extraction protocols for the determination of inorganic As in ash, sediment and soil were also proposed [32,72,77]. Accordingly, Gonzalez et al. [72] employed a two-step, sequential extraction procedure using water and a 1.0 mmol L^{-1} phosphate buffer to identify As oxidation states in soluble and exchangeable As fractions, respectively, in fly ash and sediment samples. Similarly, Shi et al. [32] introduced a four-step extraction procedure, where deionized water, 0.6 mol L^{-1} KH_2PO_4, 1% HCl and 1% NaOH were sequentially used to leach extractable As forms from soils. Solubilized inorganic As compounds were speciated in these extracts by HG-AAS [72] and HG-AFS [32]. Shortened procedures were also proposed to assess the potential bioavailability of toxic As species. Accordingly, Petrov et al. [77] selectively extracted all four As species with EDTA (0.05 mol L^{-1}, pH 6–7) from soils and sediments prior to analysis of the EDTA-extractable As fraction by HG-AAS on the content of the sum of As(III,V), DMA and MMA. In another two works [56,57], to understand the solubility, mobility and transport of As in herbaceous plant samples for traditional Chinese medicines (TCMs), as well as commercially available TCMs (herbs), extraction with low concentration HCl [57] and water [56] was recommended. In cases of plants, powdered samples were extracted overnight (RT) with 1% HCl, then filtered and diluted with water. To investigate leachable As species from commercial TCMs, samples were placed in water, heated on a hot-plate to boil for 30 min, and then cooled and filtered. Concentrations of As(III) and total i-As in the obtained sample solutions were measured directly by HG-AFS. Because water is an attractive extracting solution for As, simple brewing, i.e., an everyday culinary process of tea infusion preparation, was also proposed for the speciation of As in tea [29]. For this purpose, a portion of tea was extracted with water for 20 min at 100 °C. Next, a sample suspension was filtered, diluted with water and analyzed for its As(III,V) content by HG-AFS.

Adequate extraction procedures are crucial for speciation analyses of specific solid samples too. Thus, critical studies were performed by do Nascimento et al. [35] for speciation analysis of As in glass. The authors compared four procedures for glass decomposition in order to determine inorganic As species in various commercial clear, green or amber glass ampoules, bottles and containers. Alkaline fusion with Na_2CO_3 (2 h at 900 °C) and glass dissolution in 40% NaOH or diluted HF (24%) for 48 h with or without microwave irradiation (10-50 min at 174 W) were tested. Problems related to either losses of As or stability of As(III) and As(V) forms, as well as incomplete real sample dissolution, were observed using alkaline fusion and the treatment with NaOH/HF supported by heating. Satisfactorily, a cold diluted HF-based strategy was the procedure of choice, and complete recoveries of both As species were obtained.

4. Non-Chromatographic As Speciation by Selective HG

The separation of As species in (prepared) sample solutions or sample extracts by selective HG and subsequent spectrometric detection is the most widely applied strategy for non-chromatographic As speciation, combined with the HG technique. Generally, using HG, different reactivities of As(III), As(V), DMA and MMA at various HG reaction conditions are used for the selective generation of individual hydrides. These As hydrides can be generated either selectively at different chemical conditions, or together with other species in various reaction media. Despite these essential advantages, As speciation by HG is focused mainly on i-As species and procedures discriminating between As(III) and As(V) [2,6,12,15,17,20,27,29,32–36,38,47,49,50,56,57,59,72,80,83] or ensuring the selective determination of i-As (as As(III)) [44–46,53,54] are those dominating.

4.1. Case of Inorganic Arsenic (As(III) and As(V))

4.1.1. Individual Speciation, i.e., Separately Determined As(III) and As(V)

Several approaches provided by selective HG can be used for the differentiation of As(III) and As(V). Accordingly, species selective HG from As(III,V) is based on:

(a)   Different reaction media, i.e., the acidity-dependent reduction reaction between As species and $NaBH_4$ to generate hydrides (affected strongly by the type and the concentration of acid or buffer solutions used, as well as the pH of the reacting medium).

(b)   The reaction rate with $NaBH_4$, i.e., differences in the reduction efficiency between As(III) and As(V) at different $NaBH_4$ concentrations (especially low) in the acid medium.

(c)   The absence or presence of additives like pre-reduction agents or other specific organic substances including chelating and masking reagents. The addition of various pre-reductants ensures that any As(V) is in the As(III) form before the reaction with $NaBH_4$ takes place; this provides the highest sensitivity in the HG technique for As, and hence, makes it possible to quantify the total As content.

Considering the approaches mentioned above, the determination of As(III) and As(V) is quite easy. Generally, a three-step scheme is needed, comprising: (i) the selective determination of As(III) in the presence of As(V); (ii) the determination of total i-As as As(III) after a pre-reduction step for As(V) with a suitable pre-reductant; and finally, (iii) the calculation of the As(V) content as the difference [As(V) = i-As–As(III)]. Accordingly, different reaction media (acids, buffers) [2,6,29,32–36,70,72] and variable concentrations of $NaBH_4$ in both soft and highly acidic conditions [12,15,20,27,38,47,72,80,83] were used for the selective determination of As(III) and total i-As. Concurrently to the subtracting method, since responses of As(III) and As(V) in different HG reaction conditions are not the same, linear independent equations relating their analytical signals versus concentrations were also used for quantification [17,29,49,50,56]. The optimum conditions for inorganic As speciation (individual) in various matrices by selective HG and atomic spectrometry detection with LODs of $\leq 1 \mu g \ L^{-1}$ are summarized in Table 2.

Table 2. Non-chromatographic procedures for individual speciation, i.e., selective determination of As(III) and As(V).

Matrix	Species	Sample Preparation [a]	Speciation Approach	Detection	LOD [b], $\mu g\ L^{-1}$	Ref.
			Redox Property of NaBH₄ (kinetic-dependent reduction reaction between NaBH₄ and As(III,V) forms)			
natural waters	As(III) As(V)	direct analysis	I. Selective As(III): S: not acidified, A: 10 M HCl, R: 0.05% NaBH₄ II. i-As (as As(III)): S: L-cysteine, A: 10 M HCl, 0.6% NaBH₄	HG-ICP-OES	1.0	[83]
water CRMs (TMDA-54.3, CASS-4)	As(III) As(V)	direct analysis	I. Selective As(III): S: 4 M HCl, A: 6 M HCl, 0.2% NaBH₄ II. i-As (as As(III)): S: KI-ascorbic acid–4 M HCl, A: 6 M HCl, 0.2% NaBH₄	HG-AFS	0.05	[80]
natural waters	As(III) As(V)	direct analysis	I. Selective As(III): S: 0.1 M HCl, 4.0% NaBH₄ II. i-As (as As(III)): S: KI-ascorbic acid–1 M HCl, 4.0% NaBH₄	HG-AAS	1.2	[15]
natural waters	As(III) As(V)	direct analysis	I. Selective As(III): A: 1.5 M HCl, R: 0.5% NaBH₄ II. i-As (as As(III)): A: 9.0 M HCl 3.0% NaBH₄	HG-AAS	As(III): 0.1 As(V): 0.06	[20]
ground water	As(III) As(V)	direct analysis	I. Selective As(III): S: 0.1 M HCl, A: 2 M HCl, 0.035% NaBH₄, II. i-As (as As(III)): S: KI-ascorbic acid–2.6 M HCl, A: 2 M HCl, R: 0.2% NaBH₄,	HG-AAS	As(III): 1.4 As(V): 0.6	[12]
rice	As(III) As(V)	MAE (0.14 M HNO₃)	I. Selective As(III): S: 0.14 M HNO₃, A: 10 M HCl, 0.1% NaBH₄ II. i-As (as As(III)): S: 0.14 M HNO₃–KI-ascorbic acid–1.2 M HCl, 10 M HCl, R: 0.1% NaBH₄	HG-AAS	As(III): 0.07 As(V): 0.14	[38]
rice products	As(III) i-As	HP (0.28 M HNO₃)	I. Selective As(III): S: 0.28 M HNO₃, A: 10 M HCl, 0.1% NaBH₄ II. i-As (as As(III)): S: 0.28 M HNO₃–KI-ascorbic acid–1.9 M HCl, 10 M HCl, R: 0.1% NaBH₄	HG-AAS	As(III): 0.08 i-As: 0.14	[47]

Table 2. Cont.

Matrix	Species	Sample Preparation [a]	Speciation Approach	Detection	LOD [b], µg L^{-1}	Ref.
			Specific HG from As(III) and As(V) under controlled pH conditions or different reaction media			
drinking water	As(III) As(V)	direct analysis	I. Selective As(III): S: Tris-HCl buffer (pH 7.2), R: 3.0% NaBH$_4$ II. i-As (as As(III)): S: Tris-HCl (pH 7.2)–TGA, R: 3.0% NaBH$_4$	HG-AFS	As(III): 0.027 As(V): 0.036	[2]
ground water	As(III) As(V)	direct analysis	I. Selective As(III): S: not acidified, A: 0.1 M citric acid, R: 0.6% NaBH$_4$ II. i-As (as As(III)): S: KI-6.5 M HCl, A: 10 M HCl, R: 0.6% NaBH$_4$	HG-AAS	0.4	[6]
ash and soil CRMs (NIST 1633b GBW07302 GBW07311)	As(III) As(V)	two-step sequential LE (H$_2$O, 1 mM phosphate buffer)	I. Selective As(III): S: citric buffer (pH 4.5), A: 10% HCl, R: 0.3% NaBH$_4$; optionally: S: 2% HCl, A: 2% HCl, R: 0.3% NaBH$_4$ II. i-As (as As(III)): S: KI-ascorbic acid–HCl, A: 10% HCl, R: 0.3% NaBH$_4$; optionally: S: 0.46% TGA in sample, A: 10% HCl, R: 0.3% NaBH$_4$	HG-AAS	As(III): 0.07 As(V): 0.06	[72]
soil	As(III) As(V)	four-step sequential LE (H$_2$O, 0.6 M KH$_2$PO$_4$, 1% HCl, 1% NaOH)	I. Selective As(III): S: 0.1 M citric acid, R: 1.0% NaBH$_4$ II. i-As (as As(III)): S: 0.1 M citric acid–L-cysteine, 1.0% NaBH$_4$	HG-AFS	As(III): 0.11 As(V): 0.07	[32]
glass	As(III) As(V)	dissolution (24% HF)	I. Selective As(III): S: citric buffer (pH 4.5), R: 1.0% NaBH$_4$ II. i-As (as As(III)): S: 6.0 M HCl, R: 1.0% NaBH$_4$	HG-AAS		[35]
airborne particulate matter	As(III) i-As	SS (4 M HCl)	I. Selective As(III): S: citric buffer (pH 7.1), A: 2 M HCl, R: 2.15% NaBH$_4$ II. i-As (as As(III)): S: KI-ascorbic acid–1.8 M HCl, A: 2 M HCl, R: 2.15% NaBH$_4$	HG-AAS	0.1	[34]
phosphate fertilizer and rock	As(III) i-As	SS (6 M HCl)	I. Selective As(III): S: citric buffer (pH 7.1), A: 2 M HCl, R: 2.15% NaBH$_4$ II. i-As (as As(III)): S: KI-ascorbic acid–2 M HCl, A: 2 M HCl, R: 2.15% NaBH$_4$	HG-AAS	0.1	[33]

Table 2. *Cont.*

Matrix	Species	Sample Preparation [a]	Speciation Approach	Detection	LOD [b], μg L^{-1}	Ref.
phosphate fertilizer	As(III), i-As	US (0.35% Triton X-114 + 6.5 M HNO$_3$)	*I. Selective As(III):* S: citric buffer (pH 4.5), A: 10% HCl, R: 0.4% NaBH$_4$. *II. i-As (as As(III)):* S: thiourea, A: 10% HCl, R: 0.4% NaBH$_4$	HG-AAS	As(III): 0.029, i-As: 0.022	[36]
wine	As(III), i-As	EtOH evaporation	*I. Selective As(III):* S: 8 M HCl, A: 9 M HCl, R: 0.2% NaBH$_4$, *II. i-As (as As(III)):* S: KI– 8 M HCl, A: 9 M HCl, R: 0.2% NaBH$_4$	HG-AAS	0.1	[27]
plankton	As(III), As(V)	MAE (H$_2$O)	*I. Selective As(III):* S: citric buffer (pH 4.5), A: 2% HCl, R: 0.1% NaBH$_4$. *II. i-As (as As(III)):* S: citric buffer (pH 4.5)–L-cysteine, A: 2% HCl, R: 0.1% NaBH$_4$	HG-MF-AAS	2.0	[70]
Species-selective respond toward two different reducing conditions (As speciation based on systems of linear independent equations)						
milk	As(III), As(V)	SS (*aqua regia*)	1.2% NaBH$_4$–3.5 M HCl without and after pre-reduction with KI-ascorbic acid-hydroxylamine hydrochloride in 10.8 M HCl	HG-AAS	As(III): 0.0081, As(V): 0.0103	[49]
mushrooms	As(III), As(V)	US (1 M H$_3$PO$_4$-0.1% Triton X-100 + 0.1% EDTA)	0.7% NaBH$_4$–3.5 M HCl without and after pre-reduction with KI-ascorbic acid in 9 M HCl	HG-AFS	6.5 ng g^{-1} c	[50]
TCM (herbs)	As(III), As(V)	LE (H$_2$O)	1.0% KBH$_4$–1 M HCl without and after pre-reduction with KI-thiourea in 2 M HCl	HG-AFS	0.0797	[56]
tea	As(III), As(V)	brewing (H$_2$O)	*I. Selective As(III):* S: 0.1 M citric acid–5% HCl, A: 5% HCl, R: 5% KBH$_4$. *II. i-As (as As(III)):* S: thiourea-ascorbic acid–5% HCl, A: 5% HCl, R: 5% KBH$_4$	HG-AFS	As(III): 0.0070, As(V): 0.0095	[29]
natural waters	As(III), As(V)	direct analysis	*I.* S: 0.7 M HCl, R: 0.7% NaBH$_4$, *II.* S: L-cysteine–0.1 M HCl, R: 0.7% NaBH$_4$	HG-AFS	As(III) 0.013, As(V): 0.015	[17]

Table 2. *Cont.*

Matrix	Species	Sample Preparation [a]	Speciation Approach	Detection	LOD [b], µg L^{-1}	Ref.
			Specific HG from As(III) and As(V) in the presence of masking reagents			
herbaceous plant	As(III) As(V)	LE (1% HCl)	I. *Selective As(III):* S: 8-hydroxyquinoline–10% HCl, R: 2.5% NaBH$_4$ II. *i-As (as As(III)):* 8-hydroxyquinoline–KI–ascorbic acid–10% HCl, R: 2.5% NaBH$_4$	HG-AFS		[57]
dietary suplements	As(III) As(V)	SS (50% HCl)	I. *Selective As(III):* S: 8-hydroxyquinoline–1% HCl, R: 1.0% KBH$_4$ II. *i-As (as As(III)):* S: 8-hydroxyquinoline–KI–5% HCl, R: 1.0% KBH$_4$	HG-AAS	As(III): 0.080 As(V): 0.089	[59]

EtOH: ethanol. HG-AAS: hydride generation atomic absorption spectrometry. HG-AFS: hydride generation atomic fluorescence spectrometry. HG-MF-AAS: metallic furnace hydride generation atomic absorption spectrometry. HG-ICP-OES: hydride generation inductively-coupled plasma optical emission spectrometry. HP: heating in water bath on hot plate. i-As: the inorganic tri- and pentavalent As species (As(III) and As(V)). LE: solvent extraction. M: mol L^{-1}. MAE: microwave-assisted extraction. S: sample solution. A: additional acid (carrier) solution. R: reductant solution. SS: slurry sampling. TCM: traditional Chinese medicines. TGA: thioglycolic acid. US: ultrasonication in an ultrasound water bath. WB: water bath. [a] Detailed information about sample preparation procedures described in point 4. [b] For As(V) or i-As as As(III) without or after previous pre-reduction step. [c] In the original sample, taking into account the amount of sample and the final dilution employed in the proposed procedure.

It can be concluded that, on the one hand, As(III) reacts selectively with $NaBH_4$ in buffered media (with citrates at pH 4.5 [35,36,70,72] and at pH 7.1 [33,34] or Tris-HCl at pH 7.2 [2]) or low concentrated organic acids such as citric acid (0.1 mol L^{-1} [6,32]) or citric acid combined with HCl (0.1 mol L^{-1}-5% [29]). On the other hand, selective AsH_3 formation for As(III) without interference from As(V) can be also achieved via the HG reaction at (very) low $NaBH_4$ concentrations (0.035-0.2%) [12,27,38,47,72,80,83], using 0.65–10 mol L^{-1} HCl as the reaction medium. Interestingly, using a batch type HG system, 4% $NaBH_4$ combined with 0.1 mol L^{-1} HCl was found to be optimal for direct measurements of As(III) coexisting with As(V) [15]. It should be noted that the As signal increases with the concentration of $NaBH_4$; this effect is evident, particularly for As(V). Therefore, the use of low or even very low concentrations of $NaBH_4$ decreases the sensitivity of As(III) determination, and hence, the method LOD of As(III) would be affected too [72]. Generally, for this approach, the selected $NaBH_4$ concentration is a compromise between the error due to the As(V) interference and the LOD value, making it possible to reliably determine the As(III) content in examined samples [12,83]. Typically, total i-As is determined at higher $NaBH_4$ concentrations (0.2–3%) [2,6,12,29,33,34,36,72,83] and 2–10 mol L^{-1} HCl for the HG reaction after the offline pre-reduction of As(V) to As(III) using KI with ascorbic acid [6,12,34,72], thiourea (alone [36] or in combination with ascorbic acid [29]) and L-cysteine [83]. Total i-As can be also determined under the same HG reaction conditions as for As(III) after online [2,32,80] or offline [6,15,27,38,47,70] pre-reduction of As(V) to As(III) with KI alone [6,27], KI-ascorbic acid [15,38,47,80], L-cysteine [32,70] or thioglycolic acid (TGA) [2].

Potassium iodide is one of the most widely used reagents in the pre-reducing step; however, it works effectively only under strong acidic conditions for sample acidification, i.e., 1–11 mol L^{-1} HCl (see e.g., [6,12,13,15,27,34,38,39,45–47,49,51,52,71,80]). Sample acidifications of 2–3 mol L^{-1} HCl were the most common [12,13,34,39,45,46,51,52]; nevertheless, lower HCl concentrations in sample solutions (0.1-0.7 mol L^{-1}) were also used [3,28]. Typically, ascorbic acid is added along with KI in order to avoid the self-oxidation of I^- ions to free I_2 by the O_2 that is present in solutions [12,15,46,77,80]. Importantly, the ascorbic acid concentration has no effect on the As signals; thus, its role is primarily to stabilize the pre-reduction medium [80]. Despite this, samples need to be analyzed within 6 h after the addition of reagents to avoid oxidation by atmospheric O_2, leading to the formation of triiodide and the eventual re-oxidation of As(III) to As(V) [46]. Optionally, to transgress I_2, sample solutions treated with KI alone can be heated to boil for 5–10 min [59]. Pre-reduction with KI runs commonly for 30–60 min [28,33,34,38,39,44,46,50–53,59,77], but it can be achieved in a shorter time (i.e., 5 min) by increasing the reagent concentration [80]. By contrast, reagents with thiol (–SH) functional groups such as L-cysteine, thiourea and TGA were found to be more advantageous. Besides its pre-reduction properties, thiourea can also be used as a sensitization reagent, improving the As intensity [56]. L-cysteine pre-reduces all pentavalent As species to their trivalent oxidation state forms with similar responses at a relatively low and narrow optimum HCl concentration range (0.01–0.1 mol L^{-1}) [28,71,81]. This reagent was used also to level off the responses of different As species [28,32]. The reaction time for the complete pre-reduction at RT, i.e., 30–60 min [28,52], can be shortened to less than 60 s at 100 °C [52]. Interestingly, pre-reduction with TGA is fast and measurements can be made just after adding this reagents to sample solutions [2,72]. Additionally, it can be used as a pre-reductant for all four As species [2], or can be treated as a reaction medium, making it possible to achieve the same responses with As(III) and As(V) [72].

Gonzalez et al. [72] compared several speciation procedures for the selective determination of As(III) in the presence of As(V) and total i-As by HG-AAS in water-soluble and phosphate-exchangeable extracts of sediments and fly ash CRMs (NIST 1633b, GBW 07311 and GBW 07302). This included (i) AsH_3 generation under soft HG conditions, i.e., low $NaBH_4$ and HCl concentrations (0.05%–2.0%, respectively) and from different reaction media, i.e., citric acid at pH 4.5 and acetic acid for lonely As(III) determination, and (2) total i-As determination in the TGA medium or after pre-reduction with a KI-ascorbic acid mixture. Except for the acetic acid based procedure, all of them could be used to distinguish between these two forms. Acetic acid was not selective enough for the determination of

As(III) coexisting with As(V). The best analytical performance was achieved using the procedure with the citric buffer and KI-ascorbic pre-reduction for As(III) and total i-As, respectively. The obtained LODs were two times better than those achieved with the two remaining procedures. On the other hand, in contrast to the pre-reduction of As(V) with KI-ascorbic, the determination of As(III) and As(V) in the presence of TGA medium, with virtually the same responses of both species, could be achieved just after adding this acid to sample solutions. With the KI-ascorbic acid mixture, the conversion of As(V) into As(III) was completed within 1 h. Citric acid was also found to be the most effective reagent in the determination of As(III) alone in soil extracts by HG-AFS [32]. Among the other tested media, including low concentration acetic and tartaric acids, HNO_3 and HCl, only when citric acid was used, the unwanted presence of As(V) could be virtually eliminated. Similarly, Lehmann et al. [70] succeeded in speciating inorganic As in plankton samples using a metallic furnace atomizer with HG-AAS by controlling the reaction medium and avoiding the reduction of As(V) to As(III). Importantly, measurements of As(III) and total i-As (as As(III) after pre-reduction) were carried out using the same mild conditions for HG, such as a slightly acidic media (a citrate buffer at pH 4.5 was used) and a low $NaBH_4$ concentration (0.1%). Among the various pre-reducing agents tested, including KI-ascorbic acid, $Na_2S_2O_3$, L-cysteine and thiourea, it was found that L-cysteine was able to reduce As(V) to As(III) at pH 4.5. This was probably due to the strong affinity of the As(III) to –SH group present in the L-cysteine structure, which may change the mechanism of reduction and promote the conversion of As(V) to As(III).

Interestingly, in two works [57,59], AsH_3 was generated selectively for As(III) in the presence of 8-hydroxyquinoline under HCl acidity. Possibly, an ion associate of As(V) with 8-hydroxyquinoline was likely formed, making As(V) unreactive. On the other hand, 8-hydroxyquinoline had an enhancing effect on the responses of both As(III) and As(V) pre-treaded with KI/KI-ascorbic acid before HG. Additionally, the presence of 8-hydroxyquinoline in the reaction medium lessened the interference from transition metals occurring in the generation of arsine. The proposed strategy was found to be attractive in analyses of herbaceous plants by HG-AFS [57] and dietary supplements by HG-AAS [59].

Regarding the different behavior of As(III) and As(V) in the HG process, speciation was also made using proportional equations corresponding to two different measurement conditions for the same sample (acid-$NaBH_4$ combination) [17,29,49,50,56]. In three cited works [49,50,56], these two different conditions referred to direct measurements of diluted sample extracts and measurements after the previous pre-reduction step with KI-ascorbic acid-hydroxylamine hydrochloride [49], KI-ascorbic acid [50] or KI-thiourea [56] mixtures. This was successfully applied to determine both i-As species in various materials, including milk [49], mushrooms [50] and TCM herbs [56] by HG-AFS/AAS. To establish these simultaneous equations, two HG reaction conditions combined with L-cysteine for pre-reduction were also used by Wang and Tyson [17] to determine As(III) and As(V) quantities in fresh water samples (sea, tap, pound) by HG-AFS. To overcome interference from Fe(III) in the case of seawater, standard addition calibration curves were employed for As quantification. Finally, Cai et al. [29] obtained a proportional dependence of As species using citric acid for As(III) and thiourea-ascorbic acid for total i-As. The methodology was used to distinguish As(III) and As(V) in tea by HG-AFS.

This subtraction-based speciation scheme could also be simplified and realized without the need of a previous pre-reduction step for total i-As. However, this was possible only for the case when As(III) and As(V) species responded similarly under applied HG conditions. Anthemidis et al. [20] reported that AsH_3 could be selectively generated either for As(III) or total i-As using different HCl and $NaBH_4$ concentrations. As a result, the authors proposed two different pairs of HCl and $NaBH_4$ concentrations, i.e., 1.5 mol L^{-1} HCl–0.5% $NaBH_4$ and 9.0 mol L^{-1} HCl–3.0% $NaBH_4$ for the selective determination of As(III) and total i-As, respectively, in natural water samples (river, lake, tap) by HG-AAS. Similarly, as shown in [35,72], As(III) and As(V) can be determined with the same sensitivity at 0.3% [72] or

1.0% [35] $NaBH_4$ in the medium of 0.48% TGA [72] or 6 mol L^{-1} HCl [35] in environmental [72] or glass [35] samples by HG-AAS; no additional pre-reduction step was required.

The main drawback of the proposed speciation schemes could be that the presence of o-As species in the sample matrix, especially MMA and DMA, which generate their respective hydrides under chosen conditions, could selectivity affect the measurements of As(III) and As(V). This is evident when a –SH group containing pre-reductants such as L-cysteine or TGA is used for As(V) pre-reduction. Both result in the same response for all four As species, leading to overestimations of As(V) concentrations [2].

In some works, the contribution of methylated As species (o-As) to the quantified i-As concentration was assessed by evaluating the interference of DMA and MMA in the As signal under selected pre-reduction and HG reaction conditions [2,12,17,27,38,47,59,83]. It was found that the increase in HCl concentration (9 mol L^{-1}) and the application of KI at the pre-reduction step successfully diminished the negative effects coming from the presence of both DMA and MMA during inorganic As speciation in wine by HG-AAS [27]. In another work [59], the selectivity of As(III) and As(V) measurements in dietary supplements by HG-AAS, i.e., with no interference from either methylated As species, was guaranteed under 1–5% HCl acidity conditions in the presence of 8-hydroxyquinoline and KI, added to pre-reduce As(V) into As(III). In two other works [38,47], using 0.1% $NaBH_4$ and 10 mol L^{-1} HCl in combination with KI-ascorbic acid for pre-reduction, it was possible to separately determine As(III) and i-As in rice and rice products by HG-AAS. However, the absence of MMA in the sample was considered; DMA remained undetected, as evidenced by its signal being close to that of the blank. Interference effects coming from DMA and MMA were also noted; therefore, methods developed for the speciation of i-As were proposed and used for samples with negligible or no methylated As compounds [2,12,17,83]. Optionally, to overcome the interference effects from DMA and MMA, solid phase extraction (SPE) with specific sorbents were used to separate i-As from o-As before HG [23,41–43,55,79] (see point 5.1). Organic As could also be removed after the HG reaction by freezing out in a liquid nitrogen trap installed between a gas/liquid separator and a detection device [83]. Occasionally, the amount of o-As was defined by subtracting the total i-As content from the total As (As_T) content determined after complete sample decomposition by wet digestion [27,29,45,50,56,57].

4.1.2. Operational Speciation, i.e., Dealing Only with i-As Determination (total As(III) and As(V))

This approach compromises the inorganic As species present in the sample, and is recommended when inter-conversion between As(III) and As(V) is not an issue. The toxicity of As depends on the presence of both inorganic forms; hence, conditions for the selective determination of i-As, i.e., total As(III) and As(V), in the presence of other organoarsenic compounds, including methylated forms and As-sugars, are of special interest.

Depending on the reagents applied at the sample preparation step, e.g., a low concentration of HNO_3 (alone [46] or combined with H_2O_2 [44,53,54]) or *aqua regia* [45]), As can be present in sample extracts in the form of As(III) and As(V) or as i-As, being exclusively As(V) after the conversion of all As(III) to As(V) during extraction. For the selective determination of i-As by HG, i-As species should be present as a single species, i.e., as As(III) or As(V), rather than a mixture of As(III) and As(V). Since As(III) reacts more effectively with $NaBH_4$ than As(V), the pre-reduction of As(V) to As(III) prior to HG is preferred. As a result, total i-As is measured in the form of As(III). Selected optimal parameters developed for the selective determination of traces of i-As in rice by HG-ICP-MS [44], HG-ICP-OES [45] or HG-AAS [46], and in more complex matrices, i.e., samples of marine origin such as seaweed, by HG-ICP-MS [44,53,54], are summarized in Table 3.

Table 3. Non-chromatographic procedures for operational speciation, i.e., i-As determination (total As(III) and As(V)).

Matrix	Species	Sample Preparation [a]	Speciation Procedures	Detection	LOD [b], $\mu g\ L^{-1}$	Ref.
rice, seaweed	i-As (as As(V))	MAE (1–2% HNO_3–3% H_2O_2)	S: 3% H_2O_2, A: 5 M HCl, R: 2% $NaBH_4$	HG-ICP-MS	0.006	[44]
rice	i-As (as As(III))	US (aqua regia)	S: 1.25 M aqua regia–KI-ascorbic acid–3 M HCl, A: 10 M HCl, R: 1% $NaBH_4$	HG-ICP-OES	0.28 ng g^{-1}	[45]
rice	i-As (as As(III))	MAE (0.28 M HNO_3)	S: 0.28 M HNO_3–KI-ascorbic acid–3 M HCl, A: 1.2 M HCl, R: 0.1% $NaBH_4$	HG-AAS	16 ng g^{-1} [c]	[46]
seafood, seaweed	i-As (as As(V))	MAE (2% HNO_3–3% H_2O_2)	S: 3% H_2O_2, A: 8 M HCl, R: 2% $NaBH_4$	HG-ICP-MS	0.01	[53]
seaweed	i-As (as As(V))	MAE (2% HNO_3–3% H_2O_2)	S: 3% H_2O_2, A: 5 M HCl, R: 2% $NaBH_4$	HG-ICP-MS	0.06	[54]

HG-AAS: hydride generation atomic absorption spectrometry. HG-ICP-OES: hydride generation inductively-coupled plasma optical emission spectrometry. HG-ICP-MS: hydride generation inductively-coupled plasma mass spectrometry. i-As: the inorganic tri- and pentavalent As species (As(III) and As(V)). M: mol L^{-1}. MAE: microwave-assisted extraction. S: sample solution. A: additional acid (carrier) solution. R: reductant solution. US: ultrasonication in an ultrasound water bath. [a] Detailed information about sample preparation procedures described in point 4. [b] For As(V) as As(III) without or after previous pre-reduction step. [c] In the original sample taking into account the amount of sample and the final dilution employed in the proposed procedure.

Generally, the selectivity of i-As determination is based on the application of a high concentration of HCl (5–10 mol L^{-1}) for HG and a KI-ascorbic acid mixture for pre-reduction of As(V). By careful selection of flow rates of solutions, a lower optimum HCl concentration (1.2 mol L^{-1}) can be achieved [46]. Unfortunately, in a variant of strong sample acidity required for selective HG, L-cysteine could not be used at the pre-reduction step [54]. The concentration of NaBH$_4$ used for the determination of i-As varied (0.1–2.0%); nevertheless, its higher concentrations had to be used when HNO$_3$ combined with H$_2$O$_2$ was used to extract both As species. At least 2% NaBH$_4$ was needed to overcome the interference effects which occurred in HG for As(V) coming from the presence of H$_2$O$_2$ in sample extracts left after extraction. Additionally, H$_2$O$_2$ interfered with pre-reduction of As(V) by KI, which made it impossible to determine the i-As content in real samples in the form of As(III). Accordingly, in these works [44,53,54], total i-As was measured as As(V) after oxidation with H$_2$O$_2$. In contrast, such an effect was not observed when *aqua regia*- [45] or HNO$_3$-based [46] sample extraction procedures were used to release As species from sample matrices.

The main goal of these optimization studies was to provide a contribution of all hydride-active methylated As species to the i-As signal as low as possible. Besides adequate reducing conditions, sample preparation in terms of the reagents used for extraction was helpful in improving the selectivity of i-As determination in the presence of coexisting o-As species. In three works [44,53,54], high concentrations of HCl for HG and H$_2$O$_2$ in samples of the same concentration, as used for extraction (3%), led to the selective conversion of i-As to volatile arsine, while HG from DMA was substantially inhibited (less than 1-3% of the i-As signal) (see Table 3). Unfortunately, these schemes introduced some errors due to a significant contribution of MMA to the i-As signal (21–43%). However, it is argued that MMA is normally absent or present in trace amounts in most samples of rice and seafood, and hence, would not affect the quantification of i-As in these materials. Interference from a rich As-sugars matrix in the determination of i-As with HG was also negligible; therefore, the described methodologies were suitable as quick reliable screening methods for i-As determination in seaweed [54]. Similar results were achieved for the i-As species measured in sample extracts containing 0.28 mol L^{-1} HNO$_3$, used to extract As species from various types of rice (paddy, brown, polished, parboiled) [46]. In one work [45], where *aqua regia* was used for extraction, interferences coming from both DMA and MMA were successfully eliminated, making it possible to reliably differentiate between i-As and o-As, and to selectively determine traces of i-As in brown rice by HG-ICP-OES. The presence of 1.25 mol L^{-1} *aqua regia* in the sample extracts and a high concentration of HCl for the HG reaction were advantageous to limit the activity of o-As during HG, but not i-As.

4.2. Case of Inorganic As(III) and As(V) and Organic Arsenic (DMA and MMA)—Speciation and Fractionation Protocols

As presented above, the response of As achievable in the HG reaction strongly depends not only on the oxidation state (III/V) and experimental conditions, but also on the nature of hydride-active As species (inorganic/organic). Accordingly, distinguishing between tri- and penta- valent arsenicals, provided by selective HG, can expand speciation analyses of i-As to its methylated forms (DMA, MMA). Undoubtedly, non-chromatographic approaches to the differentiation of four As species by means of HG are the most desirable, but also the most challenging. Their development has to be proceeded by the careful optimization of experimental parameters, being appropriate for each As species present in the sample solution.

Based on the different reactivities of all four As species under special pre-reducing and HG reaction conditions, procedures for species-selective HG of As can be evaluated using the same rules as those provided for i-As analysis, i.e., pH specific HG reaction or selective conditions (a proper acid-NaBH$_4$ combination and a pre-reductant). By combining the responses obtained for these procedures (simple mathematical subtraction or a series of independent proportional equations), protocols for the non-chromatographic speciation of As and determination of its species at a trace level in various matrices, including food [39,40,52,81], beverages [3,7,28] and environmental [3,13,75–77] samples, have been proposed. In addition to individual speciation [3,7,28,39,51,52], procedures to fractionate As were also evaluated (operational speciation) by distinguishing between species of the same nature, i.e., i-As versus o-As [71] or As_{toxic} versus $As_{non\text{-}toxic}$ [52,71]. The fraction of As_{toxic} indicates hydride-active species (As_h), i.e., the sum of As(III)+As(V)+DMA+MMA, while the content of $As_{non\text{-}toxic}$ refers to unreactive As forms toward HG (As_{nh}). Typically, the latter is determined after sample digestion by the difference between the total As content (As_T) and As_h. Variants among sensitive determinations of three- and four-species of As are detailed in Table 4.

Table 4. Non-chromatographic speciation and fractionation protocols for the determination of various As species by HG.

Matrix	Species	Sample Preparation [a]	Speciation Procedures	Detection	LOD [b], μg L⁻¹	Ref.
cereals, fish, vegetables	As(III) As(V) DMA MMA t-As$_{toxic}$ t-As$_{non-toxic}$	US (3 M HNO₃ or 1 M H₃PO₄–0.1% Triton X-114 + 0.1% EDTA)	-*Individual speciation:* I: 2 M HCl, 1.4% NaBH₄ (maximum for DMA, MMA) II: 4 M HCl, 1.4% NaBH₄ (intermediate for all species) III: 3.5 M HCl, 1.2% NaBH₄ (maximum for As(III,V) IV: KI-ascorbic acid–3.5 M HCl, 1.2% NaBH₄ (with pre-reduction As(V) and MMA to As(III)) -*Operational speciation:* t-As$_{toxic}$ and t-As$_{non-toxic}$ fractions as difference between total As content (T-As) determined after complete sample digestion and sum of As(III,V), DMA and MMA species (t-As$_{toxic}$)	HG-AFS	As(III): 0.62-3.1 ng g⁻¹ [c] As(V): 0.9-3.0 ng g⁻¹ [c] DMA: 1.5-1.8 ng g⁻¹ [c] MMA: 0.6-5.4 ng g⁻¹ [c]	[39,51,52] [52]
rice	As(III) As(V) DMA MMA	US (1% HNO₃)	I: 0.06 M citric acid (maximum for As(III)+DMA) II: K₂S₂O₈–0.06 M citric acid (maximum for DMA) III: L-cysteine-ascorbic acid–0.06 M citric acid (maximum for As(III)+As(V)+DMA) IV: thiourea-ascorbic acid–5% HCl (pre-reduction of As(V), DMA and MMA to As(III)) A: 1.6 M HCl–citrate buffer (pH 4.8), R: 2% KBH₄	HG-AFS	As(III): 0.21 μg kg⁻¹ As(V): 0.52 μg kg⁻¹ DMA: 0.65 μg kg⁻¹ MMA: 0.9 μg kg⁻¹	[40]
ground water	As(III) As(V) DMA	direct analysis	S: oxalate buffer (pH 4-4.5), R: 0.6% NaBH₄ I. As(III): II. As(III)+As(V): S: KI, A: 6 M HCl, R: 0.6% NaBH₄ III. DMA: S: L-cysteine, A: 1.5 M HCl, R: 0.6% NaBH₄	HG-AAS	As(III): 0.1 As(III,V): 0.1 DMA: 0.19	[7]
natural waters	As(III) As(V) DMA MMA	direct analysis	I. As(III)+As(V)+MMA: S: KI–0.1 M HCl, A: 1 M HCl, R: 0.6% NaBH₄ II. As(III)+DMA: S: 0.1 M HCl, A: 6 M CH₃COOH, R: 0.6% NaBH₄ III. As(III)+As(V)+DMA: S: KI–0.1 M HCl, A: 6 M CH₃COOH, R: 0.6% NaBH₄ IV. As(III)+As(V)+DMA+MMA: S: KI–0.1 M HCl, A: 1 M tartaric acid, R: 0.6% NaBH₄	HG-AAS	0.1	[3]

Table 4. *Cont.*

Matrix	Species	Sample Preparation [a]	Speciation Procedures	Detection	LOD [b], $\mu g\ L^{-1}$	Ref.
wine	As(III) As(V) DMA MMA	5-10 dilution	*I. As(III):* S: citrate buffer (pH 5.1), R: 0.6% $NaBH_4$ *II. As(III)+DMA:* S: 0.2 M CH_3COOH, R: 0.6% $NaBH_4$ *III. As(III)+As(V):* S: KI-8 mol/L HCl, R: 0.2% $NaBH_4$ *IV. As(III)+As(V)+DMA+MMA:* S: L-cysteine-0.01 M HCl, R: 0.6% $NaBH_4$	HG-AAS	As(III): 0.4 As(V): 0.3 DMA: 0.3 MMA: 0.3	[28]
fish	As(III) As(V) DMA MMA	MAE (H_2O:MeOH, 1:4 v:v))	*I. As(III):* S: citrate buffer (pH 5.2), R: 0.45% $NaBH_4$ *II. As(III)+DMA:* S: 0.2 M CH_3COOH, R: 0.45% $NaBH_4$ *III. As(III)+As(V):* S: KI-7 M HCl, R: 0.2% $NaBH_4$ *IV. As(III)+As(V)+DMA+MMA:* S: L-cysteine-0.05 M HCl, R: 0.45% $NaBH_4$	HG-AAS	As(III): 3.5 $\mu g\ kg^{-1c}$ As(V): 5.1 $\mu g\ kg^{-1c}$ DMA: 4.8 $\mu g\ kg^{-1c}$	[81]
plant, sediment	As(III) As(V) DMA	US (6 M HCl)	*I. As(III):* S: 2% HCl, R: 1.2% $NaBH_4$ *II. As(III)+As(V):* S: 15% HNO_3, R: 1.2% $NaBH_4$ *III. As(III)+DMA:* S: 10% HNO_3, R: 1.2% $NaBH_4$ S: 2% HNO_3, R: 1.2% $NaBH_4$	HG-AFS	As(III): 3.1 As(V): 5.7 DMA: 3.8	[75]
drinking water	As(III) As(V) DMA MMA i-As o-As	direct analysis	*I. As(III):* S: citrate buffer (pH 5.2), R: 1.0% $NaBH_4$ *II. As(III)+DMA:* S: acetate buffer (pH 4.5), R: 1.0% $NaBH_4$ *III. As(III)+As(V)+MMA:* S: KI-ascorbic acid-3 M HCl, A: 10 M HCl, R: 1.0% $NaBH_4$ *IV. DMA+MMA:* S: L-cysteine, A: 2 M HCl, R: 1.0% $NaBH_4$ *V. As(III)+As(V)+DMA+MMA:* S: L-cysteine, A: 10 M HCl, R: 1.0% $NaBH_4$	HG-ICP-OES		[71]

Table 4. Cont.

Matrix	Species	Sample Preparation [a]	Speciation Procedures	Detection	LOD [b], μg L⁻¹	Ref.
soil and sediment CRMs	As(III) As(V) DMA MMA	WB (0.05 M EDTA)	I. *As(III)+DMA+MMA:* S: 0.02 M EDTA (pH 5-6), II. *As(III)+As(V)+DMA+MMA:* S: 0.02 M EDTA (pH 5-6) + KI-ascorbic acid/HCl followed by neutralization to pH 5-7 A: 1.2 M HCl, R: 0.6% NaBH₄ I. *i-As (As(III)+As(V)):*	HG-AAS	0.2 mg kg⁻¹ [c]	[77]
ground water	i-As DMA	direct analysis	S: KI-ascorbic acid–2 M HCl, A: 3 M HCl, R: 2.2% NaBH₄ II. *As(III)+As(V)+DMA (UV photo-oxidation):* S: K₂S₂O₈, A: 3 M HCl, R: 2.2% NaBH₄ I. *As(III):*	HG-AFS	As(III,V): 0.09 DMA: 0.47	[13]
water, soil, sediments	As(III) As(V) DMA	direct analysis/US	S: not acidified, A: 2% HCl, R: 0.5% NaBH₄ II. *As(III)+DMA (UV photo-reduction):* S: ZnO–8% formic acid, 2% HCl, R: 0.5% NaBH₄ III. *As(III)+As(V)+DMA (UV photo-reduction):* S: ZnO–8% formic acid, 10% HCl, R: 0.5% NaBH₄	HG-AFS	As(III): 3.20 As(V): 3.86 DMA: 6.68	[76]

HG-AAS: hydride generation atomic absorption spectrometry. HG-AFS: hydride generation atomic fluorescence spectrometry. HG-ICP-OES: hydride generation inductively-coupled plasma optical emission spectrometry. HG-ICP-MS: hydride generation inductively-coupled plasma mass spectrometry. i-As: the inorganic tri- and pentavalent As species (As(III) and As(V)). M: mol L⁻¹. MAE: microwave-assisted extraction. o-As: the organic, i.e., methylated pentavalent As species (DMA and MMA). MeOH: methanol. S: sample solution. A: additional acid (carrier) solution. R: reductant solution. US: ultrasonication in an ultrasound water bath. WB: water batch. ᵃ Detailed information about sample preparation procedures described in point 4. ᵇ Pentavalent As species as As(III) without or after previous pre-reduction step. ᶜ In the original sample taking into account the amount of sample and the final dilution employed in the proposed procedure.

As shown in Table 4, to selectively generate hydrides for each As species in the reaction with $NaBH_4$, the reaction medium—i.e., the type of acid and its acidity, including inorganic (HCl [3,7,13,28, 39,51,52,71,75–77], HNO_3 [75]) and carboxylic acids (e.g., tartaric [3,77], acetic [28,71,81], citric [40,71], formic [76], oxalic [77], malic [77]) or buffers (citrate at pH 5 [28,40,71,81]), oxalate at pH 4.5 [7], acetate at pH 4.5 [71]), the concentration of $NaBH_4$ [13,28,39,51,52,71], and the kind of the pre-reducing agent employed at the step of the sample pre-treatment (mostly KI/KI-ascorbic acid [3,7,28,39,51,52,71,77,81], thiourea-ascorbic acid [40,71] and L-cysteine [7,28,40,71,81]—was typically controlled.

Unfortunately, in contrast to the speciation of inorganic As only, one uniform strategy when all four forms are speciated is difficult to establish. However, some selective reaction media for speciation purposes have been recommended. As shown in [28,71,81], which focus on As speciation in wine and fish by HG-AAS [28,81] or drinking water by HG-ICP-OES [71], the determination of only As(III) could be achieved in the presence of a citrate buffer (pH 5), while the presence of a low concentration of acetic acid [28,81] or an acetate buffer (pH 4.5) [71] ensured the generation of arsines for As(III) and DMA. The sum of i-As (As(III) and As(V)) alone [28,81] or together with MMA (i-As+MMA) [71] could be determined after the pre-reduction of V-state As to As(III) with KI/KI-ascorbic and 7–10 mol L^{-1} HCl for the HG reaction. Interestingly, the sum of o-As species, i.e., (DMA+MMA), could be selectively determined after their pre-reduction with L-cysteine, followed by HG in 2 mol L^{-1} HCl [71]. Finally, L-cysteine, used for the pre-reduction of As(V), DMA and MMA to As(III), enabled us to determine total As. Nevertheless, quantitative pre-reduction was reached under completely different HG reaction conditions in terms of the HCl concentration, i.e., 0.01–0.05 mol L^{-1} [28,81] and 10 mol L^{-1} [71]. Otherwise, all four As forms could be speciated by HG-AAS in natural waters (sea, underground, drinking) by conducting the reduction reaction at a fixed $NaBH_4$ concentration (0.6%), using different reaction media (HCl, acetic and tartaric acids) and the pre-reduction step with KI [3].

In contrast to a popular speciation method by subtraction, it seems that the strategy with HG reaction conditions under which hydrides of As species are generated with different efficiencies may be much easier to establish, as evidenced in several works cited here. In three of them, the determination of all four As forms (As(III,V), DMA and MMA) in fish [52], cereals (rice, wheat semolina) [39] and vegetables (chard, aubergine) [51] by HG-AFS was carried out using a series of independent proportional equations corresponding to four different reduction conditions (I-IV) based on various HCl and $NaBH_4$ concentrations, i.e., 2–4 mol L^{-1} HCl and 1.2–1.4% $NaBH_4$. An additional pre-treatment with KI-ascorbic acid was applied in one case (condition IV) to reduce As(V) and MMA to As(III). Importantly, non-reducible (non-hydride reactive) As species during HG, such as non-toxic AsB, remained unchanged [51,52]. In the same way, i.e., with the aforementioned linear equation approach, all four As forms were speciated and determined in rice by HG-AFS [40]. It was possible to find four different sample pre-treatment procedures to selectively generate As hydrides in the same HG reaction conditions (1.6 mol L^{-1} HCl-citrate buffer (pH 4.8)-KBH_4). Several common pre-reducing (KI, L-cysteine, thiourea, ascorbic acid) and preoxidizing (H_2O_2, $KMnO_4$, $K_2S_2O_8$) agents that would make it possible to pre-reduce As(V), DMA and MMA to As(III) or oxidize As(III) to As(V) in the sample solution acidified to 0.06 mol L^{-1} citric acid were tested. As a result, in the presence of a low concentration of citric acid alone, a maximum signal for As(III)+DMA was provided (condition I); $K_2S_2O_8$ completely oxidized As(III), but it did not degrade DMA, and hence, allowed for selective HG for DMA, importantly, without any interference effect from As(V) and MMA (condition II); the use of L-cysteine-ascorbic acid made it possible to determine the sum of As(III), As(V) and DMA, without any contribution of MMA to the overall signal (condition III); finally, the determination of all As species by HG was most effective in the presence of HCl and thiourea-ascorbic acid for pre-reduction (condition IV). It is noteworthy that the results for L-cysteine were similar to those presented in work [70], and proved the pre-reducing potential of this reagent in a citric acid medium (pH 4.5–4.8).

An interesting approach to the determination of the sum of As(III)+DMA+MMA and the sum of all toxicologically relevant hydride-active As species (As(III)+As(V)+DMA+MMA) in EDTA extracts of soil and sediment samples by HG-AAS was evaluated in [77]. An additional pre-treatment with

KI-ascorbic acid was applied to reduce As(V) to As(III), and then the As(V) concentration was calculated by the appropriate difference. Moreover, the influence of various types of carboxylic acids, their amino- and hydroxo-derivatives and monosaccharides on the efficiency of the HG process was investigated. Observations showed that EDTA, ascorbic acid, glucose and fructose leveled and equalized the responses of As(III), DMA and MMA at pH 5–7, and furthermore, that ascorbic acid, glucose and fructose maintained their leveling effect at pH 1.3–2 (0.01–0.05 mol L^{-1} HCl).

In two other works [13,76], sample pre-treatment with UV irradiation resulted in photo-oxidation or photo-reduction processes that promoted the inter-conversion of As species prior to HG. Chaparro et al. [13] proposed an approach for the selective determination of total i-As and DMA in ground water samples using an automated HG-AFS system. In the first procedure, total As (the sum of As(III), As(V) and DMA) was determined after UV irradiation of the sample in a $K_2S_2O_8$ medium and after photo-oxidation of all As species to inorganic As(V). In the second one, total i-As was measured directly (i.e., the photo-oxidation step was omitted) after the previous pre-reduction of As(V) to As(III) with KI-ascorbic acid. The DMA concentration was calculated by the difference. Pinheiro et al. [76] reported that all three As forms were speciated and determined by HG-AAS by coupling a photo-reduction system with a HG manifold. The effective photo-reduction of DMA and As(V) to inorganic As(III) was achieved by using UV treatment (catalyzed by ZnO nanoparticles) in a formic acid medium. Various reaction conditions, in terms of the HCl concentration, were used for the selective generation of individual hydrides. With 2% HCl for the HG reaction, As(III) was directly determined, while As(III)+DMA could be determined after the photo-reduction step. By applying 10% HCl and the same photo-reduction process, As(III)+As(V)+DMA were quantified. The proposed strategies were suitable for As speciation in environmental samples such as water and soil or sediments (after US-assisted extraction).

However, Pinheiro et al. [75] and Akter et al. [7] demonstrated that As(III), As(V) and DMA could be also selectively determined without UV irradiation of samples. Accordingly, the selective generation of As(III), As(V) and DMA hydrides was possible using 1.2% NaBH$_4$ and changing the acid type and its concentration [75]. It is noteworthy that no additional pre-reducing step was required. For each species, different experimental reduction conditions, based on different HCl and HNO$_3$ concentrations, were used. Four speciation procedures were developed and their reliability was verified. Employing various HCl concentrations, i.e., 2% and 10%, As(III) and i-As (As(III)+As(V)) were determined, respectively. The content of As(V) was obtained by calculating the difference. Employing various HNO$_3$ concentrations, i.e., 15% and 2%, As(III) and As(III)+DMA, were determined, respectively. Similarly to As(V), by the difference in responses found for 2 and 15% HNO$_3$, the exact content of DMA was calculated. The proposed speciation approach was applicable for As speciation in sediment and plant samples by HG-AFS. Akter et al. [7] used a well-known procedure for i-As speciation based on selective As(III) determination at pH 4.5 (oxalate buffer) and total i-As in the HCl medium (6 mol L^{-1}) after pre-reduction of As(V) to As(III) with KI. Otherwise, the reaction medium of 1.5 mol L^{-1} HCl and the sample pre-treatment with L-cysteine allowed made it possible to determine the DMA alone. The same NaBH$_4$ concentration (0.6%) was used in all three procedures. The developed methodology was applied for As speciation in ground water samples by HG-AAS.

5. Pre-Concentration and/or Separation

The concentrations of As in environmental, food and biological samples are usually very low (see Table 1). The HG technique, coupled with atomic spectrometric detectors, is a sensitive analytical tool for the determination of traces of As. However, when handling ultratrace amounts of As, additionally in complex sample matrices, the direct determination of As species is difficult, and therefore, preliminary pre-treatment comprising the separation and/or pre-concentration of these species is highly desirable. Accordingly, in the sample preparation step, one or more As species (typically inorganic) can be separated through different extraction techniques. Early evolved arsines, both inorganic and methylated, can also be isolated according to their boiling points with the cryo-trapping (CT) technique or by the pervaporation-based membrane separation technique. Both approaches provide an excellent improvement in LODs of As species to $\leq$ ng L^{-1} levels for the same detector and improve the selectivity of measurements due to the alleviation of interference from sample matrix components.

5.1. Pre-Concentration and Separation Methods before HG Process

Selective complexation–extraction techniques, selective retention on solid adsorbents, i.e., solid phase extraction (SPE), and selective co-precipitations are the most common methods to separate/pre-concentrate As species, mainly inorganic forms, used in the sample preparation stage. Accordingly, As(III) or As(V) can be selectively extracted or co-precipitated with subsequent HG and detection. Total i-As is obtained after the initial conversion of As(III)$\leftrightarrow$As(V), while the content of As(III) or As(V) can then be calculated by the difference. Non-chromatographic schemes for As speciation with the most popular pre-concentration/separation methods before HG, along with the achieved analytical performance, are summarized in Table S1.

5.1.1. Selective Complexation–Extraction

Among the various extraction techniques that could easily be adapted as the initial step in species-selective and -sensitive As determination combined with HG and spectrometric detection, cloud point extraction (CE), dispersive liquid-liquid microextraction (DLLME) and liquid–liquid extraction of aqueous two-phase systems (ATPS) are very popular.

Cloud point extraction, based on nonionic surfactants used as extracting solvents, offers many advantages like simplicity, safety, low cost and high pre-concentration factors. Different extraction procedures, making it possible to distinguish between i-As species, were proposed in methods developed to monitor As(III) and As(V) in natural water samples (drinking, tap, lake). For example, in one work [18], a selective complex of As(III) with Pyronine B in the presence of sodium dodecyl sulfate (SDS) at pH 10 was extracted using a nonionic surfactant, Triton X-114. The surfactant-rich phase with As(III) was then separated and diluted with 1 mol L^{-1} HCl prior to its determination by HG-AAS. Total i-As (As(III,V)) was extracted similarly after the pre-reduction of As(V) to As(III) with $Na_2S_2O_3$, and the As(V) content was calculated by the difference. In another work [23], As(III) and As(V) were separated by complexing with ammonium pyrrolidinedithiocarbamate (APDC) (at pH 4.6) and molybdate (at pH 2.4), respectively, followed by quantitative extraction with Triton X-114. Afterwards, the As(III) content was determined by HG-AAS after diluting the surfactant-rich phase with 5% HCl. In the case of As(V), the resulting As(V) complex was first converted to free As(V) by ultrasonication, pre-reduced to As(III) with a thiourea-ascorbic acid mixture, and finally, determined by HG-AAS.

Dispersive liquid–liquid microextraction aims to extract the analytes from an aqueous phase into an organic phase, in which a third solvent is rapidly injected to accelerate efficient dispersion. Compared to traditional LE, it presents advantages in costs, labor, solvent consumption and enrichment factors. Chen et al. [30] applied the DLLME approach to speciate and quantify As(III) and As(V) in fruit juices in the presence of o-As (DMA and MMA) by HG-AFS using APDC (complexing agent), methanol (dispersant) and CCl_4 (extractant). Samples were adjusted to pH 3 and then mixed with APDC to form the As(III)-APDC complex, followed by injection of methanol and CCl_4 to form a dispersion. After centrifugation, the organic phase with the As(III)-APDC complex was evaporated to dryness, and then the residue was dissolved in 1 mol L^{-1} HCl and subjected to analysis by HG-AFS. Total i-As was determined after the pre-reduction of As(V) to As(III) using $Na_2S_2O_3$; next, the same protocol as for As(III) was used. Finally, the As(V) content was calculated from the difference. Under the selected pre-reduction and HG reaction conditions, limitations of the method for DMA were found to be advantageous to selectively measure i-As. In contrast, MMA contributed to the As response for i-As. However, the degree of this interference was pH-dependent; at pH 1.7–1.8, the error was <10%.

A recent alternative to LE extraction is extraction by the aqueous two-phase system (ATPS). Its advantages include simplicity, low cost and high enrichment factors. These aqueous two-phase systems are primarily composed of water and other compounds of low toxicity. They can be formed by two heterogeneous phases composed of aqueous solutions of two incompatible polymers, i.e., a polymer and an electrolyte, or two types of physically incompatible electrolytes. Assis et al. [84] studied ATPS extraction comprising a polymer and an electrolyte prior to speciation of i-As forms in tap waters by HG-ICP-OES. The authors developed a procedure for the selective extraction of As(III) coexisting with As(V) using the ATPS composed of L64 (copolymers), water and Na_2SO_4 at pH 6.0, while APDC was used to extract As(III); As(V) was poorly extracted in these conditions (<18%).

5.1.2. Selective Retention—Solid Phase Extraction

Solid phase extraction is the most popular and commonly applied separation/pre-concentration technique. In SPE, analytes are extracted by sorption, eluted with a small amount of a solvent and then directly detected. Speciation analysis is realized by selective sorption or selective elution. SPE can fulfill the separation/pre-concentration requirements of As species before HG and spectrometric detection using a variety of sorbents. These could be conventional substances such as resins or gels (whose analytical capabilities could be further modified by surface modifications (functionalization) to improve the absorption performance), as well as alternative and novel ones like nanometer-sized materials or biosorbents with exceptional properties to effectively separate specific As species. Moreover, the method can easily be combined with different detection techniques in online [9–11,25,26,37,55] or offline modes [4,8,14,16,21,41,43,79].

The selection of appropriate sorbents is a key parameter in SPE. The development of alternative/novel sorbents is a general trend in pre-concentration and separation procedures for As speciation, whereas methodologies aimed at separating arsenate from arsenite are more widespread. Deng et al. [14] developed a simple and rapid method for the determination of trace amounts of total i-As in environmental water samples (ground, river, lake) by SPE on an aluminum hydroxide gel and HG-AFS detection. Trivalent arsenic was first oxidized to As(V) by $KMnO_4$; then, the sample solution was adjusted to pH 6, followed by the addition of a freshly prepared gel for extraction of As(V). After centrifugation, the resultant precipitate with adsorbed i-As was dissolved in concentrated HCl; then, As(V) was pre-reduced with a thiourea-ascorbic acid mixture, and finally, total i-As was determined as As(III) by HG-AFS. In two other works, SPE columns packed with cigarette filters [9] or PTFE particles [25] were used prior to HG and determination of the i-As content in various water samples (including tap, ground and seawater). The developed methods were based on selective online formation of the As(III)-APDC complex and its retention on SPE columns. As(V) did not form any complexes with APDC, and could not be retained in these conditions; hence, it passed through the columns. After reducing As(V) to As(III) with L-cysteine [9] or thiourea [25], the same system was applied to determine the total i-As, and As(V) was calculated by the difference. The adsorbed As(III)-APDC complex was online removed from SPE columns using HCl (1.7-2.0 mol L^{-1}) and merged with KBH_4 (2.1%) [9] or $NaBH_4$ (4.1%) [25] solutions to generate AsH_3 before entering AFS [9] or AAS [25] spectrometers for As(III) detection. Importantly, diluted HCl was used as the eluent because it also provided a favorable medium for the HG reaction.

In a few recent studies, graphene or carbon nanotubes have been used as sorbents for SPE. Considering As speciation, Khaligh et al. [4] presented an interesting approach to the speciation of i-As using nonporous graphene functionalized with carboxyl groups (G-COOH) for the ultrasound assisted, dispersive, micro-solid phase extraction (US-D-μ-SPE) of As(V) from several natural water (tap, drinking, river, waste) and biological (human serum/urine) samples prior to its determination by HG-AAS. Briefly, As(V) was selectively retained on the G-COOH sorbent at pH 3.5 by US-D-μ-SPE with the next separation of the solid phase being achieved through centrifugation. Then, the As(V) retained on the sorbent was eluted with NaOH (0.3–0.5 mol L^{-1}), pre-reduced to As(III) with a KI-ascorbic acid mixture, and determined by HG-AAS. The previous oxidation of As(III) using $KMnO_4$ made it possible to determine the total i-As. The difference between total i-As and As(V) yielded the As(III) content in the analyzed samples. The application of the carbon nanotube (CNT) sorbents for the determination of As(III) and As(V) at (ultra)trace levels in various environmental water samples (rain, snow, sea, river) is demonstrated in [10,26]. Wu et al. [10] packed a micropipette with single-walled (SW) CNTs to achieve the selective adsorption of the As(III)-APDC complex. The proposed speciation scheme involved the online formation and retention of the As(III)-APDC complex at pH 3 on a SWCNTs-packed micro-column, followed by its online elution with 20% HNO_3 and determination of As(III) by HG-AFS using a sequential flow injection manifold. Total i-As was determined by the same protocol after pre-reduction of As(V) to As(III) with thiourea; As(V) was calculated by the difference. In contrast, to effectively improve the CNT material performance for a favorable selective adsorption of As(V) in the presence of As(III), Chen et al. [26] employed multi-walled (MW) CNTs functionalized with branched cationic polyethyleneimine (BPEI) that were packed into a mini-column for online SPE of As(V) in a sequential injection system following HG-AFS detection. Adsorption of As(V) was carried out at pH 5.8, and the analyte was eluted with 0.6% NH_4HCO_3. By following the same procedure, total i-As was determined after the oxidation of As(III) to As(V) with H_2O_2.

Other nanosized adsorbents for the separation and pre-concentration of As(III,V) species from natural water samples (tap, sea, ground, underground) were also found to be useful. For example, Erdogan et al. [16] synthesized nano-zirconium dioxide-boron oxide (ZrO_2/B_2O_3), called a "hybrid sorbent", and employed it for the selective sorption of As(V) by the SPE column technique prior to its determination by HG-AAS. In this SPE procedure, As(V) ions, retained at pH 3.0, were eluted with 3 mol L^{-1} HCl and then pre-reduced to As(III) with a KI-ascorbic acid mixture and determined using HG-AAS. For total i-As, As(III) was oxidized firstly to As(V) by $KMnO_4$ prior to SPE and then determined by HG-AAS; As(III) was calculated by the difference. To enhance the adsorption efficiency, Montoro Leal et al. [11] proposed magnetic nanoparticles (MNPs), i.e., ferrite (Fe_3O_4), which were further functionalized with [1,5-bis (2-pyridyl) 3-sulfonophenylmethylene] thiocarbonohydrazine (PSTH-MNPs) and applied for the speciation of i-As by HG-ICP-MS. This procedure was based on the retention of As(III) and As(V) at pH 4 in two knotted reactors filled with PSTH-MNPs, followed by the sequential elution of As(III) and total i-As in 7% HNO_3-0.1% thiourea-2.8% L-cysteine medium before HG using different $NaBH_4$ concentration, i.e., 0.1 for As(III) and 0.5% for total i-As and measurement by ICP-MS. The concentration of As(V) was obtained by subtracting As(III) from total i-As.

Finally, strategies involving the use of biomaterials for SPE, including baker's yeasts (*Saccharomyces cerevisiae*) [8,37] or egg-shell membranes [21], were also proposed to speciate As species. Smichowski et al. [8] proposed a simple and sensitive method for the biosorption and pre-concentration of As(III) in the presence of As(V) in aqueous solutions using a batch system. A sample solution was combined with yeasts and 0.1 mol L^{-1} oxalic acid (acting as a reaction medium), adjusted to pH 7 and then placed in a water bath (60 °C, 30 min) to extract the As(III) form. After centrifugation, the solid phase was re-suspended in 4.0 mol L^{-1} HCl to form a slurry, the liquid phase (supernatant) was acidified to 3.5 mol L^{-1} HCl, and As(III) and As(V) were determined correspondingly in both phases by HG-ICP-AES. To overcome possible matrix effects, the method of the standard addition was used for the determination of As(III) in suspension. Under selected conditions, As(III) was almost completely (~97%) retained by the biomass, likely bounded through –SH groups of yeast proteins, while As(V) remained in the supernatant. This made it possible to determine both As species in separate phases. Several different ground water samples were analyzed following the proposed method. In contrast, Koh et al. [37] showed that As(V) was retained better than As(III) in a yeast-immobilized column. In the cited work [37], *S. cerevisiae* was covalently bound onto controlled pore glass (CPG), packed into the column and used to selectively pre-concentrate As(V) over As(III). As a result, a simple flow injection system using the yeast-immobilized column coupled online with HG and ICP-AES for sensitive determinations of As(III) and As(V) was proposed. The manifold consisted of the SPE column and a manual injector. While the CPG-yeast column (pH 7) was loaded with the sample solution, As(III) was passed through the column, and hence, could be determined by HG-ICP-OES. Moving the injector to an alternative position, elution with 3 mol L^{-1} HNO_3 solution took place, releasing the As(V) retained on the column. The proposed method was applied for the determination of As species in herbicide, pesticide and cigarette samples. More recently, Zhang et al. [21] used a natural egg-shell membrane (ESM) as a sorptive material for SPE combined with HG-AFS to separate and determine As(V) in environmental water samples. The retention of As(V) on the ESM surface was via anion-exchange due to the presence of positively charged functional groups such as $-NH_3^+$ and $-CO-NH_2^+$. The ESM was obtained from fresh eggs and packed into a cartridge (1 g, 6 mL) by replacing its original C18 packing material. Sample solutions were adjusted to pH 11, loaded on the ESM column, and then the cartridge was washed out with water and dried. The retained As(V) was eluted with 2 mol L^{-1} HNO_3 and measured by HG-AFS.

Regarding i-As, the application of SPE not only results in enhanced sensitivity due to the pre-concentration of the analytes; the separation of i-As also plays a crucial role in minimizing interference coming from sample matrix compounds, further improving the selectivity of i-As measurements by HG. In this way, negative effects from hydride-active methylated As species (DMA and MMA) on activity of As(III,V) under reducing conditions are overcome. Successful applications of SPE include the selective determination of total i-As using anionic exchange or nonpolar resins [41–43,55,79] as sorbents. In three cited works [41,42,79], silica-based strong anion exchange (SAX) cartridges were used for the determination of i-As in rice [42], various rice products [41] and seafood samples [79]. They were used for offline SPE separation of i-As from DMA and MMA, followed by HG and spectrometric detection of the sum of As(III) and As(V). Species such as As(III), As(V), DMA and MMA were extracted with dilute HNO_3-H_2O_2 solutions (0.06–0.1 mol L^{-1}, 1–3%) to solubilize them and oxidize all i-As to As(V). Next, sample extracts were buffered (pH 5–7.5) and loaded onto SPE cartridges. Organic As species were washed out using an acetic acid solution (0.1–0.5 mol L^{-1}), and the retained As(V) ions were back-extracted (eluted) with HCl (0.5 mol L^{-1}) [42,79] or HNO_3 (0.4 mol L^{-1}) solutions [41]. Finally, As(V) was pre-reduced with a KI-ascorbic acid mixture and total i-As was measured in SPE eluates as As(III) by HG-AFS [42] or HG-AAS [41,79]. It is worth mentioning that the successful separation of i-As (as As(V)), via this SAX sorbent, was achieved through pH adjustment based on dissociation constants. Unfortunately, these methods were useless when the target analyte and matrix species had similar dissociation constants. To solve these problems, a novel method for separating i-As from other species by SPE was developed, including the chemical conversion of polar i-As to nonpolar compounds, such as $AsCl_3$, which can be retained by a nonpolar resin. Accordingly, Huang et al. [43] employed a polystyrene (PS) resin to retain i-As from the matrix of rice as $AsCl_3$. Arsenic species were extracted with HNO_3 (0.02 mol L^{-1}), and then sample extracts were acidified with HCl to 10 mol L^{-1}, treated with thiourea to reduce As(V) into to As(III) and loaded on SPE cartridges. Then, cartridges were rinsed with HCl (10 mol L^{-1}), and the retained $AsCl_3$ was eluted by hydrolysis with water. Finally, HG-AFS was applied to quantify the concentration of total i-As as As(III) in SPE eluates. Similarly, Zhang et al. [55] reported online SPE using PS resin cartridges coupled with HG-AFS for the determination of i-As in a complicated and arsenosugar-rich algae matrix. However, to fully overcome any matrix interferences and improve the retention efficiency of i-As, Br^- ions were found to be more advantageous than Cl^- ones for As(III) halogenation. In the proposed procedure, As species were initially extracted with $HClO_4$ (1%) by a heat-vortex technique (80 °C, 20 min). Then, thiourea, KBr and HCl were added. In the presence of these reagents, As(V) was first reduced to As(III), and then i-As (as As(III)) was converted into $AsBr_3$, while the role of HCl was to maintain an acidic environment for the production of $AsBr_3$ retained on the cartridge. The retained $AsBr_3$ was eluted from the sorbent with water and total i-As was measured as As(III) by HG-AFS.

Otherwise, the separation o-As from i-As by SPE can be also applied only as a sample pre-treatment before analysis [23]. In this scenario, the examined water samples were passed through a glass column filled with small, activated Al_2O_3 to remove o-As. Adsorbed i-As species were desorbed by 0.2 mol L^{-1} HCl, made up with water to a proper volume before subsequent analysis.

5.1.3. Selective Co-Precipitation

Arsenic species can also be selectively co-precipitated for separation. Van Elteren et al. [5] proposed a method for the determination of As(III) and As(V) in bottled mineral waters at ultratrace levels by HG-AFS coupled with a flow-injection system based on selective co-precipitation of As(III) with dibenzyldithiocarbamate (DBDTC) at low pH (pH 2). The As-DBDTC precipitate was dissolved in 0.01 mol L^{-1} NaOH and 30% H_2O_2, and then As(III) was determined. The pre-reduction of As(V) to As(III) with a KI-$K_2S_2O_7$ mixture was carried out before co-precipitation; this allowed the authors to determine total i-As and As(V), calculated by the difference.

5.2. Pre-Concentration and Separation Methods after HG Process

Another category of pre-concentration/separation methods is the trapping of As species hydrides in a cryogenic trap (CT) before their detection. This system (HG-CT) is a convenient approach to speciation of all hydride-active As species, i.e., inorganic and methylsubstituted, due to its pre-concentration and separation ability. Generally, non-selectively formed hydrides for various As species are first cryofocussed/trapped at liquid-N_2 temperature in a U-shaped tube (acting as a separator of different hydrides), then sequentially released from the trap by its heating according to given boiling points (BP), i.e., AsH_3 (−55 °C, derived from As(III,V)), CH_3AsH_2 (2 °C, derived from MMA) and $(CH_3)_2AsH$ (35.6 °C, derived from DMA), and finally transported to atomic spectrometers. To separate arsines in the trap, various types of packings were used; however, chromatographic materials are typically applied [24,61].

The main advantage of the HG-CT system is that analyses can be performed directly or with only a minimum sample pre-treatment, minimizing the risk of species inter-conversion. In view of this, an interesting approach to the determination of i-As, DMA and MMA in baby food (porridge powders and baby meals) by HG-CT-AAS, based on a slurry sampling using 3 mol L^{-1} HCl, was proposed [48]. Moreover, complex biological matrices such as human urine, cells, tissue or blood were simply lysed and/or diluted with water before analysis [31,60–63]. These minimally pretreated samples were then directly introduced into the HG device in the form of suspensions.

A remarkable quality of the HG-CT technique for As speciation is its ability to determine methylated three- and penta- valent As species in addition to inorganic ones. Here, the differentiation between inorganic and methylated As(III)- and As(V)-species in the reaction with $NaBH_4$ is based on pH-dependent selective HG of arsines of As species of respective valences. Alternatively, it can be HG in the presence or absence of pre-reduction agents (usually L-cysteine), making it possible to overcome the different sensitivities of individual As species, as observed with the pH-specific HG approach [60]. To overcome the problems associated with the low and narrow range of HCl concentrations required for HG reactions in the presence of L-cysteine, the use of buffered media, i.e., Tris-HCl, was proposed. Furthermore, this approach also ensured a selective HG for the trivalent As species without its pre-reduction. This selective HG-CT approach was successfully combined with AAS and applied for analyses of different biological materials. Matousek et al. [60] and Hernandez-Zavala et al. [61] proposed oxidation state-specific speciation of inorganic and methylated arsenicals in complex biological matrices (e.g., cell cultures or tissue homogenates) by an automated HG-CT-AAS system equipped with a multiatomizer. Arsines for As(III,V), DMA(III,V) and MMA(III,V) were pre-concentrated and separated in a cryogenic chromatographic trap. To differentiate between As(III)- and As(V)-containing arsenicals, arsines for As(III)-species were generated in the presence of a Tris-HCl buffer (pH 6). Under the same conditions, a sample pre-treatment with L-cysteine allowed the authors to generate arsines for both As(III)- and As(V)-species. Finally, the content of As(V)-species was calculated as the difference. LODs of arsenicals ranged between 0.016–0.040 µg L^{-1} [61]. The proposed method was further improved by the online pre-reduction step integrated with the HG-CT system using TGA for pre-reduction [62]. The applicability of this method was also demonstrated in the case of As speciation in human urine samples. Replacing AAS with more sensitive AFS or ICP-MS detectors, and employing the identical HG-CT system, LODs of As species were significantly lowered to low ng L^{-1} and sub ng L^{-1} levels. This permitted analyses to be performed of limited-size samples (e.g., tissue) or samples with an extremely low As content. Accordingly, Musil et al. [63] carried out speciation of inorganic and methyl-substituted arsenicals. i.e., As(III), i-As, DMA and MMA, in exfoliated bladder epithelial cells isolated from human urine by selective HG-CT-AFS with extremely low LODs of these As species, i.e., 0.00044, 0.00074, 0.00015 and 0.00017 μg L^{-1}, respectively. In another work [31], the HG-CT-ICP-MS-based method was shown to be suitable for direct As speciation in whole blood and blood plasma at low exposure levels with LODs of 0 μg L^{-1} (i-As), 0.002 μg L^{-1} (MMA) and 0.001 μ L^{-1} (DMA). For comparison, these LODs were about an order of magnitude lower than those achievable with the HG-CT-AAS system that was used for the analysis of the same samples, i.e., 0.15 μg L^{-1} (i-As), 0.09 μg L^{-1} (MMA) and 0.07 μg L^{-1} (DMA).

Although a typical HG-CT system is equipped with a U-shaped tube, improvements to this trap have also been reported. For example, Hsiung and Wang [24] proposed a novel packed cold finger trap (PCFT) packed with a chromatographic material for the determination of As(III,V), DMA and MMA in fresh water and seawater by HG-AAS with a flame-heated, quartz-tube atomizer (QTA) under species-selective HG conditions. The advantage of this PCFT module over a typical U-shaped cryogenic trap lay in its better separation of collected arsines prior to their detection. In the proposed speciation scheme, As(III) was determined using a citrate buffer (pH 6.4), and As(V) was assessed by subtracting the As(III) content from the total i-As content determined after pre-reduction of As(V) to As(III) with KI, while for DMA and MMA determination, compromised levels of chemical and instrumental parameters were selected. These parameters refer to the HCl-NaBH$_4$ concentrations for effective HG reactions, the flow rate of the carrier gas (He) and the heating voltage of the PCFT device for reasonable separation between both methylated species. The achieved LODs were 0.047, 0.042, 0.0045 and 0.0063 μg L^{-1} for As(III), As(V), MMA and DMA, respectively. More recently, Maratta et al. [19] developed a novel methodology for As speciation in ground and cistern water samples based on selective HG and CT on a CNTs-packed column, followed by elution of adsorbed arsines with 5% HNO$_3$ and their quantification by ETAAS. This speciation strategy involved the selective determination of As(III) and As(V), in addition to either inorganic (hydride-active) and organic (non-hydride-active) As fractions. As(III) was determined selectively using a citric buffer (pH 4), and As(III)+As(V) was determined after the pre-reduction of As(V) with L-cysteine and thiourea. In the case of the As organic fraction, it was decomposed into hydride-active species by UV photo-oxidation catalyzed by TiO$_2$. This method yielded an enrichment factor of 60 and the LOD of As of 0.00078 μg L^{-1}.

Interestingly, by combining two complementary techniques, i.e., HG-AAS and HG-CT-AAS, a complete experimental speciation protocol for the determination of As(III), As(V), MMA, DMA and non-hydride reactive As species at ng L^{-1} levels in seawater was proposed [68]. In the first step, As was pre-concentrated by collecting the hydrides of As species into a graphite furnace before AAS detection. By selecting different HG reaction conditions, it was possible to determine As(III) alone (Tris-HCl, pH7), total hydride reactive As species (after alkaline persulfate digestion), total As (after alkaline persulfate digestion submitted to an oxidative UV irradiation treatment of non-hydride reactive As species), and by the difference, non-hydride reactive As species. In the second step, hydrides of reactive As species were cryogenically trapped on a chromatographic column, followed by their sequential release and AAS determination in a heated quartz furnace. As a consequence, this system made it possible to separate and determine i-As, DMA and MMA.

Finally, hydrides of As species can be per-evaporated (P), i.e., separated from a sample matrix by simultaneous evaporation and gas diffusion through a membrane in a single step before performing measurements. In this manifold, the reactant mixture is transported to the lower chamber of a pervaporation unit from which the generated arsines evaporate first to the air gap between the liquid and the membrane. Then, they diffuse through the membrane into an acceptor stream (located in the acceptor chamber of this pervaporation module) for subsequent detection. Pervaporation appeared to be an excellent technique for the determination of inorganic As species in complex aqueous samples, known as "dirty", i.e., matrices containing an elevated content of organic matter or solid particles in suspension. Moreover, it allows direct analyses to be made of this kind of sample, as no prior pre-treatment such as filtration is required. Caballo-Lopez and de Castro [82] developed the HG-P-AFS method with the LOD of As of 0.42 μg L^{-1} to determine As(III) and As(V) in various "dirty" samples with suspended particulate matter. Selective HG was carried out by varying the pH at which arsines were generated, i.e., pH 1.3 for As(III) and 0.2 for total i-As. Arsenic hydrides were generated in the reaction with 0.5% $NaBH_4$ and 6 mol L^{-1} HCl in the absence of the pre-reduction step for total i-As. The concentration of As(V) was calculated as the difference. Importantly, under optimal conditions, the interference caused by DMA and MMA was negligible. In another work [86], the HG-P method coupled with CCD spectrophotometric detection was proposed for the selective determination of As(III) and As(V) in turbid river water samples with a high content of organic carbon. As(III) was selectively determined, generating AsH_3 in the presence of a citrate buffer (pH 4.5). Both As(III) and As(V) were determined when HG was performed under highly acidic conditions (pH < 1). The concentrations of the As species were calculated by simultaneously solving two proportional equations.

6. Alternative HG Technique-Electrolytic Hydride Generation

Undoubtedly, traditional HG using acid-$NaBH_4$ for the species selective generation of As hydrides is the predominant method. Nevertheless, in recent years, interest in alternative techniques like electrolytic hydride generation (EcHG) has grown. The great advantage of the EcHG technique is that formation of hydride eliminates the use of the reducing $NaBH_4$ reagent. In this technique, the reduction of the analyte to its hydride takes place on the cathode surface, and only acidic electrolyte solutions are required to ensure the electric current for the reduction process. Therefore, the efficiency of EcHG depends mainly on the cathode material and electrolytic current. Accordingly, Arbab-Zavar et al. [85] proposed a EcHG spectrophotometric method for the determination of As(III) and As(V) in tap water without pre-reduction of As(V). In the method, a graphite cathode was used to perform the reduction of As(III) to AsH_3, while a Sn/Pb alloy wire cathode was used to reduce As(V) to AsH_3. Next, species were determined at 510 nm as the As(III)-SDDC complex obtained from the reaction between AsH_3 and silver diethyldithiocarbamate (SDDC). The proposed method resulted in LODs of 20 (As(III)) and 60 µg L^{-1} (As(V)). Similarly, Li et al. [78] selectively determined i-As species by EcHG-AAS using a glassy carbon cathode in a 0.06 mol L^{-1} H_2SO_4 catholyte medium. Differentiation between As(III) and As(V) was based on the different HG efficiencies attained by controlling electrolytic currents at 0.6 and 1.0 A, respectively. Concentrations of both species were calculated from slopes of calibration curves with LODs of 0.2 µg L^{-1} for As(III) and 0.5 µg L^{-1} for As(V). The method was successfully applied to speciate soluble i-As in Chinese medicines. Yang et al. [22] reported that the EcHG behavior of As(III), As(V), DMA and MMA could be changed at the modified graphite electrode (GE) by –SH modifiers including L-cysteine (Cys) and glutathione (GSH). The authors proposed a four-step analysis approach for the determination of all four As species in rice and natural water (river, lake, rain) samples by EcHG-AFS. Accordingly, As(III) was reduced on the GSH/GE at an applied current of 0.4 A. As(III), and As(V) generated AsH_3 on the Cys/GE at an applied current of 0.6 A. Measurements of As(III) and DMA were done on Cys/GE at two applied currents. Finally, the total As was determined after the reduction on Cys/GE at an applied current of 2.0 A. Under optimal conditions, LODs ranged between 0.10–0.25 µg L^{-1}. Lu et al. [58] demonstrated the application of a chemically modified carbon paste electrode (CMCPE) in EcHG to detect (ultra)trace amounts of As species. Given this, the authors developed a procedure for the selective determination of As(III) and total As in Chinese herbal medicines by EcHG-AFS. In the procedure, As(III), As(V), MMA and DMA were first selectively reduced to AsH_3 at an applied current of 1.0 A on Cys/CMCPE. Then, under the same conditions, total As was determined after the pre-reduction of As(V), DMA, and MMA to As(III) with L-cysteine. Compared to traditional GE, the CMCPE provided better sensitivity; with CMCPE-AFS, LODs were successfully reduced to 0.095 (As(III)) and 0.087 (total As) µg L^{-1}.

7. Conclusions

In terms of toxicological investigations, knowledge of As species is crucial to understanding the potentially harmful effects associated with exposure to this element. HPLC-ICP-MS is undoubtedly the most popular coupled method for speciation analyses of As, because it provides a complete and sophisticated picture of species eluted from one injection of a sample, quantified at a (ultra)trace level. Nevertheless, as shown in this review, current research' interest is more focused on developing robust and reliable alternative methods obviating chromatographic separation. Non-chromatographic approaches to As speciation, based on the use of simple instrumentation which is available in most of laboratories (like atomic absorption or fluorescence detectors), are simpler, faster and economically friendly. They additionally provide degrees of sensitivity which are of the same order of magnitude or even better (related to possible species separation improving their pre-concentration) than those with hyphenated traditional chromatographic techniques using ICP-MS detection. In this sense, speciation achieved by selective HG prior to spectrometric detection is one of the most effective tools for distinguishing among four major toxic As species, i.e., As(III), As(V), DMA and MMA, or differentiating these toxic (hydride-active) species from non-toxic (non-hydride-active) arsenicals. In this way, this non-chromatographic approach gives sufficient information about As speciation for appropriate risk evaluations. Different physical-chemical properties of hydride-active As species (e.g., their volatility, redox potential) and the possibility of controlling experimental parameters such as pH, reaction medium, reactant concentration, the presence of additives, temperature, etc., make it possible to develop species-selective non-chromatographic strategies for the speciation of trace amounts of different As species in a wide range of environmental, food and biological matrices. Furthermore, HG combined with various pre-concentration/separation approaches to accomplish As speciation serves as an excellent tool for the determination of As species at ultratrace levels. There is a belief that accurate (precise and true) methods for As speciation, which could be used for routine analyses of different samples, both with simple and complex matrices, will continue to develop.

The possibility of effectively speciating As using HG and atomic spectrometry gives rise to a willingness to further improve this technique, focused on reagent consumption, reduction, miniaturization and automation (employing various online flow systems). Simultaneously, as alternatives to chemical HG techniques, electrochemical (Ec)HG, for example, will be extensively studied and developed in the future, in an attempt to augment conventional HG for the determination and speciation of hydride forming elements.

Author Contributions: Paper concept, data analysis and interpretation, writing, M.W.; paper preparation and visualization, A.S.-M., M.W.; discussion, A.S.-M., M.W., P.P.; review and editing, A.S.-M., P.P.; supervision, P.P.; funding acquisition, P.P. All authors have read and agreed to the published version of the manuscript.

Acknowledgments: This work was financed by a statutory activity subsidy obtained by the Faculty of Chemistry, Wroclaw University of Science and Technology from the Polish Ministry of Science and Higher Education.

References

1. Caroli, S.; Torre, F.L.; Petrucci, F.; Violante, N. Arsenic speciation and Health Aspects. In *Element Speciation in Bioinorganic Chemistry*; Caroli, S., Ed.; John Wiley and Sons: New York, NY, USA, 1996; pp. 445–463.

2. Alp, O.; Tosun, G. A rapid on-line non-chromatographic hydride generation atomic fluorescence spectrometry technique for speciation of inorganic arsenic in drinking water. *Food Chem.* **2019**, *290*, 10–15. [CrossRef] [PubMed]

3. Bundaleska, J.M.; Stafilov, T.; Arpadjan, S. Direct analysis of natural waters for arsenic species by hydride generation atomic absorption spectrometry. *Int. J. Environ. Anal. Chem.* **2005**, *85*, 199–207. [CrossRef]

4. Khaligh, A.; Mousavi, H.Z.; Shirkhanloo, H.; Rashidi, A. Speciation and determination of inorganic arsenic species in water and biological samples by ultrasound assisted-dispersive-micro-solid phase extraction on carboxylated nanoporous graphene coupled with flow injection-hydride generation atomic absorption spectrometry. *RSC Adv.* **2015**, *5*, 93347–93359.

5.   van Elteren, J.T.; Stibilj, V.; Slejkovec, Z. Speciation of inorganic arsenic in some bottled Slovene mineral waters using HPLC–HGAFS and selective coprecipitation combined with FI-HGAFS. *Water Res.* **2002**, *36*, 2967–2974. [CrossRef]

6.   Maity, S.; Chakravarty, S.; Thakur, P.; Gupta, K.K.; Bhattacharjee, S.; Roy, B.C. Evaluation and standardisation of a simple HG-AAS method for rapid speciation of As(III) and As(V) in some contaminated groundwater samples of West Bengal, India. *Chemosphere* **2004**, *54*, 1199–1206. [CrossRef]

7.   Akter, K.F.; Chen, Z.; Smith, L.; Davey, D.; Naidu, R. Speciation of arsenic in ground water samples: A comparative study of CE-UV, HG-AAS and LC-ICP-MS. *Talanta* **2005**, *68*, 406–415.

8.   Smichowski, P.; Marrero, J.; Ledesma, A.; Polla, G.; Batistoni, D.A. Speciation of As(III) and As(V) in aqueous solutions using baker's yeast and hydride generation inductively coupled plasma atomic emission spectrometric determination. *J. Anal. At. Spectrom.* **2000**, *15*, 1493–1497. [CrossRef]

9.   Li, N.; Fang, G.; Zhu, H.; Gao, Z.; Wang, S. Determination of As(III) and As(V) in water samples by flow injection online sorption preconcentration coupled to hydride generation atomic fluorescence spectrometry. *Microchim. Acta* **2009**, *165*, 135–141. [CrossRef]

10.   Wu, H.; Wang, X.; Liu, B.; Liu, Y.; Li, S.; Lu, J.; Tian, J.; Zhao, W.; Yang, Z. Simultaneous speciation of inorganic arsenic and antimony in water samples by hydride generation-double channel atomic fluorescence spectrometry with on-line solid-phase extraction using single-walled carbon nanotubes micro-column. *Spectrochim. Acta B* **2011**, *66*, 74–80. [CrossRef]

11.   Montoro Leal, P.; Vereda Alonso, E.; López Guerrero, M.M.; Siles Cordero, M.T.; Cano Pavón, J.N.; García de Torres, A. Speciation analysis of inorganic arsenic by magnetic solid phase extraction on-line with inductively coupled mass spectrometry determination. *Talanta* **2018**, *184*, 251–259. [CrossRef]

12.   Sigrist, M.E.; Beldomencio, H.R. Determination of inorganic arsenic species by flow injection hydride generation atomic absorption spectrometry with variable sodium tetrahydroborate concentrations. *Spectrochim. Acta B* **2004**, *59*, 1041–4045. [CrossRef]

13.   Chaparro, L.L.; Ferrer, L.; Cerda, V.; Leal, L.O. Automated system for on-line determination of dimethylarsinic and inorganic arsenic by hydride generation–atomic fluorescence spectrometry. *Anal. Bioanal. Chem.* **2012**, *404*, 1589–1595. [CrossRef] [PubMed]

14.   Deng, F.; Dong, R.; Yu, K.; Luo, X.; Tu, X.; Luo, S.; Yang, L. Determination of trace total inorganic arsenic by hydride generation atomic fluorescence spectrometry after solid phase extraction-preconcentration on aluminium hydroxide gel. *Microchim. Acta* **2013**, *180*, 509–515. [CrossRef]

15.   Ciftci, T.D.; Henden, E. Arsenic Speciation of Waters from the Aegean Region, Turkey by Hydride Generation: Atomic Absorption Spectrometry. *Bull. Environ. Contam. Toxicol.* **2016**, *97*, 272–278. [CrossRef] [PubMed]

16.   Erdogan, H.; Yalcınkaya, O.; Turker, A.R. Determination of inorganic arsenic species by hydride generation atomic absorption spectrometry in water samples after preconcentration/separation on nano ZrO_2/B_2O_3 by solid phase extraction. *Desalination* **2011**, *280*, 391–396. [CrossRef]

17.   Wang, N.; Tyson, J. Non-chromatographic speciation of inorganic arsenic by atomic fluorescence spectrometry with flow injection hydride generation with a tetrahydroborate-form anion-exchanger. *J. Anal. At. Spectrom.* **2014**, *29*, 665–673. [CrossRef]

18.   Ulusoy, H.I.; Akcay, M.; Ulusoy, S.; Gürkan, R. Determination of ultra trace arsenic species in water samples by hydride generation atomic absorption spectrometry after cloud point extraction. *Anl. Chim. Acta* **2011**, *703*, 137–144. [CrossRef]

19.   Maratta, A.; Martinez, L.D.; Pacheco, P. Development of an on line miniaturized non-chromatographic arsenic speciation system. *Microchem. J.* **2016**, *127*, 199–205. [CrossRef]

20.   Anthemidis, A.N.; Zachariadis, G.A.; Stratis, J.A. Determination of arsenic(III) and total inorganic arsenic in water samples using an on-line sequential insertion system and hydride generation atomic absorption spectrometry. *Anal. Chim. Acta* **2005**, *547*, 237–242. [CrossRef]

21.   Zhang, Y.; Wang, W.; Li, L.; Huang, Y.; Cao, J. Eggshell membrane-based solid-phase extraction combined with hydride generation atomic fluorescence spectrometry for trace arsenic(V) in environmental water samples. *Talanta* **2010**, *80*, 1907–1912. [CrossRef]

22.   Yang, X.-A.; Lu, X.-P.; Liu, L.; Chi, M.-B.; Hu, H.-H.; Zhang, W.-B. Selective determination of four arsenic species in rice and water samples by modified graphite electrode-based electrolytic hydride generation coupled with atomic fluorescence spectrometry. *Talanta* **2016**, *159*, 127–136. [CrossRef] [PubMed]

23.   Li, S.; Wang, M.; Zhong, Y.; Zhang, Z.; Yang, B. Cloud point extraction for trace inorganic arsenic speciation analysis in water samples by hydride generation atomic fluorescence spectrometry. *Spectrochim. Acta B* **2015**, *111*, 74–79. [CrossRef]

24.   Hsiung, T.-M.; Wang, J.-M. Cryogenic trapping with a packed cold finger trap for the determination and speciation of arsenic by flow injection/hydride generation/atomic absorption spectrometry. *J. Anal. At. Spectrom.* **2004**, *19*, 923–928. [CrossRef]

25.   dos Santos, Q.O.; Silva Junior, M.M.; Lemos, V.A.; Ferreira, S.L.C.; de Andrade, J.B. An online preconcentration system for speciation analysis of arsenic in seawater by hydride generation flame atomic absorption spectrometry. *Microchem. J.* **2018**, *143*, 175–180. [CrossRef]

26.   Chen, M.; Lin, Y.; Gu, C.; Wang, J. Arsenic sorption and speciation with branch-polyethyleneimine modified carbon nanotubes with detection by atomic fluorescence spectrometry. *Talanta* **2013**, *104*, 53–57. [CrossRef]

27.   Tasev, K.; Karadjova, I.; Stafilov, T. Determination of Inorganic and Total Arsenic in Wines by Hydride Generation Atomic Absorption Spectrometry. *Microchim. Acta* **2005**, *149*, 55–60. [CrossRef]

28.   Karadjova, I.B.; Lampugnani, L.; Onor, M.; D'Ulivo, A.; Tsalev, D.L. Continuous flow hydride generation-atomic fluorescence spectrometric determination and speciation of arsenic in wine. *Spectrochim. Acta B* **2005**, *60*, 816–823. [CrossRef]

29.   Cai, L.; Xu, C.; Zhong, M.; Wu, Y.; Zheng, S. Arsenic Speciation in Drinking Tea Samples by Hydride Generation Atomic Fluorescence Spectrometry. *Asian J. Chem.* **2013**, *14*, 8169–8172. [CrossRef]

30.   Lai, G.; Chen, G.; Chen, T. Speciation of As^{III} and As^{V} in fruit juices by dispersive liquid–liquid microextraction and hydride generation-atomic fluorescence spectrometry. *Food Chem.* **2016**, *190*, 158–163. [CrossRef]

31.   Matousek, T.; Wang, Z.; Douillet, C.; Musil, S.; Styblo, M. Direct Speciation Analysis of Arsenic in Whole Blood and Blood Plasma at Low Exposure Levels by Hydride Generation-Cryotrapping-Inductively Coupled Plasma Mass Spectrometry. *Anal. Chem.* **2017**, *89*, 9633–9637. [CrossRef]

32.   Shi, J.-B.; Tang, Z.-Y.; Jin, Z.-X.; Chi, Q.; He, B.; Jiang, G.-b. Determination of As(III) and As(V) in soils using sequential extraction combined with flow injection hydride generation atomic fluorescence detection. *Anal Chim. Acta* **2003**, *477*, 139–147. [CrossRef]

33.   Macedo, S.M.; de Jesus, R.M.; Garcia, K.S.; Hatje, V.; de Queiroz, S.A.F.; Ferreira, S.L.C. Determination of total arsenic and arsenic (III) in phosphate fertilizers and phosphate rocks by HG-AAS after multivariate optimization based on Box-Behnken design. *Talanta* **2009**, *80*, 974–979. [CrossRef] [PubMed]

34.   Macedo, S.M.; dos Santos, D.C.; de Jesus, R.M.; da Rocha, G.O.; Ferreira, S.L.C.; de Andrade, J.B. Development of an analytical approach for determination of total arsenic and arsenic(III) in airborne particulate matter by slurry sampling and HG-FAAS. *Microchem. J.* **2010**, *96*, 46–49. [CrossRef]

35.   do Nascimento, P.C.; Bohrer, D.; Becker, E.; de Carvalho, L.M. Comparison of different sample treatments for arsenic speciation in glass samples. *J. Noncryst. Solids* **2005**, *351*, 1312–1316. [CrossRef]

36.   Rezende, H.C.; Coelho, N.M.M. Determination of Total Arsenic and Arsenic(III) in Phosphate Fertilizers by Hydride Generation Atomic Absorption Spectrometry After Ultrasound-Assisted Extraction Based on a Control Acid Media. *J. AOAC Int.* **2014**, *97*, 736–741. [CrossRef]

37.   Koh, J.; Kwon, Y.; Pak, Y.-N. Separation and sensitive determination of arsenic species (As^{3+}/As^{5+}) using the yeast-immobilized column and hydride generation in ICP-AES. *Microchem. J.* **2005**, *80*, 195–199. [CrossRef]

38.   Cerveira, C.; Pozebon, D.; de Moraes, D.P.; de Fraga, J.C.S. Speciation of inorganic arsenic in rice using hydride generation atomic absorption spectrometry (HGAAS). *Anal. Methods* **2015**, *7*, 4528–4534. [CrossRef]

39.   Matos Reyes, M.N.; Cervera, M.L.; Campos, R.C.; de la Guardia, M. Determination of arsenite, arsenate, monomethylarsonic acid and dimethylarsinic acid in cereals by hydride generation atomic fluorescence spectrometry. *Spectrochim. Acta B* **2007**, *62*, 1078–1082. [CrossRef]

40.   Wang, Y.; Li, Y.; Lv, K.; Chen, X.; Yu, X. A simple and sensitive non-chromatographic method for quantification of four arsenic species in rice by hydride generation–atomic fluorescence spectrometry. *Spectrochim. Acta B* **2019**, *149*, 197–202. [CrossRef]

41.   Rasmussen, R.R.; Qian, Y.; Sloth, J.J. SPE HG-AAS method for the determination of inorganic arsenic in rice—Results from method validation studies and a survey on rice products. *Anal. Bioanal. Chem.* **2013**, *405*, 7851–7857. [CrossRef]

42.   Chen, G.; Chen, T. SPE speciation of inorganic arsenic in rice followed by hydride-generation atomic fluorescence spectrometric quantification. *Talanta* **2014**, *119*, 202–206. [CrossRef]

43. Huang, Y.; Shan, J.; Fan, B.; He, Y.; Hia, S.; Sun, Y.; Lu, J.; Wang, M.; Wang, F. Determination of inorganic arsenic in rice by solid phase extraction and hydride generation atomic fluorescence spectrometry. *Anal. Methods* **2015**, *7*, 8896–8900. [CrossRef]

44. Musil, S.; Petursdottir, A.H.; Raab, A.; Gunnlaugsdottir, H.; Krupp, E.; Feldmann, J. Speciation without Chromatography Using Selective Hydride Generation: Inorganic Arsenic in Rice and Samples of Marine Origin. *Anal. Chem.* **2014**, *86*, 993–999. [CrossRef] [PubMed]

45. Welna, M.; Pohl, P.; Szymczycha-Madeja, A. Non-chromatographic Speciation of Inorganic Arsenic in Rice by Hydride Generation Inductively Coupled Plasma Optical Emission Spectrometry. *Food Anal. Methods* **2019**, *12*, 581–594. [CrossRef]

46. Schlotthauer, J.; Brusa, L.; Liberman, C.; Durand, M.; Livore, A.; Sigrist, M. Determination of inorganic arsenic in Argentinean rice by selective HGAAS: Analytical performance for paddy, brown and polished rice. *J. Food Compos. Anal.* **2020**, *91*, 103506. [CrossRef]

47. dos Santos, G.M.; Pozebon, D.; Cerveira, C.; de Moraes, D.P. Inorganic arsenic speciation in rice products using selective hydride generation and atomic absorption spectrometry (AAS). *Microchem. J.* **2017**, *133*, 265–271. [CrossRef]

48. Huber, C.S.; Vale, M.G.R.; Dessuy, M.B.; Svoboda, M.; Musil, S.; Dedina, J. Sample preparation for arsenic speciation analysis in baby food by generation of substituted arsines with atomic absorption spectrometry detection. *Talanta* **2017**, *175*, 406–412. [CrossRef]

49. Cava-Montesions, P.; de la Guardia, A.; Teutsch, C.; Cervera, M.C.; de la Guardia, M. Non-chromatographic speciation analysis of arsenic and antimony in milk hydride generation atomic fluorescence spectrometry. *Anal. Chim. Acta* **2003**, *493*, 195–203. [CrossRef]

50. Gonzalvez, A.; Llorens, A.; Cervera, M.L.; Armenta, S.; de la Guardia, M. Non-chromatographic speciation of inorganic arsenic in mushrooms by hydride generation atomic fluorescence spectrometry. *Food Chem.* **2009**, *115*, 360–364. [CrossRef]

51. Matos Reyes, M.N.; Cervera, M.L.; Campos, R.C.; de la Guardia, M. Non-chromatographic speciation of toxic arsenic in vegetables by hydride generation-atomic fluorescence spectrometry after ultrasound-assisted extraction. *Talanta* **2008**, *75*, 811–816. [CrossRef]

52. Cava-Montesinos, P.; Nilles, K.; Cervera, M.L.; de la Guardia, M. Non-chromatographic speciation of toxic arsenic in fish. *Talanta* **2005**, *66*, 895–901. [CrossRef] [PubMed]

53. Marschner, K.; Petursdottir, A.H.; Bucker, P.; Raab, A.; Feldmann, J.; Mester, Z.; Matousek, T.; Musil, S. Validation and inter-laboratory study of selective hydride generation for fast screening of inorganic arsenic in seafood. *Anal. Chim. Acta* **2019**, *1049*, 20–28. [CrossRef] [PubMed]

54. Petursdottir, A.H.; Gunnlaugsdottir, H. Selective and fast screening method for inorganic arsenic in seaweed using hydride generation inductively coupled plasma mass spectrometry (HG-ICPMS). *Microchem. J.* **2019**, *144*, 45–50. [CrossRef]

55. Zhang, W.; Qi, Y.; Qin, D.; Liu, J.; Mao, X.; Chen, G.; Wei, C.; Qian, Y. Determination of inorganic arsenic in algae using bromine halogenation and on-line nonpolar solid phase extraction followed by hydride generation atomic fluorescence spectrometry. *Talanta* **2017**, *170*, 152–157. [CrossRef]

56. Yang, L.-L.; Gao, L.-R.; Zhang, D.-Q. Speciation Analysis of Arsenic in Traditional Chinese Medicines by Hydride Generation-Atomic Fluorescence Spectrometry. *Anal. Sci.* **2003**, *9*, 897–902. [CrossRef]

57. Deng, T.; Liao, M.; Jia, M. Arsenic Speciation in Chinese Herb Ligusticum chuanxiong Hort by Hydride Generation Atomic Fluorescence Spectrometry. *Spectrosc. Lett.* **2005**, *38*, 109–119. [CrossRef]

58. Lu, X.-P.; Yang, X.-A.; Liu, L.; Hu, H.-H.; Zhang, W.-B. Selective and sensitive determination of As(III) and tAs in Chinese herbal medicine samples using L-cysteine modified carbon paste electrode-based electrolytic hydride generation and AFS analysis. *Talanta* **2017**, *165*, 258–266. [CrossRef]

59. Sun, H.; Liu, X.; Miao, Y. Speciation Analysis of Trace Inorganic Arsenic in Dietary Supplements by Slurry Sampling Hydride Generation Atomic Absorption Spectrometry. *Food Anal. Methods* **2011**, *4*, 251–257. [CrossRef]

60. Matousek, T.; Hernandez-Zavala, A.; Svoboda, M.; Langrova, L.; Adair, B.M.; Drobna, Z.; Thomas, D.J.; Styblo, M.; Dedina, J. Oxidation state specific generation of arsines from methylated arsenicals based on L-cysteine treatment in buffered media for speciation analysis by hydride generation-automated cryotrapping-gas chromatography-atomic absorption spectrometry with the multiatomizer. *Spectrochim. Acta B* **2008**, *63*, 396–406.

61. Hernandez-Zavala, A.; Matousek, T.; Drobna, Z.; Paul, D.S.; Walton, F.; Adair, B.M.; Dedina, J.; Thomas, D.J.; Styblo, M. Speciation analysis of arsenic in biological matrices by automated hydride generation cryotrapping-atomic absorption spectrometry with multiple microflame quartz tube atomizer (multiatomizer). *J. Anal. At. Spectrom.* **2008**, *23*, 342–351. [CrossRef]

62. Musil, S.; Matousek, T. On-line pre-reduction of pentavalent arsenicals by thioglycolic acid for speciation analysis by selective hydride generation–cryotrapping–atomic absorption spectrometry. *Spectrochim. Acta B* **2008**, *63*, 685–691. [CrossRef] [PubMed]

63. Musil, S.; Matousek, T.; Currier, J.M.; Styblo, M.; Dedina, J. Speciation Analysis of Arsenic by Selective Hydride Generation-Cryotrapping-Atomic Fluorescence Spectrometry with Flame-in-Gas-Shield Atomizer: Achieving Extremely Low Detection Limits with Inexpensive Instrumentation. *Anal. Chem.* **2014**, *86*, 10422–10428. [CrossRef] [PubMed]

64. Yu, H.; Li, C.; Tian, Y. Recent developments in determination and speciation of arsenic in environmental and biological samples by atomic spectrometry. *Microchem. J.* **2020**, *152*, 104312. [CrossRef]

65. Pohl, P. Hydride Generation—Recent Advances In Atomic Emission Spectrometry. *TrAC Trends Anal. Chem.* **2004**, *23*, 87–101. [CrossRef]

66. Marschner, K.; Musil, S.; Dedina, J. Demethylation of Methylated Arsenic Species during Generation of Arsanes with Tetrahydridoborate(1−) in Acidic Media. *Anal. Chem.* **2016**, *88*, 6366–6373. [CrossRef]

67. Carrero, P.; Malave, A.; Burguera, J.L.; Burguera, M.; Rondon, C. Determination of various arsenic species by flow injection hydride generation atomic absorption spectrometry: Investigation of the effects of the acid concentration of different reaction media on the generation of arsines. *Anal. Chim. Acta* **2001**, *438*, 195–204. [CrossRef]

68. Cabon, J.Y.; Cabon, N. Speciation of major arsenic species in seawater by flow injection hydride generation atomic absorption spectrometry. *Fresenius J. Anal. Chem.* **2000**, *368*, 484–489. [CrossRef]

69. Marschner, K.; Musil, S.; Miksik, I.; Dedina, J. Investigation of hydride generation from arsenosugars—Is it feasible for speciation analysis? *Anal. Chim. Acta* **2018**, *1008*, 8–17. [CrossRef]

70. Lehmann, E.L.; Fostier, A.H.; Arruda, M.A.Z. Hydride generation using a metallic atomizer after microwave-assisted extraction for inorganic arsenic speciation in biological samples. *Talanta* **2013**, *104*, 187–192. [CrossRef]

71. Welna, M.; Pohl, P. Potential of the hydride generation technique coupled to inductively coupled plasma optical emission spectrometry for non-chromatographic As speciation. *J. Anal. At. Spectrom.* **2017**, *32*, 1766–1779. [CrossRef]

72. Gonzales, J.C.; Lavilla, I.; Bendicho, C. Evaluation of non-chromatographic approaches for speciation of extractable As(III) and As(V) in environmental solid samples by FI-HGAAS. *Talanta* **2003**, *59*, 525–534.

73. Anwar, H.M. Arsenic speciation in environmental samples by hydride generation and electrothermal atomic absorption spectrometry. *Talanta* **2012**, *88*, 30–42.

74. Ramesh Kumar, A.; Riyazuddin, P. Non-chromatographic hydride generation atomic spectrometric techniques for the speciation analysis of arsenic, antimony, selenium, and tellurium in water samples—A review. *Intern. J. Environ. Anal. Chem.* **2007**, *7*, 469–500.

75. Pinheiro, B.S.; Gimenes, L.L.S.; Moreira, A.J.; Freschi, C.D.; Freschi, G.P.G. Arsenic Speciation in Environmental Samples Using Different Acid Concentrations and Ultrasonic Extraction for the Determination by HG-FAAS. *At. Spectrosc.* **2016**, *37*, 83–89.

76. Pinheiro, B.S.; Moreira, A.J.; Gimenes, L.L.S.; Freschi, C.D.; Freschi, G.P.G. UV photochemical hydride generation using ZnO nanoparticles for arsenic speciation in waters, sediments, and soils samples. *Environ. Monit. Assess.* **2020**, *192*, 331–343. [PubMed]

77. Petrov, P.K.; Serafimovska, J.M.; Arpadjan, S.; Tsalev, D.L.; Stafilov, T. Influence of EDTA, carboxylic acids, amino- and hydroxocarboxylic acids and monosaccharides on the generation of arsines in hydride generation atomic absorption spectrometry. *Cent. Eur. J. Chem.* **2008**, *6*, 216–221.

78. Li, X.; Jia, J.; Wang, Z. Speciation of inorganic arsenic by electrochemical hydride generation atomic absorption spectrometry. *Anal. Chim. Acta* **2006**, *560*, 153–158.

79. Rasmussen, R.R.; Hedegaard, R.V.; Larsen, E.H.; Sloth, J.J. Development and validation of an SPE HG-AAS method for determination of inorganic arsenic in samples of marine origin. *Anal. Bioanal. Chem.* **2012**, *403*, 2825–2834. [PubMed]

80. Leal, L.O.; Forteza, R.; Cerda, V. Speciation analysis of inorganic arsenic by a multisyringe flow injection system with hydride generation–atomic fluorescence spectrometric detection. *Talanta* **2006**, *69*, 500–508. [PubMed]

81. Serafimovski, I.; Karadjova, I.B.; Stafilov, T.; Tsalev, D.L. Determination of total arsenic and toxicologically relevant arsenic species in fish by using electrothermal and hydride generation atomic absorption spectrometry. *Microchem. J.* **2006**, *83*, 55–60.

82. Caballo-Lopez, A.; Luque de Castro, M.D. Hydride generation-pervaporation-atomic fluorescence detection prior to speciation analysis of arsenic in dirty samples. *J. Anal. At. Spectrom.* **2002**, *17*, 1363–1367. [CrossRef]

83. Müller, J. Determination of inorganic arsenic(III) in ground water using hydride generation coupled to ICP-AES (HG-ICP-AES) under variable sodium boron hydride (NaBH4) concentrations. *Fresenius J. Anal. Chem.* **1999**, *363*, 572–576. [CrossRef]

84. Assis, R.C.; de Araujo Faria, B.A.; Caldeira, C.L.; Mageste, A.B.; de Lemos, L.R.; Rodrigues, G.D. Extraction of arsenic(III) in aqueous two-phase systems: A new methodology for determination and speciation analysis of inorganic arsenic. *Microchem. J.* **2019**, *147*, 429–436. [CrossRef]

85. Arbab-Zavar, M.H.; Chamsaz, M.; Heidari, T. Speciation and Analysis of Arsenic(III) and Arsenic(V) by Electrochemical Hydride Generation Spectrophotometric Method. *Anal. Sci.* **2010**, *26*, 107–110. [CrossRef] [PubMed]

86. Boonjob, W.; Miro, M.; Kolev, S.D. On-line speciation analysis of inorganic arsenic in complex environmental aqueous samples by pervaporation sequential injection analysis. *Talanta* **2013**, *117*, 8–13. [CrossRef] [PubMed]

2

Chemometrics for Selection, Prediction and Classification of Sustainable Solutions for Green Chemistry

Marta Bystrzanowska * and Marek Tobiszewski (iD)

Department of Analytical Chemistry, Faculty of Chemistry, Gdańsk University of Technology (GUT), 80-233 Gdańsk, Poland; marektobiszewski@wp.pl
* Correspondence: marbystr@student.pg.edu.pl

Abstract: In this review, we present the applications of chemometric techniques for green and sustainable chemistry. The techniques, such as cluster analysis, principal component analysis, artificial neural networks, and multivariate ranking techniques, are applied for dealing with missing data, grouping or classification purposes, selection of green material, or processes. The areas of application are mainly finding sustainable solutions in terms of solvents, reagents, processes, or conditions of processes. Another important area is filling the data gaps in datasets to more fully characterize sustainable options. It is significant as many experiments are avoided, and the results are obtained with good approximation. Multivariate statistics are tools that support the application of quantitative structure–property relationships, a widely applied technique in green chemistry.

Keywords: multivariate statistics; sustainable chemistry; missing data; classification; grouping; solvents

1. Introduction

The term "chemometrics" was coined by the Swedish scientist Svante Wold in early 1970s while submitting a grant proposal for the application of statistical methods to chemical data [1]. It appeared as the word "kemometri," a combination of the forms "kemo-" for chemistry and "-metri" for measure [2].

Initially, chemometrics was defined as a "science of relating measurements made on a chemical system or process to the state of the system via application of mathematical or statistical methods." According to the name, the discipline of chemometrics originated from chemistry, where one of the first applications focused on improving the quantitative performance of analytical instruments, such as NIR (near infrared) calibration, HPLC (high-performance liquid chromatography) resolution, and UV–VIS deconvolution [3]. Chemometrics took the form of an interdisciplinary field that uses mathematical and statistical methods to design or select optimal measurement procedures and experiments and to provide maximum chemical information by analysing chemical data. The numerous domains that are covered by chemometrics are presented by Santos et al. on a bibliometric map generated using more repeated words in the authors' search for the period 2014–2018 performed in the Science Citation Index Expanded [4]. However, the breakthrough in chemometrics is a response to various software and new high-dimensional hyphenated equipment appearance. These devices in chromatography have been allowed for the determination of various analytes in complex matrices with high resolution and precision. On the other hand, obtained results as large datasets become more difficult to interpret.

Due to rapid technological advances, the focus on multivariate methods is visible. Therefore, the distribution of multiple variables simultaneously provides more information than what could be obtained by considering each variable individually. Then some meaningful information may be chemometrically extracted. As mentioned above, chemometrics is a very important issue in fields concerning environmental monitoring, forensics, chemical biology, food and nutrition, pharmaceutics, polymer, safety and healthcare diagnostics, fraud detection, green chemistry and sustainability, and omics sciences. The latter, together with some bioinformatics and cheminformatics, is becoming more and more popular recently (especially in an advanced data analysis).

However, the use of chemometrics is responsible not only for intelligent data analysis but more specifically for modelling, classification, selection, or searching for missing data. Due to the fact that chemical sciences are based on complex processes involving multistep chemical processes, with condition optimizations, selection of chemical reagents, and so forth, they are a great representative of a wide spectrum of chemometric utilization.

It is also worth noting that chemometric application may be an incredible approach to incorporating the green chemistry concept to chemical sciences via the usage of more environmentally friendly chemicals, analytical procedures, or chemical processes and their optimization (saving energy and materials) and prediction of properties to provide additional information and estimate environmental fate of chemical compounds and pollutants.

In the study, the application of chemometrics in green chemistry as a tool for selection (chemical substances, mainly solvents), classification (different types of organic solvents and ionic liquids), and property prediction (i.e., viscosity, density, carbon dioxide solubility, toxicity, partition coefficient, bioconcentration factor) is presented and discussed.

2. The Outline of Chemometric Tools

Chemometric tools may be divided into two groups: qualitative and quantitative methods. The first group is dedicated to solving problems of classification and pattern recognition. In other words, they allow for assigning an individual sample to a given group of samples or finding a sorting pattern in the underlying data structure of a set [5]. The idea of these methods is based on two philosophies dividing methods into unsupervised and supervised methods. The aim of unsupervised methods is to reveal the underlying data structure without the potential bias of knowing the group memberships beforehand. On the other hand, supervised methods are based on producing the best possible separation of the groups. Therefore, they maximize the capability of the classification method to predict the class membership of samples with unknown membership. Accordingly, it is worth bearing in mind that depending on the problem, one group of methods could be more suited for a given purpose. However, due to fact that it is not always an unambiguous choice, sometimes several chemometric tools are applied. In finding the connection between the detected signals and the exact concentration values, quantitative methods are used. As it is widely known, modern analytical devices generate huge datasets with thousands of spectral data (from Fourier transform infrared/near-infrared, mass spectrometry, nuclear magnetic resonance, etc.); therefore, finding a correlation is very often unclear and difficult. The quantitative analysis is based on regression techniques, whose concept involves exploration of a connection (linear or nonlinear) between one or several independent variables and one (or more, but usually one) dependent variable. If there is only one dependent and one independent variable, then the easiest case is presented—a univariate regression. However, sometimes, as in analytical chemistry problems, the situation is more complicated, including a greater number of dependent variables [6]. Taking the above into account, the selection of an appropriate chemometric tool is dictated by the purpose of the analysis and the characteristics of a given problem. Moreover, obtaining satisfactory results may require the use of several tools. The most commonly used chemometric tools in chemical analysis are briefly described below [7].

The most commonly used chemometric tools in chemical sciences are principal component analysis (PCA) [8,9] and cluster analysis (CA) [6,10]. These unsupervised techniques are very often applied for reducing the dimension of the original data [11], finding internal patterns in the dataset [12,13], or discovering the dominant factors [14,15]. In element classification, very popular are supervised techniques such as linear discriminant analysis (LDA) [16] and partial least squares (PLS) [17,18]. However, they may also be used for prediction [19,20]. An example of regression algorithms may be similar to each other: multiple linear regression (MLR) [21] and principal component regression (PCR) [22]. They are mainly used in data analysis for finding the relationship among variables that effect the prediction of variable values (e.g., chemical compounds' properties). Nevertheless, the most widely used prediction tools are mathematical models from the quantitative structure–activity relationship (QSAR) family [23,24]. They allow for finding the physicochemical, biological, and environmental fate properties of compounds in reference to the knowledge of their chemical structure (new and existing chemical compounds) without animal use in, for example, toxicological testing. Nowadays, artificial neural network (ANN) and genetic algorithm (GA) are gaining more attention in the field of chemical sciences while identifying patterns in data, even complex ones. This is due to their structures and mechanisms, because both of them are comparable to evolutionary processes in nature, namely, equivalents of genes and chromosomes in GA [25] or the biological (human or animal) central nervous system (including neurons) in ANN [26]. They can be successfully used separately [27] or often as a combined tool [28,29]. It is worth noting that these are not all of the techniques that may be used for this purpose. Other approaches, for instance, sum of ranking differences (SRD) [30], k-nearest neighbours (KNN) method [31], and support vector machine, (SVM) [32,33], may also be successfully applied for alternative data treatment in the context of green chemistry. Details of the mentioned chemometric techniques are described elsewhere (some references given in brackets); therefore, they are not fully described in this review.

3. Selection

The problem of selection can be related to the solvents and other chemical reagents (for instance, derivatization agents) used in operations, such as extraction, clean-up, and derivatization. In these cases, the selection of appropriate solvents and chemical reagents for additional chemical activities is extremely important to obtain satisfactory results. Nevertheless, it is worth looking for substitutes for those chemicals mentioned above that are less hazardous to the environment, which correspond to the 5th and 8th of the 12 principles of green chemistry for solvents and derivatization agents, respectively. Considering the above, it is not surprising that the selection of appropriate chemical reagents is a topic of interest in chemometrics.

An approach for fast selection of solvents for a given industrial application with the use of chemometric tools is proposed by García et al. [34]. First, the QSPR (quantitative structure–property relationship) model is developed to find the relationship between the molecular structure and some fundamental solvent properties. Then MLR (multiple linear regression) and PLS (partial least squares) are used for the selection of 62 glycerol-based solvents with respect to three solvent features: the behaviour of the dissolution processes (solvatochromic parameter E_T^N), mechanical aspects (viscosity), and volatility aspects (closely related to safety, toxicity, and air pollution considered through the boiling point). A comparison of applied chemometric tools shows that both of them represent good results in the E_T^N solvation parameter. MLR is only appropriate in the E_T^N solvation parameter, whereas PLS offers better fitting of two of the three properties considered simultaneously. Viscosity and boiling point do not fit well enough to lead to a fully predictive model; however, PLS provides a higher value of determination coefficient for boiling point.

A solvent selection system based on a combination of chemometrics and multicriteria decision analysis is proposed by Tobiszewski et al. in line with the concept of green chemistry [35]. CA (cluster analysis), together with the TOPSIS (the technique for order of preference by similarity to ideal solution) algorithm, allows for, first, grouping and then ranking within groups of 151 solvents in respect to physicochemical, toxicological, and hazard parameters. Three clusters, as presented in Figure 1, are obtained: nonpolar and volatile (35 solvents), nonpolar and sparingly volatile (35 solvents), and polar (81 solvents). The results are compared with another SSG (solvent selection guide) developed by Pfizer [36], GlaxoSmithKline [37], AstraZeneca [38], Sanofi [39], and CHEM21 [40], which are well known in the pharmaceutical industry, confirming a general agreement of solvent rankings within each cluster.

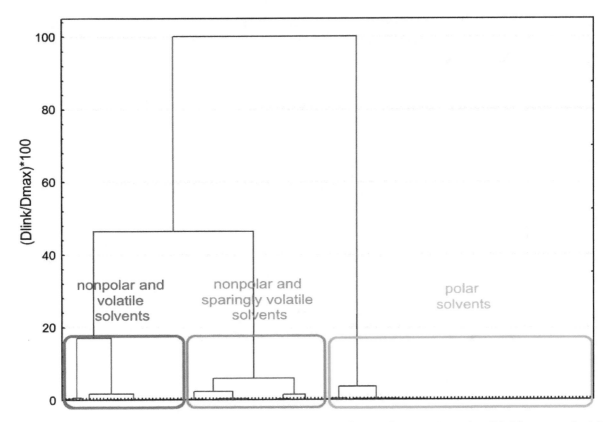

Figure 1. Clustering of the solvents based on their 9 physicochemical properties using CA (cluster analysis). Reproduced from Ref. A solvent selection guide based on chemometrics and multicriteria decision analysis (Tobiszewski et al. [35]) with permission from the Royal Society of Chemistry.

Similar results were recently presented by Sels et al. with the application of MDS (multidimensional scaling) [41]. Solvents were assigned to three groups based on their 22 physical properties according to safety, health, and environment scores: polar compounds, slightly water-soluble solvents, and hydrophobic solvents. In the MDS visualization, the solvents that were similar were plotted closer together in the 2D solvent space. However, it was noted that the relative influence of a functional group decreased with increasing chain length and molecular size. Then a straight line in the MDS visualization was not visible for homologous series from alcohols (due to drastic increase in boiling point and decrease in water solubility, vapour pressure, and relative evaporation rate). Moreover, the application of SUSSOL (Sustainable Solvents Selection and Substitution Software), a specially created software by applying artificial intelligence (AI), is presented for finding solvent replacements for N-methylpyrrolidone (NMP), toluene, and tetramethyl oxolane (TMO). The proposed alternative solvents are as follows: 10 candidate alternative solvents (including dimethyl sulfoxide, Cyrene, N-butyl pyrrolidone, pyridine, acetone, methyl acetoacetate, 1-ethyl pyrrolidone, dimethylacetamide, dimethylformamide, nicotine) for NMP; isobutylbenzene and p-cymene for toluene; and toluene, 1,1-dichloroethene, 1,1-dichloroethane, 1,1,1-trichloroethane, 1,1-dichloropropane, ethylene glycol diethyl ether (1,2-diethoxyethane), and so forth for TMO. An example of visualization dedicated to possible alternatives for NMP by SUSSOL software is presented in Figure 2.

A screening of potential PBT (persistent, bioaccumulative, and toxic) compounds (in an environment based on persistence, bioconcentration, and toxicity data) is another example of chemical selection, but different from solvents [42]. PCA is used to group chemicals representing many classes of pollutants of various chemical structures, such as dioxins, PCBs, PAHs, and pesticides, and various industrial chemicals according to their potential cumulative PBT behaviour. However, due to unavailability of experimental data, an approach combining multivariate analysis and QSAR/QSPR (quantitative structure–activity relationship) was applied, which allowed for the reduction of data gaps in the dataset. The strength of the approach is validated in two sequential steps: first, performed on the available experimental dataset, including 54 chemicals, and then performed on the dataset of 180 chemicals (developed by QSPR). In Figure 3, the analysis of the latter dataset of organic compounds using PCA is presented.

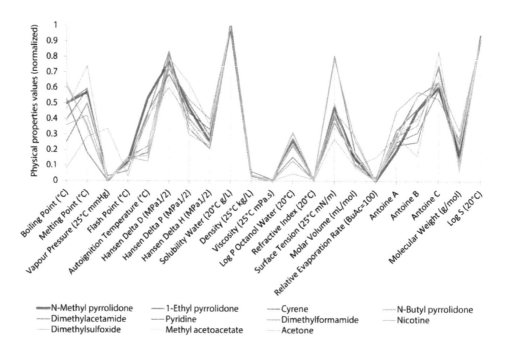

Figure 2. Visualization of the analysis results of substitution candidates for NMP in SUSSOL software. Reproduced from Ref. SUSSOL—Using Artificial Intelligence for Greener Solvent Selection and Substitution (Sels et al. [41]).

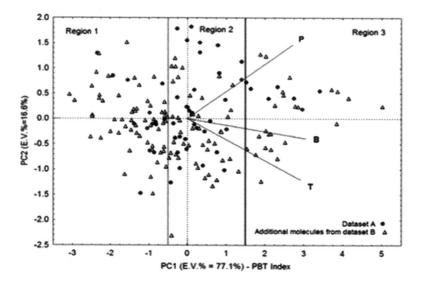

Figure 3. PCA (principal component analysis) on experimental and predicted PBT (persistent, bioaccumulative, and toxic) data for 180 organic compounds (dataset A – 54 comp. + dataset B – 126 comp.). Reproduced from Ref. QSPR as a support for the EU REACH regulation and rational design of environmentally safer chemicals: PBT identification from molecular structure (Papa and Gramatica [42]) with permission from the Royal Society of Chemistry.

According to PBT index values, chemicals are grouped into three regions: region 1—not PBT chemicals, region 2—chemicals with medium PBT properties, and region 3—PBT and vPvB (very persistent and very bioaccumulative) chemicals.

4. Classification

Classification as a systematic arrangement in groups or categories according to established criteria is sometimes very useful in designing a chemical process or reaction. It allows for recognizing some alternatives with corresponding characterization.

Translating the principle *similia similibus solvuntur* into the field of chemistry means solvents belonging to the same group demonstrate similar abilities to dissolve compounds. Therefore, chemometric classification of solvents according to the degree of polarity may provide information about possible substitutes. This kind of grouping addressed to organic solvents is one of the frequently undertaken problems in chemometrics, which is summarized in Table 1.

Table 1. Organic solvent classification according to the degree of polarity by chemometric application—summarized exemplary studies.

Classification Object	Chemometric Tool	Evaluated Parameters	Results—Groups of Solvents	Ref.
83 organic solvents	PCA	• the Kirkwood function (K) • molecular refraction (MR) • molecular dipole moment (μ) • the parameter of Hildebrand • index of refraction (n) • boiling point (bp) • energies of HOMO (Highest Occupied Molecular Orbital) and LUMO (Lowest Unoccupied Molecular Orbital)	9 groups of solvents: • aprotic dipolar: acetonitrile, acetone, ethyl acetate, dichloromethane • aprotic highly dipolar: dimethyl sulfoxide, N,N-dimethyl formamide, pyridine • aprotic highly polarizable dipolar: hexamethylphosphotriamide • aromatic apolar: toluene, benzene • aromatic polar: chlorobenzene, o-dichlorobenzene • electron-pair donor: triethylamine, diethyl ether, dioxane • hydrogen bonding: methanol, ethanol, pentan-2-ol • hydrogen bonding strongly associated: formamide, water, ethylene glycol • miscellaneous: carbon disulphide, chloroform, aniline	Chastrette et al. (1985) [43]

101 organic solvents	Parker–Reichardt classification	correlation between dielectric β parameter and empirical solvent polarity parameter E_T^N	4 groups (and 2 subgroups) of solvents:
			• weakly dipolar nonhydrogen bonding donor: ethers, carboxylic esters, tertiary amines, halogen-substituted hydrocarbons
			• dipolar nonhydrogen bonding donor: ketones, N,N-disubstituted amides, nitro-substituted hydrocarbons, nitriles, sulphoxides, sulphones, cyclic carbonates, pyridine
			• hydrogen bonding donor: water, alcohols, carboxylic acid, glycols
			  ○ nonprimary alcohols and aniline
			  ○ phenol and its derivatives
			• N-monosubstituted amides and formamide
				Dutkiewicz (1990) [44]

Table 1. *Cont.*

Classification Object	Chemometric Tool	Evaluated Parameters	Results—Groups of Solvents	Ref.
51 solvents	KNN	Empirical scale parameters: • PAC (polarity/acidity) • PBC (polarity/basicity) • PPC (polarity/polarizability)	8 groups of solvents: • Nonpolar inert solvents: aliphatic hydrocarbons) • nonpolar-polarizable: aromatic hydrocarbons, tetrachloromethane, carbon disulphide • nonpolar-basic: ethers, triethylamine • little polar-polarizable: aliphatic halogen derivatives, substituted benzenes with heteroatom-containing substituents • little polar-basic: cyclic ethers, ketones, esters, pyridine • polar-aprotic: acetanhydride, dialkylamides, acetonitrile, nitromethane, dimethyl sulfoxide, sulfolane • polar-protic: alcohols, acetic acid • exceptional solvents: water, formamide, glycol, hexamethylphosphoric triamide	Pytela (1989) [45]
152 organic solvents	KNN, CP-ANN, QSPR	4 molecular descriptors (theoretical descriptions of the molecular structure)	5 groups of solvents: • aprotic polar • aromatic apolar or lightly polar • electron-pair donors • hydrogen bonding donors • aliphatic aprotic apolar	Gramatica et al. (1999) [46]
76 solvents	ANN	9 characteristics (application in a field of C60 fullerene solubility)	9 groups of solvents: • apolar and slightly polar: n-pentane, n-hexane, n-octane, n-decane • apolar and slightly polar: n-dodecane, benzene, m/o/p-xylene, toluene, ethylbenzene, cumene • apolar and slightly polar: carbon disulphide, tetrachloroethylene • weakly polar: fluorobenzene, dichloromethane, o-cresol • weakly polar: chlorobenzene, pyridine • weakly polar: bromobenzene, bromoform • hydrogen bond donors and others: methanol, ethanol, 1-propanol, 1-butanol, acetone • hydrogen bond donors and others: 1-pentanol, 1-hexanol, 1-octanol, 1-decanol • highly polar: nitrobenzene, benzonitrile • highly polar: 1,2-ethanediol, water, *N*-methylformamide, acetonitrile, *N,N*-dimethylformamide • miscellaneous: chloroform, 1-aminobutane	Pushkarova and Kholin (2014) [47]

Table 1. *Cont.*

Classification Object	Chemometric Tool	Evaluated Parameters	Results—Groups of Solvents	Ref.
236 industrial solvents	PCA, CA	quantum and experimental parameters	10 groups of solvents: • hydrogen bond donor: short-chain alcohols, phenols, acetic acid, butyric acid • hydrogen bond donor with high polarizability: tributylamine, glycols, long-chain alcohols • hydrogen bonds acceptor/electron-pair donor: amines, pyridines, aniline, anisole, dioxane • aprotic dipolar: ethyl acetate, cyclohexanone, acetophenone, acetone • aprotic dipolar-polarizable: sulfolane, ketones with at less C7, hexamethylphosphoramide • aprotic very strongly dipolar: nitro/nitrile compounds • aprotic apolar: linear or cyclic alkanes • aprotic apolar with pi bonds: aromatics, xylenes, cyclohexane • halogenated hydrocarbons: dichloromethane, carbon disulphide, halogenated derivatives of benzene, carbon tetrachloride	Levet et al. (2016) [48]
72 solvents	FCM, FLDA	Chemical parameters connected with polarity and selectivity developed by Snyder (related to different polar interactions): • proton acceptor (x_e) • proton donor (x_d) • dipole (x_n) • chromatographic strength (P') derived from gas–liquid partition coefficient • toluene similitudes (x_t) • methylethylketone similitudes (x_m)	FCM—8 groups (selected examples): • cyclohexanone, ethylmethylketone, dioxane, acetophenone, benzonitrile, ethyl acetate, nitrobenzene • dimethyl sulfoxide, ethyleneglycol, m-cresol, m-methylpyrrolidone • p-xylene, toluene, benzene, bromobenzene • aniline, dimethylformamide, propylene carbonate, N,N-dimethyl acetamide, acetic acid • 1-propanol, 2-propanol, tetrahydrofuran, 1-butanol, tert-butanol, anisole, ethanol • fluorobenzene, 1-octanol • pyridine, triethylene glycol, benzyl alcohol, acetonitrile, methanol, acetone • formamide, water, dodecafluoroheptanol FLDA—8 groups (selected examples): • diethylether, triethylamine • propanol, 1-octanol, 2-propanol, 1-butanol, ethanol, tert-butanol, methanol • pyridine, methylformamide, triethylene glycol, N,N- dimethyl acetamide, dimethyl sulfoxide • acetic acid, ethylene glycol, formamide • methylene chloride, ethylene chloride • acetophenone, dioxane, acetonitrile, acetone, tetrahydrofuran, aniline, ethyl acetate • chlorobenzene, p-xylene, benzene, anisole, toluene, chloroform • dodecafluoroheptanol, water, m-cresol	Guidea and Sârbu(2020) [49]

Interestingly, these classifications are carried out for various objects (types of solvents) using different chemometric tools, for instance, PCA, KNN (*k*-nearest neighbours method), Parker-Reichardt classification, CP-ANN (counter-propagation artificial neural network), ANN (artificial neural network),

PCA, and CA, obtaining similar results. An example may be the study performed by Dutkiewicz [44] using the Parker–Reichardt classification, whose results highly correspond to those obtained by a more complex multivariate statistical method presented by M. Chastrette et al. [43]. Moreover, there are applications with few tools applied. The idea is to improve the results of classification, for instance, by making them more chemically interpretable, as in organic solvent classification based on molecular descriptors (theoretical descriptions of the molecular structure), where KNN application is followed by CP-ANN [46].

One of the latest works considers a classification of 72 solvents according to polarity and selectivity issues based on the Snyder approach (related to different polar interactions), performed using FCM (fuzzy *c*-means) and FLDA (fuzzy linear discriminant analysis) [49]. The used fuzzy chemometric techniques show high efficiency and information power methods in solvent characterization and classification (an approach for rationalchoosing of a good solvent). The obtained results (division into eight groups of solvents) are in good agreement with the Snyder classification, especially using FLDA (the highest value of 100% for the solvents corresponding to groups II and V and the lowest value of 66.67% for the solvents of group I).

However, the classification does not always take into account a large number of groups/classes. Salahinejad [50] proposed a division of solvents for single-walled carbon nanotube dispersion into two groups: solvents and nonsolvents (solvents with effectively zero of nanotube dispersibility). The classification is conducted separately with several tools, such as RF (random forest), SVM (support vector machine), MLP (multilayer perceptron), and QDA (quadratic discriminant analysis). According to the results of the sum of ranking difference (SRD) procedure, the RF classifier based on selected descriptors is the best classification model, while the SVM, MLP, and QDA are ranked as good models.

Moreover, another classification of solvents based on a chemical group of compounds was performed by Katritzky et al. [51] and Tobiszewski et al. [52]. In the first case, a classification of the theoretical molecular descriptors, derived from the chemical structure alone (QSPR model), according to their relevance to specific types of intermolecular interaction (including cavity formation, electrostatic polarization, dispersion, and hydrogen bonding) in liquid media is presented. According to the PCA results, 11 classes of solvents were formed: hydrocarbons; halo-hydrocarbons; saturated, unsaturated, and cyclic ethers; esters and polyesters; aldehydes, ketones, and amides; nitriles and nitro hydrocarbons; hydroxylic compounds; amines and pyridines; thiols, sulphides, sulfoxides, and thio compounds; phosphorus compounds; and compounds with vastly different chemical functionalities. In the latter case, CA and PCA were used to group around 130 potentially green organic solvents according to their similarity based on physiochemical parameters, as well as to assess and identify variables from which properties missing values such as bioconcentration factors, water–octanol, and octanol–air partitioning constants can be predicted. The CA results show that polar solvents are divided into three major groups: (a) less volatile solvents, slightly water soluble with high values of logKOW and logBCF (alcohols with ether functional groups, aromatic alcohols, and short-chain organic acids apart from formic and acetic); (b) less volatile and very highly water-soluble solvents (lactate esters, formic and acetic acids, glycerol, and some alcohols with other functional groups); and (c) highly volatile, low-boiling-point, high vapour pressure, and Henry's law constant solvents ("traditional" polar solvents, like short-chain alcohols, ketones, aldehydes, and esters). On the other hand, nonpolar solvents were divided into volatile, water-nonsoluble, and slightly water-soluble solvents. According to a chemometric analysis connected with finding the internal relationship between bioconcentration factors and physiochemical parameters, in polar solvents, the variable logBCF forms a separate latent factor not directly correlated with other variables (specific importance of this parameter as a discriminant for the dataset). Unlike in nonpolar solvents, the relationship between parameters like logBCF and logKOW and Henry's law constant and the correlation of logKOA with a whole group of physicochemical parameters, like surface tension, density, boiling, and melting point, is visible.

A different approach for the classification of 259 solvents according to the experimentally found and theoretically predicted physicochemical parameters presented by 15 specific descriptors is proposed by Nedyalkova et al. (2020) [53]. The variables involved parameters such as melting point, boiling point, density, water solubility, vapour pressure, Henry's law constant, octanol–water and octanol–air partition coefficients, and bioconcentration factor, some of which are implemented within the modules of EPI Suite or by the SMILES codes (simplified molecular input line entry system). The fuzzy hierarchical clustering methods allow for checking whether the experimental values of the respective variables correspond to the calculated ones, and the partitioning procedure could determine stable groups of similarity between the variables with highly different degrees of membership. The performed partitioning with respect to specific descriptors divides solvents into 10 classes (some examples of solvents within each class are presented in brackets) (i.e., chlorinated solvents—class 1 (iodoethane, n-butyl acetate, m-cresol, diethyl carbonate, chloroform), nonpolar and volatile solvents—class 2 (bromoethane, benzonitrile, isobutyl acetate, carbon disulphide), polar and nonpolar solvents mixed—class 3 (benzene, dichloromethane, diethyl ether, triethylene glycol, polyethyleneglycol 200), polar solvents—classes 4–7 (dioctylsuccinate, oleic acid, 2-pyrrolidone, glycerol, water, 1-octanol, nitrobenzene, methyl stearate), high molecular weight polar solvents—class 8 (ethyl laurate, anisole), large group of mostly polar solvents with some exceptions—class 9 (triethylamine, ethanol, 1-butanol, formamide, toluene, o-xylene, aniline, n-heptane, d-limonene, styrene, acetone, phenol, acetonitrile), and outlier—class 10 (perfluorooctane 20). The relationships between solvents of various natures (polar, nonpolar, volatile, etc.) and the physicochemical variables are found, despite the fact that missing data of specific descriptors are fulfilled via theoretical calculation. Moreover, applied chemometric techniques allow for partitioning solvents with more or less similar characteristics in terms of higher, smallest, or intermediate values of considered descriptors.

One of the most interesting groups of solvents are ionic liquids (ILs) due to their desired feature—designing of solvents with particular properties (within certain ranges) by a combination of selected cation and anion. Therefore, characterization of their types is very important for finding an appropriate alternative, for instance, in phases for gas chromatography. This aspect is discussed by González-Álvarez et al. in the classification of three ILs with hexacationic imidazolium, polymeric imidazolium, and phosphonium as cations and halogens, thiocyanate, boron anions, triflate, and bistriflimide as anions [54]. The application of CA, LDA (linear discriminant analysis), D-PLS (discriminant partial least squares), and MLR shows that two main groups of phases may be distinguished: ILs with acidic and basic characterization. After the identification of the two natural groups of ILs by CA, several supervised chemometric techniques, such as LDA, D-PLS, and MLR were used to construct models of pattern recognition and classification rules for ILs. All tools showed high prediction capacity and were successfully used for characterizing IL classes. The best results were obtained via LDA with >96% for classification and >92% for prediction, followed by MLR with 96.7% and 92% in the prediction for classes A and B, respectively.

In another study, 227 ionic liquids and their related salts were also classified based on their toxicities towards rat cell lines [55]. Regardless of the used chemometric method (LDA, CA, SVM (support vector machine), or CP-ANNs (counter-propagation artificial neural networks)), ILs were classified into four categories: low, moderate, high, and very high toxicity. In this study, CP-ANN turned out to be more favourable over other methods in terms of accuracy of classification, underlining that CP-ANNs may extract actual information and knowledge from the dataset.

An interesting approach with a classification map called the Σpider diagram was proposed by Lesellier [56]. Solvents were classified based on physiochemical properties encountered with other visual presentations, such as Snyder triangle, Hansen parameters, LSER (linear solvation energy relationships), Abraham descriptors, COSMO-RS (Conductor like Screening Model for Real Solvents) parameters, and solvatochromic solvent selectivity. Visualization of the last solvent classification is presented in Figure 4.

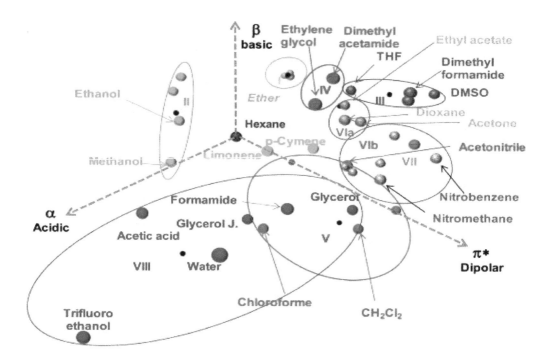

Figure 4. Spider diagram based on solvatochromic parameters π^*, α, β. Reprinted from Journal of Chromatography A, 1389, E. Lesellier, Σpider diagram: A universal and versatile approach for system comparison and classification:

This diagram shows many advantages of solvent classification through a better view of solvents having no acidic character (for the solvatochromic solvent selectivity), easier usage due to the "flattening" of the spherical view down to a single plane (for Hansen parameters), more subtle classification due to the use of five parameters instead of three (for COSMO-RS), and simple view of the solvent groups having similar or different properties (for Abraham descriptors). An approach may be useful not only for selecting suitable solvents for extraction, separation, or purification approaches and for solubility studies but also for choosing greener solvents.

There are also other fields of interest apart from solvents, for instance, pharmaceutical excipients in reference to their solubility parameters [57]. PCA is used to predict a behaviour of materials in a multicomponent system (e.g., for the selection of the best materials to form stable pharmaceutical liquid mixtures or stable coating formulation). It is significantly important because similarity between the values of the respective components of the solubility parameter allows for the estimation of the compatibility between different materials (solvents, colorants, lubricants, coating components, and powder blends).

5. Properties (Prediction and Correlation)

Knowledge of the physicochemical properties of compounds is necessary to predict their behaviour under various conditions or factors during chemical reactions, and their behaviour in various media or compartments in the environment (environmental fate). Therefore, this explains the need to obtain information on the solvents' and other chemical reagents' properties. Unfortunately, sometimes there are missing points in chemical characteristics. Thus, some prediction and computational methods for filling the gaps are highly required and successfully applied.

An example of the most popular advanced and computational modelling approaches may be QSAR (quantitative structure–activity relationship) and EPI Suite (Estimation Programs Interface Suite). QSAR models allow for the prediction of the physicochemical, biological, and environmental fate properties of compounds in reference to knowledge of their chemical structure. The concept is based on establishing quantitative relationships between descriptors (referring to the chemical structure) and the target property capable of predicting activities of novel compounds [58]. On the other hand, EPI Suite may estimate physical/chemical and environmental fate properties such as water solubility, octanol–water partition coefficient, Henry's law constant, melting point, boiling point, and aquatic toxicity, taking into account chemical structure as input data (depending on the chosen estimation model program) [59]. However, the easiest manner is chemical predictive modelling, which is based on an observation of some patterns, correlations between variables in dataset. In this respect, the chemometric tools play an important role.

As mentioned in Section 3, the use solvents in chemistry is one of the most important issues with respect to environmental aspects. In this manner, the type of solvent and its amount are of great importance. ILs are very often described in the context of solvents with incredible features, such as negligible vapour pressure, high chemical and thermal stability, low flammability, large liquidus range, high ionic conductivity, large electrochemical window, excellent solvation ability of a wide range of compounds, and most of all, possibility of designing for specific demands (due to an appropriate selection of cation and anion). However, there are also numerous studies where the authors pay attention to the environmental problem due to poor biodegradability, toxicity, and methods of preparation and degradation after use [60–65]. Nevertheless, the lack of data for IL characterization in the context of greenness assessment is a serious problem. It may make the evaluation difficult and in some sense inaccurate and inappropriate in flat assertions on ILs as alternative green solvents [66]. Hence, a large number of publications on predicting the properties of ionic liquids have been performed, as shown in Table 2.

Table 2. Prediction of ionic liquid properties by applying chemometric tools—summarized exemplary studies.

Predicted Property	Chemometric Tools	Evaluated Objects	Way of Estimation	Ref.
Carbon dioxide solubility	RB, MLP, MQR, MPE	• [emim][PF6] • [hmim][PF6] • [bmim][BF4] • [hmim][BF4] • [omim][BF4]	experimental thermodynamic data and molecular structure information	Torrecilla et al. (2008) [67]
Melting point	ANN	97 imidazolium salts with varied anions	14 molecular descriptors	Torrecilla et al. (2008) [68]
Viscosity	ANN	58 ionic liquids at several temperatures	molecular mass of the anion and cation, the mass connectivity index, and the density at 298 K	Valderrama et al. (2011) [69]
Electric conductivity	MLR, BP-ANN	35 ILs at different temperatures	structural descriptors	Cao et al. (2013) [70]
Density	ER, ANN	mixtures of ionic liquids and molecular solvents (water, alcohols, ketones, ethers, hydrocarbons, esters, and acetonitrile)	molar mass, critical volume, temperature, acentric factor of each component of the IL mixtures	Huang et al. (2014) [71]
Design of ionic liquids	PCA, CA	172 ILs	structural similarity and identification of structure aspects responsible for a given IL physicochemical properties (viscosity, n-octanol–water partition coefficient, solubility and enthalpy of fusion via ILPC predictor)	Barycki et al. (2016) [72]
Lipophilicity	QSPR, PCA	selected ionic liquid (only imidazolium-based cations)	comparison of hydrophobic or hydrophilic character according to some methods: chromatographic analysis, statistical, and chemometric approach	Studzińska et al. (2007) [73]
Toxicity	PCR, PLS, decision tree(s) model	various combinations of cations (imidazole, pyridinium, quinolinium, ammonium, phosphonium) and anions (BF_4, Cl, PF_6, Br, CFNOS, NCN_2, $C_6F_{18}PBF_4$, $C_6F_{18}P$)	molecular descriptors and EC_{50} concentrations for inhibition of acetylcholinesterase	Ž. Kurtanjek (2014) [74]

Table 2. *Cont.*

Predicted Property	Chemometric Tools	Evaluated Objects	Way of Estimation	Ref.
Toxicity	PCA	375 ILs with six different types of cations namely, imidazolium, ammonium, phosphinium, pyridinium, pyrolidinium, and sulfonium	multiple endpoints for various organisms based on WHIM descriptors	Sosnowska et al. (2014) [75]
Toxicity	QSAR, MLR, ELM	160 ILs with 57 cations and 21 anions	toxicity towards AChE based on theSEP area and the screening charge density distribution area (S σ) descriptors	Zhu et al. (2019) [76]
Toxicity	QSPR, MLR	304 ILs of different combinations of 8 cations (ammonium, imidazolium, morpholinium, phosphonium, piperidinium, pyridinium, pyrrolidinium, quinolinium) and 12 anions (chloride, bis(trifluoromethylsulfonyl) amide, bromide, iodide ion, sulfonate, borate, phosphate, fatty acid, dicyanamide, formate, thiocyanate, acetate, etc.)	toxicity against leukaemia rat cell line IPC-81 (logEC$_{50}$) based on 33 descriptors describing the structural features of ionic liquids related to toxicity (i.e., chain length of the cationic head group)	Wu et al. (2020) [77]

Abbreviations: AChE—Acetylcholinesterase; BP-ANN—Back Propagation Artificial Neural Network; ELM—Extreme Learning Machine; ILPC—Ionic Liquid PhysicoChemical; MPE—Mean Prediction Error; SEP—Surface Electrostatic Potential; WHIM descriptors—Weighted Holistic Invariant Molecular descriptors

The prediction of IL properties may be successfully conducted using different chemometric tools. It is mostly proved by a comparison of predicted values with experimental/literature ones, such as in estimation melting point [68] or viscosity [69]. Moreover, it sometimes happens that one technique is applied to select appropriate descriptors; then another one is used for the prediction of a particular feature. In some cases, the applications of several chemometric methods are compared, as presented with the example of carbon dioxide solubility [67], electric conductivity [70], density [71], and toxicity [74]. In first case, nonlinear models, such as RB (radial basis network) and MLP (multilayer perceptron) turned out to be more adequate when the mathematical complexity of the model is not important or a high accuracy is necessary. On the other hand, MQR (multiple quadratic regression) is recommended for faster computation if the operating conditions are stable. Prediction of electric conductivity using an ANN model is more favourable than using an MLR model due to more rational nonlinear modelling. An interesting approach is presented for the latter case—toxicity prediction based on molecular descriptors and EC$_{50}$ concentrations for the inhibition of acetylcholinesterase using a decision tree(s) model. Decision tree(s) models (R = 0.992) significantly outperform other models, such as PCR (principal component regression) and PLS (R = 0.62 and 0.64), for numerical predictions of EC$_{50}$ concentrations and the classification of ILs into four levels of toxicity. The visualization of this division into four classes is presented in Figure 5.

It is not always the rule that one of the models used is clearly better than the others. Very often, all of them or some of them lead to satisfactory results, which is described by Huang et al. [71] for density prediction. ER (extended Riedel) and ANN proved to be accurate in a wide range of compositions and temperatures. However, the ER model is a better alternative because it can be used directly without any adjustable parameter and computer-aided program. Sometimes satisfactory results may be obtained by the application several chemometric tools, one by one. Barycki et al. proposed the application of PCA for the definition of the distribution trends of four IL properties dependently on their structures. Then CA is used to provide some detailed information concerning IL distribution [72]. It is also worth noting that chemometrics may be the basis for developing other tools. According to the observed strong relationship between the variance in the observed toxicity and the cations' descriptors, a toxicity ranking index based on the structural similarity of cations (TRIC) for initial toxicity screening studies of ILs has been developed [75]. However, the use of TRIC cannot be individual. It is limited to the prediction of toxicity endpoints used in its development.

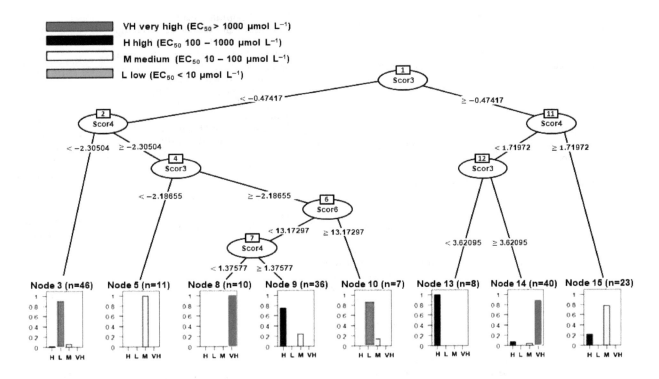

Figure 5. Decision tree model for predictions based on the IL classification of $\ln(EC_{50})$ into four toxicity categories. Reproduced from Ref. Chemometric versus Random Forest Predictors of Ionic Liquid Toxicity (Kurtanjek [74]).

One of the most frequently predicted environmental parameters is toxicity, which may be noticed due to the visible trend in IL properties' prediction analysis as summarized above [74–77]. It is expressed by different endpoints towards various organisms. Toxicity assessment is very important from green chemistry's point of view. Some examples of studies concerning the prediction of toxicity for selected chemicals as potential pollutants are summarized in Table 3.

Table 3. Examples of toxicity prediction for different groups of chemical compounds by applying chemometric tools.

Chemical Compound	Chemometric Tool	Organism	Toxicity Results	Ref.
Metals as: Tl, Cd, and Ag	RSM	growth of cabbage seedlings	Ag is observed to be the most toxic, while Tl and Cd, although toxic, exhibited fairly similar effects.	Allus et al. (1988) [78]
Nitrobenzenes	LS-SVM, QSPR, PLS, PCA, GA-PLS, MLR	*Tetrahymena pyriformis* [79]	n/a	Niazi et al. (2007) [80]
Organic compounds (including some pharmaceuticals)	QSTR, PLS	human (human lethal concentration)	The ETA models suggest that the toxicity increases with bulk, chloro (hydrophobic) functionality, presence of heteroatoms within a chain or a ring and unsaturation, and decreases with hydroxyl (polar) functionality and branching.	Roy and Ghosh (2008) [81]
Chemical compounds	SVM, ANN	*Pimephales promelas*	n/a	Tan et al. (2010) [82]

Table 3. *Cont.*

Chemical Compound	Chemometric Tool	Organism	Toxicity Results	Ref.
Organic chemicals	QSAR, MLR, PLS, GFA, G/PLS	*Daphnia magna*	Higher lipophilicity and electrophilicity, less negative charge surface area and presence of ether linkage, hydrogen bond donor groups and acetylenic carbons are responsible for greater toxicity of chemicals. Diversity in chemically different compounds in mechanisms of toxic actions is observed.	Kar and Roy (2010) [83]
Per- and polyfluorinated (PFCs) chemicals	PCA, QSAR, MLR, GA,	rodents (oral)	The importance of negative hydrophobicity and positive electronegativity for the overall toxicity of PFCs for rodents.	Bhhatarai and Gramatica (2011) [84]
Herbicides	ANN, QSAR	rat (oral)	n/a	Hamadache et al. (2016) [85]
Agrochemicals (fungicides, herbicides, insecticides, and microbiocides)	QSAR	*Daphnia magna*	The toxicity increases with lipophilicity and decreases with polarity.	Khan et al. (2019) [86]
Silver nanoparticles	CA, PCA	links between ecotoxicity and physicochemical features (*Daphnia magna, Thamnocephalus platyurus,* and *Daphnia galeata*)	n/a	Nedyalkova et al. (2017) [87]
Silver nanoparticles	PCA, CA, *k*-means clustering, MLR	*Daphnia magna, Thamnocephalus platyurus, Escherichia coli, Pseudomonas fluorescens, Pseudokirchneriella subcapitata, Pseudomonas putida, Pseudomonas aeruginosa, Staphylococcus aureus,* mammalian cells, algae, yeast, and fungi	The relation AT/ZP (acute toxicity measure, EC_{50}/LC_{50}/zeta potential of nanomaterial in the test) is not very indicative for the toxic impact of the AgNPs studied.	Nedyalkova et al. (2019) [88]

Abbreviations: ETA—Extended Topochemical Atom; GA-PLS—Genetic Algorithm-Partial Least Square; GFA—Genetic Function Approximation; G/PLS—Genetic Partial Least Squares; QSTR—Quantitative Structure Toxicity Relationship; RSM—Response Surface Methodology

Based on the above studies, the methods from the family of QSAR models are willingly used for toxicity prediction. They allow for the achievement of good results and provide more than 95% predictions for agrochemical toxicity towards *Daphnia magna* [86]. QSAR models are often supported by chemometrics; however, there is no dominant chemometric tool that ensures the best prediction ability. In nitrobenzene toxicity prediction, LS-SVM (least squares-support vector machines) turned out to be the more powerful method than the rest [80]. The reason is fact that LS-SVM (for quantum chemical descriptors) drastically enhances the ability of prediction in QSAR (prediction of IGC_{50} toxicity) studies superior to MLR and PLS.

Other parameters of great importance for the assessment of the environmental risk associated with the use of chemical compounds are the partition coefficients towards different media. They allow for the estimation of the affinity of a particular chemical compound to a selected phase system. Octanol–air or octanol–water partition coefficients may be applied as the predictors of the partitioning of semivolatile organic chemicals to aerosols or a chemical compound to dissolve in fats, oils, lipids, and nonpolar solvents, respectively. Moreover, the value of the latter coefficient could provide information on the potential for bioaccumulation as well as in persistent compounds undergoing biomagnification [89,90]. In Table 4, a list of studies on the chemometric prediction of partition coefficients in presented.

Table 4. Prediction of partition coefficients by applying chemometric tools—summarized exemplary studies.

Partition Coefficient	Chemometrics Tool	Evaluated Objects	Way of Estimation	Ref.
n-octanol-ir partition coefficient	QSAR/QSPR, PCA, PCR	chloronaphthalene congeners	190 different quantum-chemical, thermodynamical, and topological characteristics of chloronaphthalenes as descriptors	Puzyn and Falandysz (2005) [91]
Water-polydimethylsiloxane partition coefficient	QSPR, GA, MLR, ANN	organic compounds	molecular descriptors: minimum atomic orbital electronic population, Kier shape index, polarity parameter/square distance, and complementary information content	Golmohammadi and Dashtbozorgi (2010) [92]
n-octanol-water partition coefficient	LS-SVM, QSPR, MLR, SVR, ANN	organic compounds (derivative phenolic compounds)	n/a	Goudarzi and Goodarzi (2008) [21]
n-octanol-water partition coefficient	QSPR, mRMR-GA-SVR	aromatic compounds	68 molecular descriptors derived solely from the structures of the aromatic compounds	Yang et al. (2008) [93]
n-octanol-water partition coefficient	QSPR, MLR/PLS/RBF-PLS	organic compounds		Goudarzi and Goodarzi (2010) [94]
n-octanol-water partition coefficient	QSAR, CoMFA, CoMSIA	21 polychlorinated naphthalenes (PCNs) congener	3D descriptors according to the experimental values of logKOW for 21 PCNs	Gu et al. (2017) [95]
polyurethane foam-air partition coefficients	QSPR, MLR, ANN, SVM	170 organic compounds comprising 9 distinct classes (PAHs, benzenes, esters, aliphatic and cyclic hydrocarbons, polychlorinated biphenyls, musk, nitrogen and sulphur compounds, pesticides, other compounds)	368 molecular descriptors	Zhu et al. (2020) [96]

The information summarized in Table 4 shows that the application of the combination the QSPR model and chemometric methods is common. In the estimation of the water–polydimethylsiloxane [92] and n-octanol–water [21] partition coefficients of organic compounds, the best techniques turned out to be ANN and LS–SVM, respectively. This results in a significant improvement in prediction quality. Two years later, Goudarzi and Goodarzi [94] conducted a prediction of the n-octanol–water partition coefficient for the same dataset of organic compounds but using different techniques, namely, MLR, PLS, and RBF-PLS (radial basic function-partial least squares). This time, due to flexible mapping of the selected features by manipulating their functional dependence implicitly unlike regression analysis, RBF-PLS is considered to be better than MLR and PLS models.

An interesting approach for the n-octanol–water partition coefficient for polychlorinated naphthalenes (PCNs) congener is proposed by Gu et al. [95], where QSAR is combined with comparative molecular field analysis (CoMFA) and comparative molecular similarity indices analysis (CoMSIA). These two models are dedicated to 3D-QSAR approaches, where the 3D conformation property of compounds has to be taken into account (possibility of exploring, visualizing a structural information, and designing new compounds with particular properties). Although the results of both models show good prediction ability, the CoMSIA model is better in designing new types of compound molecules due to the higher number of descriptors. The readiness of chemicals to concentrate in organisms when the compounds are present in the environment may also be defined by bioconcentration factor (BCF).

Prediction of this environmental property for some organic compounds using QSAR combined with GA-ANN (for the selection of appropriate descriptors) is proposed by Fatemi et al. [29].

6. Conclusions

There are various chemometric tools that can give benefits in terms of green chemistry. Application of even the simplest and well-known techniques for dimensionality reduction and grouping of objects or variables, such as CA or PCA, may result in significant advantages. These are the treatments for missing data, so chemical parameters are predicted without performing problematic, time-consuming, and material-demanding measurements. Even finding correlations in the dataset can give clues on the selection of proper materials. In this way, there is a possibility of estimation of the environmental fate of chemical compounds if the predicted datapoints refer to their behaviour in the environment. Reducing the number of elements in the dataset by grouping objects according to similarities leads to a preselection of objects for further consideration by more detailed studies. Selection of chemical compounds with similar characteristics by chemometric techniques is helpful in finding greener alternatives, compounds that are less problematic but retain their desired features. Multivariate statistics are successfully applied in green chemistry studies, and their significance is expected to be growing.

Author Contributions: Conceptualization, M.B. and M.T.; data curation, M.B.; writing—original draft preparation, M.B. and M.T.; writing—review and editing, M.T.; supervision, M.T. All authors have read and agreed to the published version of the manuscript.

References

1. Brereton, R.G. Chemometrics in analytical chemistry. A review. *Analyst* **1987**, *112*, 1635–1657. [CrossRef]
2. Kiralj, R.; Ferreira, M.M.C. The past, present, and future of chemometrics worldwide: Some etymological, linguistic, and bibliometric investigations. *J. Chemom.* **2006**, *20*, 247–272. [CrossRef]
3. Brereton, R.G.; Jansen, J.; Lopes, J.; Marini, F.; Pomerantsev, A.; Rodionova, O.; Roger, J.M.; Walczak, B.; Tauler, R. Chemometrics in analytical chemistry—Part I: History, experimental design and data analysis tools. *Anal. Bioanal. Chem.* **2017**, *409*, 5891–5899. [CrossRef] [PubMed]
4. Santos, M.C.; Nascimento, P.; Guedes, W.N.; Pereira-Filho, E.R.; Filletti, É.R.; Pereira, F.V. Chemometrics in analytical chemistry—An overview of applications from 2014 to 2018. *Eclética Química J.* **2019**, *44*, 11–25. [CrossRef]
5. Defernez, M.; Kemsley, E. The use and misuse of chemometrics for treating classification problems. *TrAC Trends Anal. Chem.* **1997**, *16*, 216–221. [CrossRef]
6. Rácz, A.; Bajusz, D.; Héberger, K. Chemometrics in Analytical Chemistry. In *Applied Chemoinformatics: Achievements and Future Opportunities*; Engel, T., Gasteiger, J., Eds.; Wiley-VCH: Weinheim, Germany, 2018; pp. 471–499. [CrossRef]
7. Hastie, T.; Tibshirani, R.; Friedman, J. *The Elements of Statistical Learning: Data Mining, Inference, and Prediction*; Springer: New York, NY, USA, 2009.
8. Vončina, D.B. Chemometrics in analytical chemistry. *Nova Biotechnol.* **2009**, *211*, 211–216.
9. Camacho, J.; Picó, J.; Ferrer, A. Data understanding with PCA: Structural and Variance Information plots. *Chemom. Intell. Lab. Syst.* **2010**, *100*, 48–56. [CrossRef]
10. Huang, B.K.; Huang, L.Q.; Qin, L.P. Cluster analysis of NIR fingerprint of four species plants in *Valeriana officinalis* L. *J. Chin. Med. Mater.* **2008**, *31*, 1494–1496.
11. Mohsin, G.F.; Schmitt, F.-J.; Kanzler, C.; Hoehl, A.; Hornemann, A. PCA-based identification and differentiation of FTIR data from model melanoidins with specific molecular compositions. *Food Chem.* **2019**, *281*, 106–113. [CrossRef]
12. Guo, Y.; Ni, Y.; Kokot, S. Evaluation of chemical components and properties of the jujube fruit using near infrared spectroscopy and chemometrics. *Spectrochim. Acta Part A Mol. Biomol. Spectrosc.* **2016**, *153*, 79–86. [CrossRef]
13. Massart, D.L. *The Interpretation of Analytical Chemical Data by the Use of Cluster Analysis*; John Wiley & Sons: New York, NY, USA, 1983.

14. Liang, Y.Z.; Xie, P.; Chan, K. Quality control of herbal medicines. *J. Chromatogr. B* **2004**, *812*, 53–70. [CrossRef]
15. Wesołowski, M.; Konieczynski, P. Thermoanalytical, chemical and principal component analysis of plant drugs. *Int. J. Pharm.* **2003**, *262*, 29–37. [CrossRef]
16. Bansal, A.; Chhabra, V.; Rawal, R.K.; Sharma, S. Chemometrics: A new scenario in herbal drug standardization. *J. Pharm. Anal.* **2014**, *4*, 223–233. [CrossRef]
17. Li, S.; Ng, T.-T.; Yao, Z.P. Quantitative analysis of blended oils by matrix-assisted laser desorption/ionization mass spectrometry and partial least squares regression. *Food Chem.* **2020**, *334*, 127601. [CrossRef] [PubMed]
18. Wold, S.; Kettaneh, N.; Tjessem, K. Hierarchical multiblock PLS and PC models for easier model interpretation and as an alternative to variable selection. *J. Chemom.* **1996**, *10*, 463–482. [CrossRef]
19. Denham, M.C. Choosing the number of factors in partial least squares regression: Estimating and minimizing the mean squared error of prediction. *J. Chemom.* **2000**, *14*, 351–361. [CrossRef]
20. Wold, S.; Sjöström, M.; Eriksson, L. PLS-regression: A basic tool of chemometrics. *Chemom. Intell. Lab. Syst.* **2001**, *58*, 109–130. [CrossRef]
21. Goudarzi, N.; Goodarzi, M. Prediction of the logarithmic of partition coefficients (log P) of some organic compounds by least square-support vector machine (LS-SVM). *Mol. Phys.* **2008**, *106*, 2525–2535. [CrossRef]
22. Lee, H.; Park, Y.M.; Lee, S. Principal Component Regression by Principal Component Selection. *Commun. Stat. Appl. Methods* **2015**, *22*, 173–180. [CrossRef]
23. Tong, W.; Hong, H.; Xie, Q.; Shi, L.; Fang, H.; Perkins, R. Assessing QSAR Limitations—A Regulatory Perspective. *Curr. Comput. Aided Drug Des.* **2005**, *1*, 195–205. [CrossRef]
24. Tropsha, A.; Golbraikh, A. Predictive QSAR Modeling Workflow, Model Applicability Domains, and Virtual Screening. *Curr. Pharm. Des.* **2007**, *13*, 3494–3504. [CrossRef] [PubMed]
25. Hibbert, D.B. Genetic algorithms in chemistry. *Chemom. Intell. Lab. Syst.* **1993**, *19*, 277–293. [CrossRef]
26. Mehrotra, K.; Mohan, C.K.; Ranka, S. *Elements of Artificial Neural Networks*; MIT Press: Cambridge, MA, USA, 2000.
27. Attia, K.A.; Nassar, M.W.; El-Zeiny, M.B.; Serag, A. Effect of genetic algorithm as a variable selection method on different chemometric models applied for the analysis of binary mixture of amoxicillin and flucloxacillin: A comparative study. *Spectrochim. Acta Part A Mol. Biomol. Spectrosc.* **2016**, *156*, 54–62. [CrossRef] [PubMed]
28. Golmohammadi, H.; Dashtbozorgi, Z. Quantitative structure–property relationship studies of gas-to-wet butyl acetate partition coefficient of some organic compounds using genetic algorithm and artificial neural network. *Struct. Chem.* **2010**, *21*, 1241–1252. [CrossRef]
29. Fatemi, M.H.; Jalali-Heravi, M.; Konuze, E. Prediction of bioconcentration factor using genetic algorithm and artificial neural network. *Anal. Chim. Acta* **2003**, *486*, 101–108. [CrossRef]
30. Gere, A.; Rácz, A.; Bajusz, D.; Károly, H. Multicriteria decision making for evergreen problems in food science by sum of ranking differences. *Food Chem.* **2020**, 128617. [CrossRef]
31. Cover, T.; Hart, P. Nearest neighbor pattern classification. *IEEE Trans. Inf. Theory* **1967**, *13*, 21–27. [CrossRef]
32. Cao, D.-S.; Dong, J.; Wang, N.-N.; Wen, M.; Deng, B.-C.; Zeng, W.-B.; Xu, Q.-S.; Liang, Y.-Z.; Lu, A.-P.; Chen, A.F. In silico toxicity prediction of chemicals from EPA toxicity database by kernel fusion-based support vector machines. *Chemom. Intell. Lab. Syst.* **2015**, *146*, 494–502. [CrossRef]
33. Li, X.; Kong, W.; Shi, W.; Shen, Q. A combination of chemometrics methods and GC–MS for the classification of edible vegetable oils. *Chemom. Intell. Lab. Syst.* **2016**, *155*, 145–150. [CrossRef]
34. García, J.I.; Garcia-Marin, H.; Mayoral, J.A.; Pérez, P. Quantitative structure–property relationships prediction of some physico-chemical properties of glycerol based solvents. *Green Chem.* **2013**, *15*, 2283–2293. [CrossRef]
35. Tobiszewski, M.; Tsakovski, S.; Simeonov, V.; Namieśnik, J.; Pena-Pereira, F. A solvent selection guide based on chemometrics and multicriteria decision analysis. *Green Chem.* **2015**, *17*, 4773–4785. [CrossRef]
36. Alfonsi, K.; Colberg, J.; Dunn, P.J.; Fevig, T.; Jennings, S.; Johnson, T.A.; Kleine, H.P.; Knight, C.; Nagy, M.A.; Perry, D.A.; et al. Green chemistry tools to influence a medicinal chemistry and research chemistry based organisation. *Green Chem.* **2008**, *10*, 31–36. [CrossRef]
37. Henderson, R.K.; Jiménez-González, C.; Constable, D.J.C.; Alston, S.R.; Inglis, G.G.A.; Fisher, G.; Sherwood, J.; Binks, S.P.; Curzons, A.D. Expanding GSK's solvent selection guide–embedding sustainability into solvent selection starting at medicinal chemistry. *Green Chem.* **2011**, *13*, 854–862. [CrossRef]
38. Hargreaves, C.R.; Manley, J.B. ACS GCI Pharmaceutical Roundtable–Collaboration to Deliver a Solvent Selection Guide for the Pharmaceutical Industry. 2008. Available online: http://www.acs.org/content/dam/acsorg/greenchemistry/industriainnovation/roundtable/solvent-selection-guide.pdf (accessed on 3 August 2020).

39. Prat, D.; Pardigon, O.; Flemming, H.-W.; Letestu, S.; Ducandas, V.; Isnard, P.; Guntrum, E.; Senac, T.; Ruisseau, S.; Cruciani, P.; et al. Sanofi's Solvent Selection Guide: A Step Toward More Sustainable Processes. *Org. Process. Res. Dev.* **2013**, *17*, 1517–1525. [CrossRef]

40. Prat, D.; Hayler, J.; Wells, A. A survey of solvent selection guides. *Green Chem.* **2014**, *16*, 4546–4551. [CrossRef]

41. Sels, H.; De Smet, H.; Geuens, J. SUSSOL—Using Artificial Intelligence for Greener Solvent Selection and Substitution. *Molecules* **2020**, *25*, 3037. [CrossRef] [PubMed]

42. Papa, E.; Gramatica, P. QSPR as a support for the EU REACH regulation and rational design of environmentally safer chemicals: PBT identification from molecular structure. *Green Chem.* **2010**, *12*, 836–843. [CrossRef]

43. Chastrette, M.; Rajzmann, M.; Chanon, M.; Purcell, K.F. Approach to a general classification of solvents using a multivariate statistical treatment of quantitative solvent parameters. *J. Am. Chem. Soc.* **1985**, *107*, 1–11. [CrossRef]

44. Dutkiewicz, M. Classification of organic solvents based on correlation between dielectric β parameter and empirical solvent polarity parameter ENT. *J. Chem. Soc. Faraday Trans.* **1990**, *86*, 2237–2241. [CrossRef]

45. Pytela, O. A new classification of solvents based on chemometric empirical scale of parameters. *Collect. Czechoslov. Chem. Commun.* **1990**, *55*, 644–652. [CrossRef]

46. Gramatica, P.; Navas, N.; Todeschini, R. Classification of organic solvents and modelling of their physico-chemical properties by chemometric methods using different sets of molecular descriptors. *TrAC Trends Anal. Chem.* **1999**, *18*, 461–471. [CrossRef]

47. Pushkarova, Y.; Kholin, Y.V. A procedure for meaningful unsupervised clustering and its application for solvent classification. *Cent. Eur. J. Chem.* **2014**, *12*, 594–603. [CrossRef]

48. Levet, A.; Bordes, C.; Clément, Y.; Mignon, P.; Chermette, H.; Forquet, V.; Morell, C.; Lantéri, P. Solvent database and in silico classification: A new methodology for solvent substitution and its application for microencapsulation process. *Int. J. Pharm.* **2016**, *509*, 454–464. [CrossRef] [PubMed]

49. Guidea, A.; Sârbu, C. Fuzzy characterization and classification of solvents according to their polarity and selectivity. A comparison with the Snyder approach. *J. Liq. Chromatogr. Relat. Technol.* **2020**, *43*, 336–343. [CrossRef]

50. Salahinejad, M. Application of classification models to identify solvents for single-walled carbon nanotubes dispersion. *RSC Adv.* **2015**, *5*, 22391–22398. [CrossRef]

51. Katritzky, A.R.; Fara, D.C.; Kuanar, M.; Hür, E.; Karelson, M. The Classification of Solvents by Combining Classical QSPR Methodology with Principal Component Analysis. *J. Phys. Chem. A* **2005**, *109*, 10323–10341. [CrossRef]

52. Tobiszewski, M.; Nedyalkova, M.; Madurga, S.; Pena-Pereira, F.; Namieśnik, J.; Simeonov, V. Pre-selection and assessment of green organic solvents by clustering chemometric tools. *Ecotoxicol. Environ. Saf.* **2018**, *147*, 292–298. [CrossRef]

53. Nedyalkova, M.; Sârbu, C.; Tobiszewski, M.; Simeonov, V. Fuzzy Divisive Hierarchical Clustering of Solvents According to Their Experimentally and Theoretically Predicted Descriptors. *Symmetry* **2020**, *12*, 1763. [CrossRef]

54. González-Álvarez, J.; Mangas-Alonso, J.J.; Arias-Abrodo, P.; Gutiérrez-Álvarez, M.D. A chemometric approach to characterization of ionic liquids for gas chromatography. *Anal. Bioanal. Chem.* **2014**, *406*, 3149–3155. [CrossRef]

55. Izadiyan, P.; Fatemi, M. Chemometric classification of 227 Ionic Liquids and their related salts according to their toxicities to Rat Cell Lines. In Proceedings of the Iranian Biennial Chemometrics Seminar, Tabriz, Iran, 9–10 November 2011.

56. Lesellier, E. Σpider diagram: A universal and versatile approach for system comparison and classification: Application to solvent properties. *J. Chromatogr. A* **2015**, *1389*, 49–64. [CrossRef]

57. Adamska, K.; Voelkel, A.; Héberger, K. Selection of solubility parameters for characterization of pharmaceutical excipients. *J. Chromatogr. A* **2007**, *1171*, 90–97. [CrossRef] [PubMed]

58. Sild, S.; Piir, G.; Neagu, D.; Maran, U. Storing and Using Qualitative and Quantitative Structure–Activity Relationships in the Era of Toxicological and Chemical Data Expansion. In *Big Data in Predictive Toxicology*; Neagu, D., Richarz, A.N., Eds.; Royal Society of Chemistry: London, UK, 2020; pp. 185–213. [CrossRef]

59. EPA Website. Available online: https://www.epa.gov/tsca-screening-tools/epi-suitetm-estimation-program-interface (accessed on 30 January 2020).

60. Gerrity, D.; Stanford, B.D.; Trenholm, R.A.; Snyder, S.A. An evaluation of a pilot-scale nonthermal plasma advanced oxidation process for trace organic compound degradation. *Water Res.* **2010**, *44*, 493–504. [CrossRef] [PubMed]

61. Coleman, D.; Gathergood, N. Biodegradation studies of ionic liquids. *Chem. Soc. Rev.* **2010**, *39*, 600–637. [CrossRef] [PubMed]

62. Siedlecka, E.M.; Czerwicka, M.; Neumann, J.; Stepnowski, P.; Fernández, J.F.; Thöming, J. Ionic Liquids: Methods of Degradation and Recovery. In *Ionic Liquids: Theory, Properties, New Approaches*; Kokorin, A., Ed.; IntechOpen: Rijeka, Croatia, 2011; pp. 701–722. [CrossRef]

63. Matzke, M.; Thiele, K.; Müller, A.; Filser, J. Sorption and desorption of imidazolium based ionic liquids in different soil types. *Chemosphere* **2009**, *74*, 568–574. [CrossRef] [PubMed]

64. Stepnowski, P.; Mrozik, W.; Nichthauser, J. Adsorption of Alkylimidazolium and Alkylpyridinium Ionic Liquids onto Natural Soils. *Environ. Sci. Technol.* **2007**, *41*, 511–516. [CrossRef]

65. Stolte, S.; Arning, J.; Bottin-Weber, U.; Müller, A.; Pitner, W.-R.; Welz-Biermann, U.; Jastorff, B.; Ranke, J. Effects of different head groups and functionalised side chains on the cytotoxicity of ionic liquids. *Green Chem.* **2007**, *9*, 760–767. [CrossRef]

66. Bystrzanowska, M.; Pena-Pereira, F.; Marcinkowski, Ł.; Tobiszewski, M. How green are ionic liquids?—A multicriteria decision analysis approach. *Ecotoxicol. Environ. Saf.* **2019**, *174*, 455–458. [CrossRef]

67. Torrecilla, J.S.; Palomar, J.; García, J.; Rojo, E.; Rodríguez, F. Modelling of carbon dioxide solubility in ionic liquids at sub and supercritical conditions by neural networks and mathematical regressions. *Chemom. Intell. Lab. Syst.* **2008**, *93*, 149–159. [CrossRef]

68. Torrecilla, J.S.; Rodríguez, F.; Bravo, J.L.; Rothenberg, G.; Seddon, K.R.; López-Martin, I. Optimising an artificial neural network for predicting the melting point of ionic liquids. *Phys. Chem. Chem. Phys.* **2008**, *10*, 5826–5831. [CrossRef]

69. Valderrama, J.O.; Muñoz, J.M.; Rojas, R.E. Viscosity of ionic liquids using the concept of mass connectivity and artificial neural networks. *Korean J. Chem. Eng.* **2011**, *28*, 1451–1457. [CrossRef]

70. Cao, Y.; Yu, J.; Song, H.; Wang, X.; Yao, S. Prediction of electric conductivity for ionic liquids by two chemometrics methods. *J. Serbian Chem. Soc.* **2013**, *78*, 653–667. [CrossRef]

71. Huang, Y.; Zhao, Y.; Zeng, S.; Zhang, S.; Zhang, S. Density Prediction of Mixtures of Ionic Liquids and Molecular Solvents Using Two New Generalized Models. *Ind. Eng. Chem. Res.* **2014**, *53*, 15270–15277. [CrossRef]

72. Barycki, M.; Sosnowska, A.; Piotrowska, M.; Urbaszek, P.; Rybińska, A.; Grzonkowska, M.; Puzyn, T. ILPC: Simple chemometric tool supporting the design of ionic liquids. *J. Cheminform.* **2016**, *8*, 40. [CrossRef] [PubMed]

73. Studzińska, S.; Stepnowski, P.; Buszewski, B. Application of Chromatography and Chemometrics to Estimate Lipophilicity of Ionic Liquid Cations. *QSAR Comb. Sci.* **2007**, *26*, 963–972. [CrossRef]

74. Kurtanjek, Ž. Chemometric versus Random Forest Predictors of Ionic Liquid Toxicity. *Chem. Biochem. Eng. Q.* **2014**, *28*, 459–463. [CrossRef]

75. Sosnowska, A.; Barycki, M.; Zaborowska, M.; Rybińska-Fryca, A.; Puzyn, T. Towards designing environmentally safe ionic liquids: The influence of the cation structure. *Green Chem.* **2014**, *16*, 4749–4757. [CrossRef]

76. Zhu, P.; Kang, X.; Zhao, Y.; Latif, U.; Zhang, H. Predicting the Toxicity of Ionic Liquids toward Acetylcholinesterase Enzymes Using Novel QSAR Models. *Int. J. Mol. Sci.* **2019**, *20*, 2186. [CrossRef]

77. Wu, T.; Li, W.; Chen, M.; Zhou, Y.; Zhang, Q. Estimation of Ionic Liquids Toxicity against Leukemia Rat Cell Line IPC-81 based on the Empirical-like Models using Intuitive and Explainable Fingerprint Descriptors. *Mol. Inform.* **2020**, *39*, 2000102. [CrossRef]

78. Allus, M.A.; Brereton, R.G.; Nickless, G. Chemometric studies of the effect of toxic metals on plants: The use of response surface methodology to investigate the influence of Tl, Cd and Ag on the growth of cabbage seedlings. *Environ. Pollut.* **1988**, *52*, 169–181. [CrossRef]

79. Dearden, J.C.; Cronin, M.T.D.; Schultz, T.W.; Lin, D.T. QSAR Study of the Toxicity of Nitrobenzenes toTetrahymena pyriformis. *Quant. Struct. Relatsh.* **1995**, *14*, 427–432. [CrossRef]

80. Niazi, A.; Jameh-Bozorghi, S.; Nori-Shargh, D. Prediction of toxicity of nitrobenzenes using ab initio and least squares support vector machines. *J. Hazard. Mater.* **2008**, *151*, 603–609. [CrossRef] [PubMed]

81. Roy, K.; Ghosh, G. QSTR with Extended Topochemical Atom Indices. 10. Modeling of Toxicity of Organic Chemicals to Humans Using Different Chemometric Tools. *Chem. Biol. Drug Des.* **2008**, *72*, 383–394. [CrossRef] [PubMed]

82. Tan, N.X.; Li, P.; Rao, H.B.; Li, Z.-R.; Li, X.-Y. Prediction of the acute toxicity of chemical compounds to the fathead minnow by machine learning approaches. *Chemom. Intell. Lab. Syst.* **2010**, *100*, 66–73. [CrossRef]

83. Kar, S.; Roy, K. QSAR modeling of toxicity of diverse organic chemicals to Daphnia magna using 2D and 3D descriptors. *J. Hazard. Mater.* **2010**, *177*, 344–351. [CrossRef] [PubMed]

84. Bhhatarai, B.; Gramatica, P. Oral LD50 toxicity modeling and prediction of per- and polyfluorinated chemicals on rat and mouse. *Mol. Divers.* **2011**, *15*, 467–476. [CrossRef]

85. Hamadache, M.; Hanini, S.; Benkortbi, O.; Amrane, A.; Khaouane, L.; Moussa, C.S. Artificial neural network-based equation to predict the toxicity of herbicides on rats. *Chemom. Intell. Lab. Syst.* **2016**, *154*, 7–15. [CrossRef]

86. Khan, P.M.; Roy, K.; Benfenati, E. Chemometric modeling of Daphnia magna toxicity of agrochemicals. *Chemosphere* **2019**, *224*, 470–479. [CrossRef]

87. Nedyalkova, M.; Donkova, B.V.; Simeonov, V. Chemometrics Expertise in the Links Between Ecotoxicity and Physicochemical Features of Silver Nanoparticles: Environmental Aspects. *J. AOAC Int.* **2017**, *100*, 359–364. [CrossRef]

88. Nedyalkova, M.; Dimitrov, D.; Donkova, B.; Simeonov, V. Chemometric Evaluation of the Link between Acute Toxicity, Health Issues and Physicochemical Properties of Silver Nanoparticles. *Symmetry* **2019**, *11*, 1159. [CrossRef]

89. Waring, M.J. Lipophilicity in drug discovery. *Expert Opin. Drug Discov.* **2010**, *5*, 235–248. [CrossRef]

90. Chen, M.; Borlak, J.; Tong, W. High lipophilicity and high daily dose of oral medications are associated with significant risk for drug-induced liver injury. *Hepatology* **2013**, *58*, 388–396. [CrossRef] [PubMed]

91. Puzyn, T.; Falandysz, J. Computational estimation of logarithm of n-octanol/air partition coefficient and subcooled vapor pressures of 75 chloronaphthalene congeners. *Atmos. Environ.* **2005**, *39*, 1439–1446. [CrossRef]

92. Golmohammadi, H.; Dashtbozorgi, Z. Prediction of water-to-polydimethylsiloxane partition coefficient for some organic compounds using QSPR approaches. *J. Struct. Chem.* **2010**, *51*, 833–846. [CrossRef]

93. Yang, S.-S.; Lu, W.C.; Gu, T.-H.; Yan, L.-M.; Li, G.-Z. QSPR Study of n-Octanol/Water Partition Coefficient of Some Aromatic Compounds Using Support Vector Regression. *QSAR Comb. Sci.* **2008**, *28*, 175–182. [CrossRef]

94. Goudarzi, N.; Goodarzi, M. QSPR study of partition coefficient (Ko/w) of some organic compounds using radial basic function-partial least square (RBF-PLS). *J. Braz. Chem. Soc.* **2010**, *21*, 1776–1783. [CrossRef]

95. Gu, W.; Chen, Y.; Zhang, L.; Li, Y. Prediction of octanol-water partition coefficient for polychlorinated naphthalenes through three-dimensional QSAR models. *Hum. Ecol. Risk Assess. Int. J.* **2017**, *23*, 40–55. [CrossRef]

96. Zhu, T.; Gu, L.; Chen, M.; Sun, F. Exploring QSPR models for predicting PUF-air partition coefficients of organic compounds with linear and nonlinear approaches. *Chemosphere* **2020**, 128962. [CrossRef]

3

Untargeted Metabolomic Profile for the Detection of Prostate Carcinoma—Preliminary Results from PARAFAC2 and PLS–DA Models

Eleonora Amante [1,2], Alberto Salomone [1,2], Eugenio Alladio [1,2], Marco Vincenti [1,2,*], Francesco Porpiglia [3] and Rasmus Bro [4]

[1] Dipartimento di Chimica, Università degli Studi di Torino, Via P. Giuria 7, 10125 Torino, Italy
[2] Centro Regionale Antidoping e di Tossicologia "A. Bertinaria", Regione Gonzole 10/1, 10043 Orbassano, Italy
[3] Division of Urology, San Luigi Gonzaga Hospital and University of Torino, 10043 Orbassano, Italy
[4] Department of Food Science, Faculty of Science, University of Copenhagen, Rolighedsvej 30, 1958 Frederiksberg, Denmark
* Correspondence: marco.vincenti@unito.it

Abstract: Prostate-specific antigen (PSA) is the main biomarker for the screening of prostate cancer (PCa), which has a high sensibility (higher than 80%) that is negatively offset by its poor specificity (only 30%, with the European cut-off of 4 ng/mL). This generates a large number of useless biopsies, involving both risks for the patients and costs for the national healthcare systems. Consequently, efforts were recently made to discover new biomarkers useful for PCa screening, including our proposal of interpreting a multi-parametric urinary steroidal profile with multivariate statistics. This approach has been expanded to investigate new alleged biomarkers by the application of untargeted urinary metabolomics. Urine samples from 91 patients (43 affected by PCa; 48 by benign hyperplasia) were deconjugated, extracted in both basic and acidic conditions, derivatized with different reagents, and analyzed with different gas chromatographic columns. Three-dimensional data were obtained from full-scan electron impact mass spectra. The PARADISe software, coupled with NIST libraries, was employed for the computation of PARAFAC2 models, the extraction of the significative components (alleged biomarkers), and the generation of a semiquantitative dataset. After variables selection, a partial least squares–discriminant analysis classification model was built, yielding promising performances. The selected biomarkers need further validation, possibly involving, yet again, a targeted approach.

Keywords: untargeted metabolomics; PARAFAC2; alignment; gas chromatography–mass spectrometry (GC–MS); prostate carcinoma

1. Introduction

Prostate cancer (PCa) is the most common non-skin cancer in men [1,2] and the second most frequently diagnosed malignancy in males worldwide [3]. The first biomarker for PCa detection was prostatic acid phosphatase (PAP), which was introduced in the 1930s [1]. In the 1980s, PAP was replaced by prostate-specific antigen (PSA) [1,4], a secreted protein encoded by a prostate-specific gene and member of the tissue kallikrein family [1], which is produced almost exclusively in the prostate [5,6]. After the introduction of PSA, more men were diagnosed with PCa, with the majority having the early-stage, clinically indolent form of the disease. However, a large number of patients affected by a benign pathology, such as inflammation or hyperplasia, exhibited abnormal PSA values, which lead to the execution of useless biopsies and demonstrate the low specificity of this biomarker [1,6]. This phenomenon was generally designated as "overdiagnosis" or "overtreatment" [1,4,7,8].

Considerable effort has been devoted to improving the PSA-test performance, including the introduction of PSA density, PSA velocity (and doubling time), the dosage of free or complexed PSA, and the quantitation of its isoforms [1,2]. A combination of these parameters yields the Prostate Health Index (PHI) [3].

Meanwhile, intensive research has been devoted to the search for different biomarkers, mainly by applying omics methods (e.g., genomics, proteomics, transcriptomics, and metabolomics) [1], and several authors have reviewed the emerging biomarkers, among which the most prominent are the urinary prostate cancer antigen 3 (PCA3) [1,3,5] and transmembrane protease, serine 2 (TMPRSS2-ERG) (sometimes combined together) [1,3,5]. Alpha-methylacyl-CoA Racemase (AMACR) demonstrated high sensitivities and specificities on prostate biopsy, but it is not suitable for non-invasive detection in urine [1,5]. Increased diagnostic performances were obtained by the serum dosage of human kallikrein-related peptidase 2 (KLK2) in combination with total and free PSA [5]. An evolution of the application of kallikreins consists in a blood measurement of the four existing isoforms which, combined with clinical information, allows the probability calculation of PCa incidence [3]. Significantly increased levels of prostasomes were found in blood samples from patients with PCa [9], while elevated levels of urinary sarcosine were found to be associated with aggressive forms of prostate cancer [1].

Studies conducted in the 1970s and 1980s highlighted the correlation between increased urinary excretion of polyamines (i.e., spermine, spermidine, and putrescine) and several types of cancer [10,11]. However, anomalous oxidative degradation reactions of these polyamines resulted in low concentrations of these biomarkers in approximately 20% of the patients, leading to false-negative prediction and consequently limiting their application as diagnostic biomarkers [10].

The correlation between altered steroidal biosynthesis and PCa is well known [12–14]. For this reason, in a previous study, we carried out a targeted analysis of urine samples, addressed to a large panel of androgens, including testosterone and its principal phase I metabolites. The multivariate statistical interpretation of these steroid profiles produced satisfactory results in terms of sensitivity, specificity, and area under the curve (AUC) [15].

In this study, the search for new urinary biomarkers was undertaken by using untargeted methods. In perspective, emerging biomarkers could possibly be combined with the most discriminating steroid biomarkers to improve their screening performances further, without altering the inherent simplicity of the instrumental procedure. In fact, the ideal biomarker should be cheap to determine, non-invasive, easily accessible, and quickly quantifiable [1,2]. Taking into account the abovementioned considerations, gas chromatography–electron impact mass spectrometry (GC–EIMS) would give a more suitable solution than the other commonly used analytical techniques to provide a three-dimensional pattern for untargeted analysis. Urine was chosen as the election matrix, as it is easily available in large volumes and involves non-invasive sampling.

2. Materials and Methods

2.1. Chemicals and Reagents

Tert-butyl methyl ether (TBME), ethyl acetate, dithioerythritol, ammonium iodide (NH$_4$I), N-Methyl-N-(trimethylsilyl)trifluoroacetamide (TMSTFA), and trifluoroacetic anhydride (TFAA) were provided by Sigma-Aldrich (Milan, Italy). β-glucuronidase from *Escherichia coli* was purchased from Roche Life Science (Indianapolis, IN, USA). Ultra-pure water was obtained using a Milli-Q® UF-Plus apparatus (Millipore, Bedford, MA, USA).

2.2. Samples Collection

The subjects involved in this study were recruited in the ambulatory of the Department of Urology at the San Luigi Hospital of Orbassano (TO, Italy), after approval of the protocol by the reference Ethical

Committee (protocol number 0019267). A total of 91 subjects were enrolled, including 43 affected by prostate carcinoma (PCa, confirmed by a positive biopsy) and 48 diagnosed with benign prostatic hyperplasia (BPH, with a PSA lower than the European cut-off of 4 ng/mL or with a PSA above the threshold but a negative biopsy result). In a previous study, the progressive modification of the urinary steroidal profile with age was investigated [16]. From this study, we decided to enroll only individuals older than 60 years, when the bias effect due to aging became negligible [16]. Moreover, since ethnicity represents another important bias factor, only Caucasian individuals were recruited. Finally, diabetes, other carcinoma, metabolic diseases, and therapies suspected to alter the urinary steroid profile (such as steroid therapy) were considered as exclusion criteria.

Body mass index (BMI), alcohol consumption, medical therapy, digital rectal examination, PSA value, and biopsy Gleason Score (GS) were recorded. In detail, the group's mean age and standard deviation was 70 ± 10 years for BPH and 70 ± 8 years for PCa. BMI was within the range of normality for all individuals (between 18.5 and 25), and PSA was 3.8 ± 2.3 ng/mL for BPH and 11.0 ± 9.5 ng/mL for PCa. The PCa class was distributed as low risk (GS = 3 + 3, 15 patients), middle risk (GS = 3 + 4 and 4 + 3, 21 patients), and high risk (GS = 4 + 4 and 4 + 5, seven patients).

2.3. Sample Treatment and GC–MS Analysis

Firstly, the protein components of the urinary samples were precipitated by centrifugation at 4000 rpm for 5 min. Two aliquots (A and B) of 5 mL each were taken from each sample. The urine pH was adjusted between 6.8 and 7.4 by adding 2 mL of phosphate buffer and a few drops of NaOH 1M or HCl 1M whenever necessary. Enzymatic hydrolysis of the glucuronide metabolites was conducted with 100 µL of β-glucuronidase from *Escherichia coli* (equivalent to 83 enzymatic units) by heating it in the oven for 1 h. After cooling to room temperature, the two aliquots were subjected to different liquid–liquid extraction (LLE) with 5 mL of TBME each, at basic (pH ≥ 10) and acid (pH ≤ 1) conditions, respectively, obtained by the introduction of some drops of NaOH 1M and HCl 1M. Both aliquots were dried under a gentle nitrogen stream at room temperature. The dried aliquot A was derivatized using 50 µL TFAA at 65 °C for 1 h. Then, the solvent was dried and the residue was dissolved in 50 µL TBME and injected into the GC–MS. The chromatographic separation was achieved with a J&W Scientific HP-5, 17 m × 0.2 mm (i.d.) × 0.33 µm (f.t.) capillary column. The oven temperature was programmed as follows: The starting temperature of 90 °C was held for 1 min. Then, the temperature of 180 °C was reached with a rate of 30 °C/min and held for 7 min. A final heating rate of 15 °C/min was applied until the temperature of 325 °C was reached (held for 3 min). The chromatographic run lasted 22.20 min.

Aliquot B was derivatized using 50 µL of TMSTFA/NH$_4$I/dithioerythritol (1.000:2:4 *v/w/w*), at 70 °C for 30 min and then injected into a GC–MS equipped with a J&W Scientific HP-1, 17 m × 0.2 mm (i.d.) × 0.11 µm (f.t.) capillary column. The oven temperature was programmed to heat up from 120 to 177 °C at a rate of 70 °C/min, and from 177 to 236 °C at a rate of 5 °C/min. A final heating rate of 30 °C/min was applied until the temperature of 315 °C was reached. The chromatographic run lasted 18.25 min. Both the runs were performed in full-scan mode, in the interval 40–650 *m/z* at a scan rate of 2.28 scans/s.

Because the samples were analyzed in five analytical sections performed on different days, it was important to monitor the occurrence of a data structure due to the different analytical sections. The exploratory unsupervised data analysis can serve to this scope, and principal component analysis (PCA) was employed. No clustering or trend related to the day of the analysis was detected.

2.4. Statistical Analysis

The main steps of the statistical analysis are reported in Figure 1.

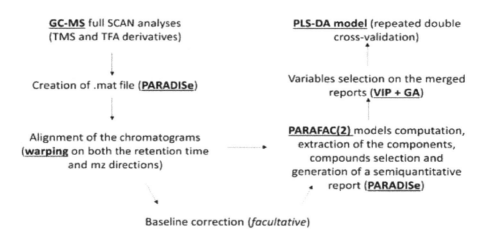

Figure 1. Statistical analysis workflow.

2.5. Pre-Treatment of the Raw Data

The .AIA files of the chromatographic runs were downloaded using the software ChemStation®. The PARADISe version 3 software was employed to convert the files in a form suitable for MATLAB (extension .mat). The alignment procedure, both propaedeutic and mandatory for the following steps of data analysis, was executed over the three-way (samples × retention time × m/z) array of size $91 \times 3099 \times 612$ (over 172 million data) and $91 \times 1640 \times 652$ (over 97 million data) for the trifluoroacetyl (TFA) and trimethylsilyl (TMS) derivatives, respectively. The correlation optimized warping (COW) was performed along the retention time and the m/z dimensions [17]. The two matrices were segmented along the retention time dimension to improve the performances of the COW algorithm, and for each slice the computation was iterated until a visually satisfying result was obtained. Lastly, to improve the visualization of the data, the baseline was subtracted. It is important to highlight that the latter computational step only served to improve the data visualization by the operator, because PARAFAC2 is able to recognize the baseline and the noise components, allowing their automatic exclusion [18–21].

2.6. PARAFAC2 Models Computation and Molecular Identification

The aligned dataset was analyzed in the PARADISe software, to proceed with the computation of PARAFAC2 models; the operating procedure consists in the manual identification of intervals along the chromatogram (with each interval ideally containing approximately one peak). The PARAFAC2 models were built introducing the non-negativity constraint and performing 10,000 iterations for interval [18–21].

Within the software, the operator can label the components as (i) baseline, (ii) noise, or (iii) compounds. All the mass spectra of the components belonging to the third category are automatically compared with the NIST database. A report is produced, including the relative concentrations of the detected compounds (assuming a uniform response factor of 1), and the n (number subjectively chosen by the user) most likely identifications for each compound. Finally, the relative concentrations were normalized using the urinary creatinine values.

2.7. Classification Models

The dataset composed by the relative concentrations of the detected metabolites for each sample was used to perform partial least squares–discriminant analysis (PLS–DA) [22], classifying the samples into having prostate carcinoma or not. Firstly, the dataset was log10 transformed (with the aim of achieving a more even distribution of each of the variables) and autoscaled. Then, the variable importance in projection (VIP) method (using a threshold of 1) [23] and genetic algorithms (GAs) [24] were run to select the most relevant variables. The reduced dataset was finally used to build the

PLS–DA classification model. The model was then validated using the repeated double cross-validation (dCV) approach [25]. The PLS_Toolbox version 8.5 (Eigenvector Research, Inc., Manson, WA, USA) was used to perform this part of the analysis [26].

3. Results and Discussion

The preliminary PARAFAC2 model extracted a total of 329 relevant compounds (184 from the chromatographic run after TMS derivatization and 145 after TFA derivatization). Of these, 89 were selected using the VIP algorithm, and a further 58 substances were discarded by one cycle of GAs. The final dataset, consisting in a 91 × 32 (subjects × variables) matrix, was employed to build a PLS–DA classification model. Due to the heterogeneity of the patients enrolled (in terms of pathology staging, prostatic volume, and PSA values), the model was validated using repeated double cross-validation (30 repetitions were performed) [25] instead of the standard external validation, in which the use of a limited and heterogeneous population may result in significant bias. The plot reporting the Y-value predicted in cross-validation (CV) in one of the several classification models produced during the repeated double cross-validation process is shown in Figure 2A. The corresponding receiver operating characteristic (ROC) curve is depicted in Figure 2B. The high values of the area under the curve (AUC) for both the estimated and cross-validated ROCs are an indicator of high performances and robustness of the model. In detail, using a discriminating Y-value of 0.5, the model provides 92.5 ± 2.2% sensitivity and 88.7 ± 3.9% specificity for the cancer-affected population. On average, misclassification occurred on about 3 ± 2 patients affected by carcinoma out of 43 and 5 ± 2 patients with hyperplasia out of 48.

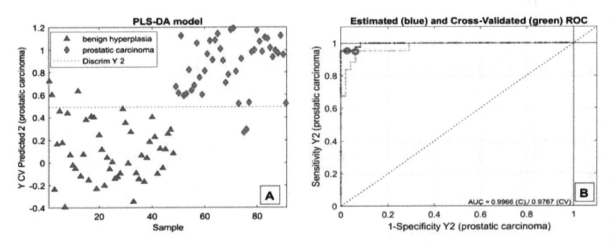

Figure 2. Y-value predicted in cross-validation (CV) (**A**) and receiver operating characteristic (ROC) curves (**B**) of one of the several classification models built during the repeated double cross-validation procedure.

Of the 32 compounds selected by the dedicated algorithms to build the model, 17 were not found in the available NIST libraries, while for seven other compounds, the identification provided by automatic spectral matching was deemed incorrect. On the other hand, manual mass spectra interpretation was made difficult by the structural similarity of many candidate biomarkers, as well as the effect of the derivatizations, that introduced functional groups (e.g., $-Si(CH_3)_3$) yielding prevalent fragment ions in the spectrum. The mass spectra of the 32 metabolites are provided as Supplementary Materials (Supplementary Figure S1). Table 1 reports the eight identified compounds, accompanied by their Human Metabolome Database (HMDB) and Kyoto Encyclopedia of Genes and Genomes (KEGG) identification numbers, when available. Since the PARADISe output provides only a rough semiquantitative report based on the total ion current (TIC) without any external calibration, the real physiological concentration of each metabolite could not be evaluated. However, these absolute TIC values can be evaluated in relative terms to provide an averaged qualitative comparison between the two populations for all the analytes. The overexpression and underexpression of these metabolites allegedly linked to the occurrence of PCa are reported in Table 1.

Table 1. List of the 32 selected metabolites. The kind of derivatization, retention time, and expression in prostate cancer (PCa)-affected individuals are reported. Moreover, metabolites with a putative identification are accompanied by the match score with NIST library and the relative identification (ID) number in the Human Metabolome Database (HMDB) and the Kyoto Encyclopedia of Genes and Genomes (KEGG) database. The mass spectra of the 32 metabolites are reported in Supplementary Materials—Figure S1.

		Compound	Derivatization	Retention Time (min)	Match with NIST	HMDB ID	KEGG ID	Expression in PCa Patients
TMS derivatives	1	5-Hydroxyindoleacetic acid	TMS	5.26	893	HMDB0000763	C05635	overexpression
	2	Unknown 1	TMS	5.86	-	-	-	overexpression
	3	Unknown 2	TMS	7.44	-	-	-	underexpression
	4	Androsterone	TMS	8.14	912	HMDB0000031	C00523	overexpression
	5	16-Hydroxydehydroisoandrosterone	TMS	9.23	888	HMDB0000352	C05139	overexpression
	6	Unknown 3	TMS	9.84	-	-	-	comparable
	7	Unknown 4	TMS	10.31	-	-	-	underexpression
	8	Unknown 5	TMS	10.61	-	-	-	underexpression
	9	Unknown 6	TMS	11.29	-	-	-	underexpression
	10	Unknown 7	TMS	11.32	-	-	-	comparable
	11	Enterodiol	TMS	12.19	826	HMDB0005056	C18166	underexpression
	12	5β-pregnanediol	TMS	12.53	853	HMDB0005943	Not available	underexpression
	13	Unknown 8	TMS	13.6	-	-	-	overexpression
	14	Unknown 9	TMS	13.67	-	-	-	comparable
	15	Pregnanetriol	TMS	13.73	904	HMDB0006070	Not available	underexpression
	16	Unknown 10	TMS	14.03	-	-	-	underexpression
	17	Unknown 11	TMS	14.50	-	-	-	underexpression
	18	Unknown 12	TMS	14.53	-	-	-	underexpression
	19	Unknown 13	TMS	14.6	-	-	-	overexpression
	20	Unknown 14	TMS	14.66	-	-	-	underexpression
	21	Unknown 15	TMS	15.04	-	-	-	underexpression
TFA derivatives	22	Unknown 16	TFA	1.63	-	-	-	underexpression
	23	Unknown 17	TFA	1.71	-	-	-	comparable
	24	Vanillyl alcohol	TFA	3.37	860	HMDB0032012	C06317	overexpression
	25	Unknown 18	TFA	4.97	-	-	-	comparable
	26	Unknown 19	TFA	5.71	-	-	-	underexpression
	27	Unknown 20	TFA	3.32	-	-	-	underexpression
	28	Epiandrosterone	TFA	15.61	925	HMDB0000365	C07635	comparable
	29	Unknown 21	TFA	16.32	-	-	-	underexpression
	30	Unknown 22	TFA	17.87	-	-	-	underexpression
	31	Unknown 23	TFA	18.11	-	-	-	overexpression
	32	Unknown 24	TFA	18.24	-	-	-	underexpression

It is interesting to note that among the eight identified compounds, five (63%) are involved in steroidal biosynthesis, confirming their potential in the detection and diagnosis of PCa. Similarly, Choi et al. found elevated levels of 16-hydroxy-dehydroepiandrosterone, epiandrosterone, etiocholanolone, and pregnanetriol in patients diagnosed with papillary thyroid carcinoma [27]. The first steroid appears to also be overexpressed in the present case for patients with PCa, but pregnanetriol was underexpressed in the same patients and epiandrosterone was found in comparable concentrations in the two populations. Dehydroepiandrosterone is involved in the expression of insulin-like growth factor 1, whose dysregulation is implicated in certain colon, liver, prostate, and breast cancers [28]. This observation may justify the inclusion of 16-hydroxy-dehydroepiandrosterone among the potential biomarkers for PCa. Pregnanetriol, together with 5 β-pregnanediol, is also known to be dysregulated in adrenal syndromes, such as adrenal tumors or Cushing's syndrome [29,30]. Increased androsterone levels were found in a cohort of PCa-affected individuals within a multivariate investigation of the urinary steroidal profile, and the present findings are in accordance with our previous study [16]. Other steroids that proved useful to discriminate PCa from BPH [27] were possibly overlooked in the present untargeted selection because of their low concentration in urine.

The overexpression of serotonin and its biomarkers (among which, 5-hydroxyindoleacetic acid) represents a potential urinary biomarker for neuroblastic and carcinoid tumors [31]. While there is no evidence in the literature of an association between 5-hydroxyindoleacetic acid and PCa, the present data suggest such a hypothesis, as its overexpression is clearly evident in the PCa-affected population considered.

Phytoestrogens are a class of substances accredited to prevent the onset of cancer [32,33]. In accordance with this hypothesis, enterodiol is underexpressed in the present PCa population.

Among the 32 selected biomarkers, different contributions to the overall discrimination achieved by the PLS–DA model (Figure 1) were expected. A rough estimation of the relative importance of these biomarkers is expressed by their selectivity ratios [34], reported in Figure 3. Nine biomarkers exhibit selectivity ratios higher that 0.1, while, for five others, values between 0.07 and 0.1 were found. Interestingly, out of the nine biomarkers with the highest selectivity ratio, eight are underexpressed in the PCa patients, apparently suggesting them as protective substances. The expression of the 14 metabolites is represented in the form of boxplots in Figure 4. Tentative PLS–DA models were built with only these 9 and 14 biomarkers, but their overall efficiency significantly dropped with respect to the model of 32 biomarkers, demonstrating that the relative contribution of the remaining biomarkers is not negligible. In particular, the specificity index was considerably reduced in the models of 9 biomarkers and 14 biomarkers, while the sensitivity score remained relatively high.

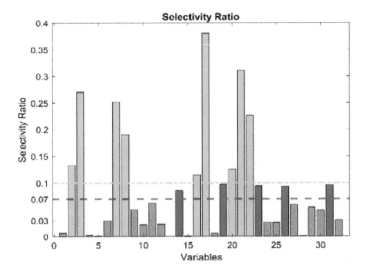

Figure 3. Selectivity ratio of the 32 selected features. The variables above the threshold of 0.1 are reported in green, and the ones between the thresholds 0.07–0.1 are reported in red.

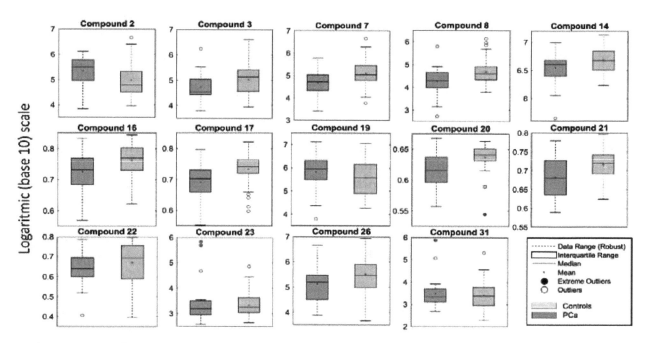

Figure 4. Boxplot, in logarithmic (base-10) scale, of the 14 compounds above the selectivity ratio threshold of 0.07 (see Figure 3).

Further testing was also conducted on the six biomarkers showing comparable mean intensity for the two populations. One variable at a time was removed, and a new classification model was computed using a simple cross-validation with each reduced dataset. Five of the seven new models yielded decreased sensitivity and specificity, while the other two models provided comparable performance, substantially confirming the choice of the 32 biomarker model.

4. Conclusions

The preliminary results reported in the present study support the premise that GC–MS tridimensional data can be profitably exploited in untargeted metabolomics studies devoted to prostatic carcinoma diagnosis. Compared to the more resource-demanding ultra-high-performance liquid chromatography–tandem mass spectrometer (UHPLC–MS/MS) and ultra-high-performance liquid chromatography–high-resolution mass spectrometry (UHPLC–HRMS) approaches frequently presented in the literature, GC–MS offers comparable chromatographic resolution and structured spectroscopic information, as is generated by the fragment ion pattern typical of electron impact ionization. On these complex data arrays, the ultimate performance in the extraction of crucial information relies on the software purposely adopted and PARAFAC2 combined with VIP and GA methods of variables selection proved to produce highly efficient models of class discrimination, allowing us to distinguish prostatic carcinoma from benign hyperplasia with good sensitivity and specificity scores.

A common limitation of untargeted metabolomics methods, including the present one, is that the most abundant components of the screened samples are preferentially isolated as potential biomarkers with respect to minor constituents, possibly present at trace levels. This explains the differences in the selected biomarkers with respect to the targeted approach that we previously tested [27]. On the other hand, complementary sets of biomarkers are extracted and then evaluated from targeted and untargeted approaches, to be subsequently combined to achieve optimal performance. More work has to be done on large populations of PCa-affected patients and controls to confirm the present findings, and further effort is necessary to reveal the identity of the most valuable biomarkers and possibly confirm their real value as interesting biomarkers by univariate statistics. Despite these limitations to be overcome in the subsequent investigations, the strategy adopted in the present study, based on non-invasive urine sampling, cheap instrumentation, and advanced data treatment by PARADISe software, appears to be extremely promising in PCa screening.

Author Contributions: All the Authors participated to the preliminary study design and planning. Recruitment, participation criteria, clinical evaluations, F.P.; Development of the analytical method, E.A. (Eleonora Amante) and A.S.; Sample processing, E.A. (Eleonora Amante); Method validation, E.A. (Eleonora Amante), E.A. (Eugenio Alladio), and M.V.; Chemometrics and statistical analysis, R.B., E.A. (Eleonora Amante), and E.A. (Eugenio Alladio); Mass spectra interpretation, M.V., E.A. (Eleonora Amante); Writing—Original Draft Preparation, E.A. (Eleonora Amante); Writing—Review & Editing, M.V. All Authors read and approved the final manuscript.

References

1.   Prenser, J.R.; Rubin, M.A.; Wei, J.T.; Chinnaiyan, A.M. Beyond PSA: The next generation of prostate cancer biomarkers. *Sci. Transl. Med.* **2012**, *4*, 127rv3.

2.   Velonas, V.M.; Woo, H.H.; Dos Remedios, C.G.; Assinder, S.J. Current status of biomarkers for prostate cancer. *Int. J. Mol. Sci.* **2013**, *14*, 11034–11060. [CrossRef] [PubMed]

3.   Hendriks, R.J.; Van Oort, I.M.; Schalken, J.A. Blood-based and urinary prostate cancer biomarkers: A review and comparison of novel biomarkers for detection and treatment decisions. *Prostate Cancer Prostatic Dis.* **2017**, *20*, 12–19. [CrossRef] [PubMed]

4.   Etzioni, R.; Penson, D.F.; Legler, J.M.; Di Tommaso, D.; Boer, R.; Gann, P.H.; Feuer, E.J. Overdiagnosis due to prostate-specific antigen screening: Lessons from U.S. prostate cancer incidence trends. *J. Natl. Cancer Inst.* **2002**, *94*, 981–990. [CrossRef] [PubMed]

5.   Sardana, G.; Dowell, B.; Diamandis, E.P. Emerging biomarkers for the diagnosis and prognosis of prostate cancer. *Clin. Chem.* **2008**, *54*, 1951–1960. [CrossRef] [PubMed]

6.   Kramer, B.S.; Brown, M.L.; Prorok, P.C.; Potosky, A.L.; Gohagan, J.K. Prostate cancer screening: What we know and what we need to know. *Ann. Intern. Med.* **1993**, *119*, 914–923. [CrossRef] [PubMed]

7.   Moyer, V.A. Screening for Prostate Cancer: U.S. Preventive Services Task Force Recommendation Statement. *Ann. Intern. Med.* **2012**, *157*, 120–134. [CrossRef]

8.   Gigerenzer, G.; Mata, J.; Frank, R. Public knowledge of benefits of breast and prostate cancer screening in Europe. *J. Natl. Cancer Inst.* **2009**, *101*, 1216–1220. [CrossRef]

9.   Tavoosidana, G.; Ronquist, G.; Darmanis, S.; Yan, J.; Carlsson, L.; Wu, D.; Conze, T.; Ek, P.; Semjonow, A.; Eltze, E.; et al. Multiple recognition assay reveals prostasomes as promising plasma biomarkers for prostate cancer. *Expert Rev. Anticancer Ther.* **2011**, *11*, 1341–1343. [CrossRef]

10. Bachrach, U. Polyamines and cancer: Minireview article. *Amino Acids* **2004**, *26*, 307–309. [CrossRef]

11. Schipper, R.G.; Romijn, J.C.; Cuijpers, V.M.; Verhofstad, A.A. Polyamines and prostatic cancer. *Biochem. Soc. Trans.* **2003**, *31*, 375–380. [CrossRef] [PubMed]

12. Lévesque, E.; Huang, S.P.; Audet-Walsh, E.; Lacombe, L.; Bao, B.Y.; Fradet, Y.; Laverdière, I.; Rouleau, M.; Huang, C.Y.; Yu, C.C.; et al. Molecular markers in key steroidogenic pathways, circulating steroid levels, and prostate cancer progression. *Clin. Cancer Res.* **2013**, *19*, 699–709. [CrossRef] [PubMed]

13. Gnanapragasam, V.J.; Robson, C.N.; Leung, H.Y.; E Neal, D. Androgen receptor signalling in the prostate. *BJU Int.* **2000**, *86*, 1001–1013. [CrossRef] [PubMed]

14. Kelloff, G.J.; Lieberman, R.; Steele, V.E.; Boone, C.W.; Lubet, R.A.; Kopelovich, L.; Malone, W.A.; Crowell, J.A.; Higley, H.R.; Sigman, C.C. Agents, biomarkers, and cohorts for chemopreventive agent development in prostate cancer. *Urology* **2001**, *57*, 46–51. [CrossRef]

15. De Luca, S.; Fiori, C.; Manfredi, M.; Amante, E. Preliminary results of prospective evaluation of urinary endogenous steroid profile and prostatic carcinoma-induced deviation. *J. Urol.* **2019**, *201*, e263–e264. [CrossRef]

16. Amante, E.; Alladio, E.; Salomone, A.; Vincenti, M.; Marini, F.; Alleva, G.; De Luca, S.; Porpiglia, F. Correlation between chronological and physiological age of males from their multivariate urinary endogenous steroid profile and prostatic carcinoma-induced deviation. *Steroids* **2018**, *139*, 10–17. [CrossRef] [PubMed]

17. Tomasi, G.; Berg, F.V.D.; Andersson, C. Correlation optimized warping and dynamic time warping as preprocessing methods for chromatographic data. *J. Chemom.* **2004**, *18*, 231–241. [CrossRef]

18. Johnsen, L.G.; Skou, P.B.; Khakimov, B.; Bro, R. Gas chromatography—Mass spectrometry data processing made easy. *J. Chromatogr. A* **2017**, *1503*, 57–64. [CrossRef]

19. Amigo, J.M.; Skov, T.; Bro, R.; Coello, J.; Maspoch, S. Solving GC-MS problems with PARAFAC2. *TrAC Trends Anal. Chem.* **2008**, *27*, 714–725. [CrossRef]

20. Bro, R.; Andersson, C.A.; Kiers, H.A.L. PARAFAC2—Part II. Modeling chromatographic data with retention time shifts. *J. Chemom.* **1999**, *13*, 295–309. [CrossRef]

21. Amigo, J.M.; Popielarz, M.J.; Callejón, R.M.; Morales, M.L.; Troncoso, A.M.; Petersen, M.A.; Toldam-Andersen, T.B.; Morales, M.L. Comprehensive analysis of chromatographic data by using PARAFAC2 and principal components analysis. *J. Chromatogr. A* **2010**, *1217*, 4422–4429. [CrossRef] [PubMed]

22. Ballabio, D.; Consonni, V. Classification tools in chemistry. Part 1: Linear models. PLS-DA. *Anal. Methods* **2013**, *5*, 3790–3798. [CrossRef]

23. Wold, S.; Johansson, E.; Cocchi, M. PLS: Partial Least Squares Projections to Latent Structures. In *3D QSAR in Drug Design: Theory, Methods and Applications*; KLUWER ESCOM Science Publisher: Heidelberg, Germany, 1993; pp. 523–550.

24. Zou, W.; Tolstikov, V.V. Probing genetic algorithms for feature selection in comprehensive metabolic profiling approach. *Rapid Commun. Mass Spectrom.* **2008**, *22*, 1312–1324. [CrossRef] [PubMed]

25. Filzmoser, P.; Liebmann, B.; Varmuza, K. Repeated double cross validation. *J. Chemom.* **2009**, *23*, 160–171. [CrossRef]

26. Wise, B.; Gallagher, N.; Bro, R. *PLS_Toolbox 8.5*; Eigenvector Research, Inc.: Manson, WA, USA, 2017; Available online: http://eigenvector.com/software/pls-toolbox/ (accessed on 22 August 2019).

27. Choi, M.H.; Moon, J.-Y.; Cho, S.-H.; Chung, B.C.; Lee, E.J. Metabolic alteration of urinary steroids in pre- and post-menopausal women, and men with papillary thyroid carcinoma. *BMC Cancer* **2011**, *11*, 342. [CrossRef] [PubMed]

28. Miller, K.K.M. The Biological Actions of Dehydroepiandrosterone. *Drug Metab. Rev.* **2006**, *38*, 89–116.

29. Arlt, W.; Biehl, M.; Taylor, A.E.; Hahner, S.; Libé, R.; Hughes, B.A.; Schneider, P.; Smith, D.J.; Stiekema, H.; Krone, N.; et al. Urine steroid metabolomics as a biomarker tool for detecting malignancy in adrenal tumors. *J. Clin. Endocrinol. Metab.* **2011**, *96*, 3775–3784. [CrossRef] [PubMed]

30. Arlt, W.; Stewart, P.M. Adrenal corticosteroid biosynthesis, metabolism, and action. *Endocrinol. Metab. Clin. N. Am.* **2005**, *34*, 293–313. [CrossRef] [PubMed]

31. Lionetto, L.; Lostia, A.M.; Stigliano, A.; Cardelli, P.; Simmaco, M. HPLC-mass spectrometry method for quantitative detection of neuroendocrine tumor markers: Vanillylmandelic acid, homovanillic acid and 5-hydroxyindoleacetic acid. *Clin. Chim. Acta* **2008**, *398*, 53–56. [CrossRef]

32. Stephens, F.O.; Unit, O. Phytoestrogens and prostate cancer: Possible preventive role. *Med. J. Aust.* **1997**, *167*, 138–140. [CrossRef]

33. Hedelin, M.; Klint, Å.; Chang, E.T.; Bellocco, R.; Johansson, J.E.; Andersson, S.O.; Heinonen, S.M.; Adlercreutz, H.; Adami, H.O.; Grönberg, H.; et al. Dietary phytoestrogen, serum enterolactone and risk of prostate cancer: The Cancer Prostate Sweden Study (Sweden). *Cancer Causes Control* **2006**, *17*, 169–180. [CrossRef] [PubMed]

34. Rajalahti, T.; Arneberg, R.; Berven, F.S.; Myhr, K.-M.; Ulvik, R.J.; Kvalheim, O.M. Biomarker discovery in mass spectral profiles by means of selectivity ratio plot. *Chemom. Intell. Lab. Syst.* **2009**, *95*, 35–48. [CrossRef]

Geographical Authentication of *Macrohyporia cocos* by a Data Fusion Method Combining Ultra-Fast Liquid Chromatography and Fourier Transform Infrared Spectroscopy

Qin-Qin Wang [1,2], Heng-Yu Huang [2,*] and Yuan-Zhong Wang [1,*]

[1] Institute of Medicinal Plants, Yunnan Academy of Agricultural Sciences, Kunming 650200, China; wqq6501@163.com

[2] College of Traditional Chinese Medicine, Yunnan University of Traditional Chinese Medicine, Kunming 650500, China

* Correspondence: hhyhhy96@163.com (H.-Y.H.); boletus@126.com (Y.-Z.W.)

Academic Editor: Marcello Locatelli

Abstract: *Macrohyporia cocos* is a medicinal and edible fungi, which is consumed widely. The epidermis and inner part of its sclerotium are used separately. *M. cocos* quality is influenced by geographical origins, so an effective and accurate geographical authentication method is required. Liquid chromatograms at 242 nm and 210 nm (LC_{242} and LC_{210}) and Fourier transform infrared (FTIR) spectra of two parts were applied to authenticate the geographical origin of cultivated *M. cocos* combined with low and mid-level data fusion strategies, and partial least squares discriminant analysis. Data pretreatment involved correlation optimized warping and second derivative. The results showed that the potential of the chromatographic fingerprint was greater than that of five triterpene acids contents. LC_{242}-FTIR low-level fusion took full advantage of information synergy and showed good performance. Further, the predictive ability of the FTIR low-level fusion model of two parts was satisfactory. The performance of the low-level fusion strategy preceded those of the single technique and mid-level fusion strategy. The inner parts were more suitable for origin identification than the epidermis. This study proved the feasibility of the data fusion of chromatograms and spectra, and the data fusion of different parts for the accurate authentication of geographical origin. This method is meaningful for the quality control of food and the protection of geographical indication products.

Keywords: *Macrohyporia cocos*; data fusion; liquid chromatography; fourier transform infrared spectroscopy; partial least squares discriminant analysis; authentication

1. Introduction

The dried sclerotium of *Macrohyporia cocos*, belonging to Polyporaceae, is an herbal medicine (called Poria) that can be used as food, and has been approved by the National Health Commission of the People's Republic of China. It plays an indispensable role in numerous drugs, such as the liquid oral formulation of *Poriacocos* polysaccharides, Sijunzi Tang, Liuwei Dihuang Wan and Chuanbei Pipa Gao. Various kinds of Poria-based foods and skin cosmetics such as sleep-friendly tea, Tuckahoe pie, Guiling jelly (drinks made from turtle shell and medicinal herbs), Guiling jelly soft candy and the Poria facial mask, are pretty popular. Present phytochemical investigation suggests that this fungus contains terpenes and polysaccharides, which present beneficial biological properties, such as a prebiotic effect, through the modulation of gut microbiota composition [1], anti-hyperlipidemic [2], anti-cancer [3] hepatoprotective [4] and affecting adipocyte and osteoblast differentiation effects [5].

Generally, the sclerotium of *M. cocos* is peeled and processed into two products, the epidermis and the inner part. The epidermis is called Poriae Cutis in Chinese, and the inner part is still called Poria. The epidermis and inner part have similar types of compounds and different secondary metabolites contents [6], which are often used and studied separately. Both Poria and Poriae Cutis are officially recorded in the Chinese Pharmacopoeia.

The provenance of *M. cocos* is mainly distributed in the Dabie mountains area and Yunnan Province of China. Yunnan is suggested as the most satisfactory habitat because the quality of Yunnan *M. cocos* is being highly recommended all the time. Due to the large demand for it, and the knowledge of cultivation mastered easily by common people, this fungus is cultivated in large quantities. Although *M. cocos* is cultivated in Yunnan, the chemical profiles influencing biological activities may be uneven owing to various cultivation sites and different management techniques. It was reported in a previous study that the contents of pachymic acid of *M. cocos* in different regions of Yunnan varied significantly [7]. Consequently, customers are increasingly demanding some sort of proof of the geographical origin. For the sake of response to the demand, it is necessary to conduct research with respect to the authentication of geographical origin, which can also provide basic technology for the protection of specific geographical indication products [8].

To date, various analytical technologies that respond to the different chemical information of samples have been implemented for the origin identification of *M. cocos* [9–11]. Although these methods proved promising for the discrimination of provenance, they were separately applied. Nowadays, data fusion has been applied in the fields of food and medicine [12,13]. For example, Ni et al. [14] discovered that, based on high-performance liquid chromatography (HPLC) and Fourier transform infrared spectroscopy (FTIR) data fusion, the type and geographical origin of *Radix Paeoniae* samples could be successfully discriminated, and the fused data matrix showed a prominent result compared with the independent technique.

Data fusion strategies, which fuse the outputs of multiple complementary information to provide rich knowledge about a sample, are hoped to achieve a more accurate characterization than single pieces of information [15]. In addition to the fusion of several datum regarding one sample, the fusion of information regarding different parts was reported. For instance, Casale et al. [16] combined the near-infrared information obtained by the three parts (pileipellis, flesh and hymenium) of each individual to check the authenticity of dried *porcini* mushrooms. Studies mentioned above demonstrated that although time and effort would be taken to collect multiple complementary data, data fusion was suggested as an alternative strategy to show accurate characterization.

Infrared spectroscopy can provide the molecular functional group structure of metabolites. Liquid chromatography can characterize the exist of compounds and determinate the special compounds. Both techniques present different and complementary information, which were used for data fusion in this study. To the best of our knowledge, infrared spectroscopy was widely used for geographical classification because of the features of simplicity and rapidity [17,18]. Liquid chromatography was almost used for determining the contents of compounds [19,20]. Multiple chromatographic data fusion has been merely reported in the authentication of the geographical origin of palm oil [21], predicting antioxidant activity of *Turnera diffusa* [22], authentication of *Valeriana* species [23] as well as a comparison of *Salvia miltiorrhiza* and its variety [24]. Actually, a wealth of information was contained in the chromatographic data, and due to extensive automation, a stable and reliable result could be obtained.

In this study, two data fusion strategies including low and mid-level fusion as well as two data combinations including the fusion of complementary information regarding a single part, and the fusion of information regarding two medicinal parts from one sclerotium were performed for the geographical authentication of *M. cocos*. Liquid chromatograms at two wavelengths (242 nm and 210 nm) and FTIR spectra of two medicinal parts (Poria and Poriae Cutis) of *M. cocos* were analyzed. Contents of five triterpene acids were measured. Chromatographic data fusion, spectral data fusion as well as chromatography and spectroscopy data fusion were implemented, combined with partial least squares discriminant analysis (PLS-DA).

2. Results and Discussion

2.1. Spectral Analysis

FTIR is an auxiliary method in the structural elucidation of organic compounds, which is also employed to assess the quality attributes of a product and authenticate geographic location [17]. With the characteristics of easy operation and rapid acquisition, it was applied to the identification of cultivation location of *M. cocos*. The second derivative spectra of samples from each geographic origin were given in Figure 1, and absorption peaks were observed in the form of negative peaks. Because a 2600–1750 cm^{-1} signal was caused by ATR crystal material [25], it was discarded and did not present in the Figure.

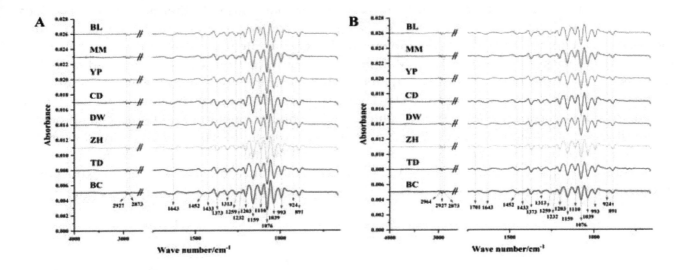

Figure 1. Second derivative spectra of Poria (**A**) and Poriae Cutis (**B**) samples from eight geographic origins.

Absorption bands at 2964 and 1704 cm^{-1} were just observed in Poriae Cutis samples. A disparity of absorption intensity exhibited in samples from different cultivation locations. Relatively high absorbance values were at around 1200–950 cm^{-1}, which were mainly caused by C-O stretching vibration, C-C stretching vibration and C-OH bending vibration of polysaccharides [26,27]. Peaks located at 2964 and 2873 cm^{-1} correspond to C-H antisymmetric and symmetrical stretching vibration of methyl group respectively, while the peak at 2927 cm^{-1} is assigned to C-H antisymmetric stretching vibration of methylene. The absorption at 1452 cm^{-1} and 1373 cm^{-1} belonged to C-H antisymmetric and symmetrical bending vibration of methyl [11]. The peak at 1643 cm^{-1} was assigned to C=O antisymmetric stretching vibration, which was related to triterpenes [28]. The band at 1704 cm^{-1} was associated with C=O group of esters [29,30]. The band at 891 cm^{-1} was assigned to the bending vibration of the C=CH$_2$ functional group [28]. The peak at 1259 cm^{-1} may be related to the amide III band [31]. In total, FTIR spectrum reflected comprehensive structural information of components in *M. cocos* samples, like triterpenes, polysaccharides, and so on.

2.2. Quantitative Analysis of Five Triterpene Acids

The content of each triterpene acid was calculated by their calibration curves and result of the validation of quantitative method was presented in Tables S1 and S2. The calibration curves of five compounds showed good linearity ($R^2 \geq 0.99$). Recovery rates calculated by the standard addition method varied from 96.32% to 106.4%. Values of relative standard deviation (RSD) of intra-day and inter-day precision were less than 1.24% and 5.68%, respectively. RSDs of repeatability did not exceed 5.95% after analyzing six solutions from the same sample in parallel. RSDs of stability were less than 0.71% after detecting a sample solution at 0, 6, 12, 17, 21 and 24 h, respectively. The method validation above indicated that the quantitative analysis was feasible. In particular, due to the obvious difference in the contents of poricoic acid A in Poria and Poriae Cutis samples, the calibration curves in two concentration ranges were prepared separately.

Contents of five triterpene acids were displayed as box-plot given in Figure 2. One-way analysis of variance was computed by SPSS 21.0 software (IBM Corporation, Armonk, NY, USA) to display the difference among eight cultivated locations. A value of $p < 0.05$ was considered significant. Poricoic acid A contents of Mengmeng were significantly different from those of Beicheng, Tuodian and Zhanhe in inner parts, and Yongping in cutis samples. Contents of dehydropachymic acid and pachymic acid in inner parts from Caodian were higher than those of other geographical origins except for Baliu. Inner parts from Baliu showed higher contents of dehydropachymic acid than those from Beicheng, Dawen and Mengmeng, and higher contents of pachymic acid than those from Tuodian, Yongping, Beicheng and Mengmeng. Inner parts from Dawen contained fairly low contents of dehydrotrametenolic acid compared with those from others with the exception of Baliu. Compared with epidermis samples from Dawen, Beicheng and Yongping showed higher contents of dehydrotumulosic acid, and Caodian and Baliu presented higher amount of pachymic acid. From the results, it was found that it was difficult to distinguish *M. cocos* samples from eight cultivation origins just in terms of contents of several target compounds. Therefore, it was necessary to take full advantage of the chromatographic fingerprint, namely, the intensity data for each retention time, to extract more information related to cultivation location.

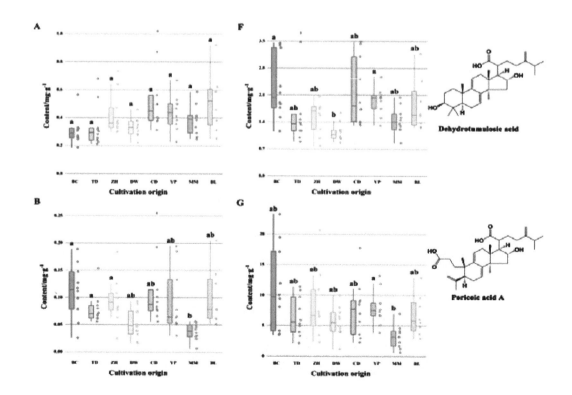

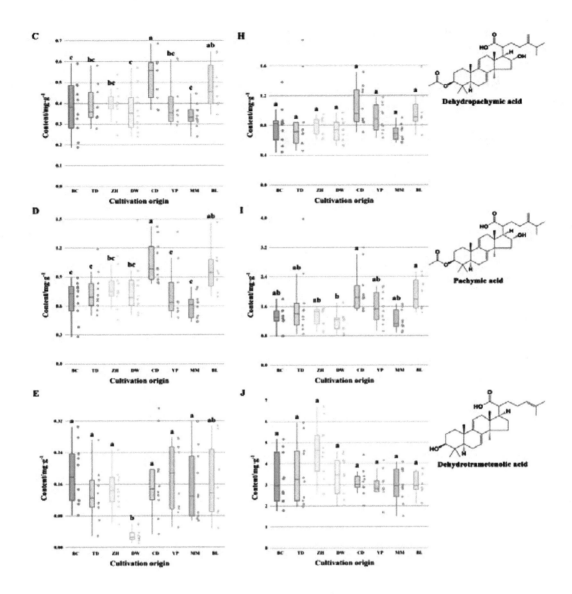

Figure 2. Box-plots of contents of five triterpene acids of Poria (**A–E**) and Poriae Cutis (**F–J**) samples from eight geographical origins. Note: Different letters show significant difference ($p < 0.05$).

2.3. Chromatographic Data Preprocessing

The chromatograms recorded at 242 nm in Figure S1 were obtained by analyzing the solution from the same sample five times successively within a day and on two consecutive days. Obviously, the retention time of each peak shifted in two days, which was inconvenient for the qualitative results of chemometric analyses. Hence, all of the chromatographic data should be aligned prior to further analysis.

The correlation optimized warping algorithm proposed by Skov et al. [32] was used to correct the retention time shifts among samples. The chromatogram that was most similar to all others was selected to be the reference chromatogram for alignment. The global search space was set to a combination of segment length from 10 to 200 and a slack size from 1 to 20 according to the observed peak widths and shifts on the chromatograms. Then the optimal combination of segment length slack size was automatically selected according to the criterion of well alignment while at the same time considering the preservation in peak shape and area. The theory for the algorithms with respect to the automated alignment of chromatographic data can be consulted in [32].

As a result, suitable combinations of segment length and slack size were achieved for chromatographic data at 242 nm of Poria (197 and 11), 210 nm of Poria (105 and 16), 242 nm of Poriae Cutis (105 and 11) and 210 nm of Poriae Cutis (198 and 16), respectively. Figure 3 presented the aligned *M. cocos* chromatographic fingerprints using these warping parameters, which displayed that the retention time shifts were properly corrected. What's more, it was observed that chromatograms of the same medicinal part recorded at 242 nm and 210 nm showed complementary information, i.e., some peaks obviously presented in liquid chromatograms at 242 nm (LC_{242}) and some compounds just displayed in liquid chromatograms at 210 nm (LC_{210}). Further, chromatograms of two parts were appreciably different. In other words, multiple chromatographic profiles presented abundant chemical information of *M. cocos* that probably facilitated to confirm cultivation areas.

The chromatographic data of one Poria sample and one Poriae Cutis sample could be represented as 7201 and 7801 data points, respectively. In order to save the time for calculation, the number of data points in the retention time dimension of the matrix was reduced by taking one in every three points without affecting the chromatographic features. Therefore, 2401 and 2601 data points were obtained after reducing data, respectively. It was proved that this method was feasible by comparing the PLS-DA results since reducing data had little influence on identifying different groups (Table S3). Additionally, the first 11 min data in the chromatogram that mainly comprised unseparated peaks and baseline shift (Figure 3), which were discarded to obtain representative fingerprints and accurate results. In this way, the final data points were 1960 and 2160, respectively.

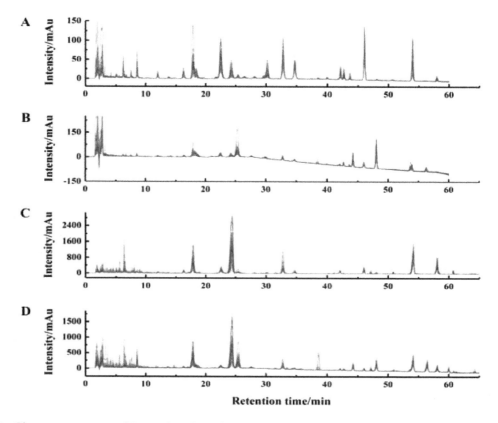

Figure 3. Chromatograms of Poria (**A,B**) and Poriae Cutis (**C,D**) recorded at 242 (**A,C**) and 210 nm (**B,D**) after the transformation of correlation optimized warping.

2.4. PLS-DA Using Chromatograms and FTIR Spectra

Partial least squares discriminant analysis is a widely-used linear classification method [33–36]. The selection of the optimal number of latent variables was an essential question for PLS-DA model, which was implemented on the basis of 7-fold cross validation procedure in present study. Unit variance scaling, which could give all variables of the same or different measurements equal importance, was performed by default when developing each PLS-DA model. The parameters of classification models were shown in Table 1 and Tables S4–S6 in detail.

Based on the preprocessing of chromatograms and FTIR spectra, a model of PLS-DA was established using the single dataset (Table 1 and Table S4). The LC_{210} dataset of Poriae Cutis samples did not build model successfully, so results of classification were not listed. FTIR and LC_{242} datasets showed better performance with higher accuracy not only in calibration set but in validation set than LC_{210} dataset. The sensitivity values of class 2 and class 8 in the validation set were 1 for Poria LC_{242} model and were smaller values for the Poria FTIR model, which indicated that LC_{242} model had stronger ability to correctly recognizing samples of class 2 and class 8. While the sensitivity of class 1 and 7 in calibration set was 0.8571 for Poria LC_{242} model smaller than that of Poria FTIR model, indicating that FTIR model had stronger ability to correctly recognizing samples of class 1 and class 7. Moreover, LC models of Poriae Cutis samples presented poorer results than those of Poria samples, which reflected the difference of two medicinal parts of *M. cocos*.

Table 1. The major parameters of PLS-DA model.

Fusion Approach	Data Matrix		Calibration Set			Validation Set
			R^2(cum)	Q^2(cum)	Accuracy	Accuracy
single technique	Poria	FTIR	0.8883	0.7268	100%	92.31%
		LC_{242}	0.6634	0.5277	96.15%	100%
		LC_{210}	0.5174	0.4012	90.38%	76.92%
	Poria Cutis	FTIR	0.9292	0.6981	100%	96.15%
		LC_{242}	0.2874	0.2204	65.38%	34.62%
low-level data fusion	Poria	FTIR-LC_{242}	0.9599	0.7917	100%	100%
		FTIR-LC_{210}	0.9468	0.7663	100%	100%
		$LC_{242-210}$	0.8097	0.6547	98.08%	92.31%
		FTIR-$LC_{242-210}$	0.8823	0.7566	100%	100%
	Poria Cutis	FTIR-LC_{242}	0.9016	0.7032	100%	100%
		FTIR-$LC_{242-210}$	0.905	0.698	100%	100%
	combination data of two medicinal parts	FTIR	0.9548	0.8064	100%	100%
		LC_{242}	0.8147	0.6495	100%	100%
		LC_{210}	0.6489	0.4806	94.23%	88.46%
mid-level data fusion	Poria	FTIR-LC_{242}	0.8266	0.5745	100%	100%
		FTIR-LC_{210}	0.7453	0.5053	96.15%	96.15%
		FTIR-$LC_{242-210}$	0.8286	0.5882	100%	100%
	Poria Cutis	FTIR-LC_{242}	0.7386	0.5493	100%	92.31%
		FTIR-LC_{210}	0.7518	0.4991	100%	96.15%
		$LC_{242-210}$	0.4617	0.228	76.92%	73.08%
		FTIR-$LC_{242-210}$	0.7607	0.5558	100%	96.15%
	combination data of two medicinal parts	FTIR	0.7564	0.5982	98.08%	88.46%
		LC_{242}	0.7761	0.4973	98.08%	100%
		LC_{210}	0.676	0.3756	96.15%	88.46%

Variable importance for the projection (VIP) plot [37] was used for assessing the significance of variable, and that the VIP score of retention time was greater than one means the compound separated at the time was important on distinguishing different cultivation origins. As an example of the Poria LC_{242} model, there were lots of variables whose VIP were higher than one including the corresponding retention time of poricoic acid A and dehydrotrametenolic acid (Figure 4). It indicated that the potential of the chromatographic fingerprint from the aspect of origin identification was greater than that of the contents of several compounds. However, all single technique models did not achieve a perfect performance, so it was necessary to carry out the data fusion strategy that was expected to enhance the classification and prediction ability of the model.

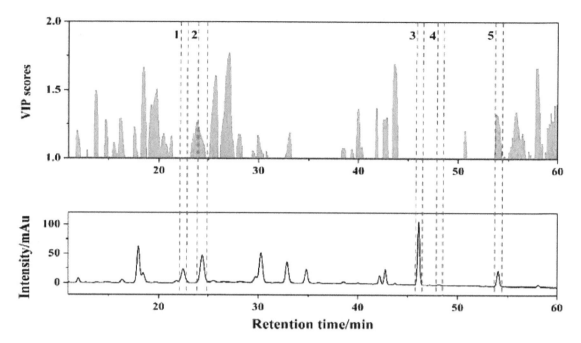

Figure 4. VIP scores of PLS-DA using LC_{242} chromatogram data of Poria samples. Note: 1, dehydrotumulosic acid; 2, poricoic acid A; 3, dehydropachymic acid; 4, pachymic acid; 5, dehydrotrametenolic acid.

2.5. Low-Level Data Fusion

2.5.1. PLS-DA of Poria

Figure 5 was the workflow of geographical authentication using data fusion, which was helpful to understand how data was combined. As shown in Table 1, accuracy rates of low-level data fusion datasets about Poria samples were 100% and higher than those of single technique models except for the model using $LC_{242\text{-}210}$ data, which implied that these models had strong classification performance. The highest R^2(cum) (0.9599) and Q^2(cum) (0.7917) were observed in FTIR-LC_{242} model, indicating a high goodness of fit for the established model in the data and good predictive ability. Therefore, the combination of FTIR and LC_{242} datasets was deemed a suitable strategy, and the fusion of three datasets was unnecessary and verbose. Furthermore, compared with the $LC_{242\text{-}210}$ model, the accuracy of FTIR-LC_{210} model was higher both in calibration and validation sets. It could be interpreted that FTIR dataset provided more helpful information to identify eight geographical origins than LC_{242} dataset in data fusion model of Poria samples. By analogy, it was found that FTIR data showed more contribution for origin discrimination than LC_{210} data when compared $LC_{242\text{-}210}$ model with FTIR-LC_{242} model.

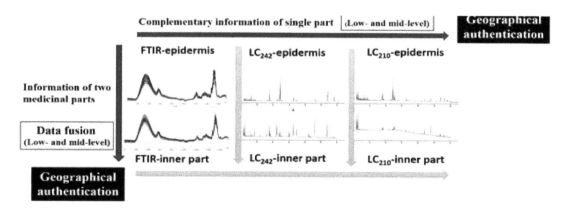

Figure 5. The workflow of geographical authentication using data fusion.

2.5.2. PLS-DA of Poriae Cutis

The accuracy of FTIR-LC$_{242}$ and FTIR-LC$_{242-210}$ models was 100%, which was greater than that of the models using the independent technique. It indicated the effectiveness of low-level data fusion. The similar Q^2(cum) of FTIR-LC$_{242}$ and FTIR-LC$_{242-210}$ models was observed. Accordingly, FTIR-LC$_{242}$ was considered as a preferred combination, and the fusion of three datasets was superfluous. Furthermore, the Q^2(cum) values of low-level fusion models about Poriae Cutis samples ($\leq$ 0.7032) were less than those of corresponding models about Poria samples (> 0.75), indicating that Poria samples were more suitable for origins identification than Poriae Cutis species. In the developing LC$_{242-210}$ and FTIR-LC$_{210}$ low-level models, latent variables could not be calculated so the models were not successfully built. It was in consistent with the state that epidermis LC$_{210}$ dataset did not built PLS-DA model, which was probably attributed by a lot of irrelevant classification information included in LC$_{210}$ dataset of epidermis.

2.5.3. PLS-DA of Combination Data of Two Medicinal Parts

Both FTIR and LC$_{242}$ datasets of two parts samples showed better performance than LC$_{210}$ dataset, which was in accordance with the results of single technique mentioned above. Compared with the single spectrum or chromatogram, data fusion of two medicinal parts proved more advantageous with greater sensitivity, specificity and efficiency. Therein, the FTIR fusion model of two part samples presented the best prediction performance from the Q^2(cum) point of view. What's more, compared with FTIR-LC$_{242}$ model of Poria samples, the Q^2(cum) of LC$_{242}$ fusion model of two parts was smaller. It could be interpreted that Poria FTIR dataset provided more helpful information to predict different geographical origins than Poriae Cutis LC$_{242}$ dataset in data fusion model. By analogy, it was found that the contribution of FTIR dataset was always more than that of LC$_{242}$ and LC$_{210}$ datasets in low-level data fusion. The low-level data fusion strategy has achieved a good classification result, but the mid-level data fusion could spend less computation time compared to the low level. Therefore, mid-level fusion was performed.

2.6. Mid-Level Data Fusion

2.6.1. The Extraction of Feature Variables

Mid-level fusion needed to first extract relevant features from each dataset independently and then concatenated them into a new matrix employed for origins identification. Principal component analysis (PCA) is a dimension reduction technique that creates a small number of new variables called principal components (PCs) from a large number of original variables, which would be applied to extract features. These PCs almost retain the same information as the original variables [38]. The optimal number of PCs was determined by 7-fold cross-validation procedure. The results of feature extraction were shown in Table S7. As an example of LC$_{210}$ dataset of Poria samples, the first thirteen PCs were extracted, which account for 90.92% of the information concerning the original variables. Then the scores of the thirteen PCs were used for data fusion.

2.6.2. PLS-DA of Poria

In agreement with the results of low-level data fusion, the accuracy rates of FTIR-LC$_{242}$ and FTIR-LC$_{242-210}$ of Poria samples were 100% not only in calibration set but in validation set. And they had stronger recognition performance with higher sensitivity, specificity, efficiency than corresponding single dataset. Nonetheless, all Q^2(cum) values of mid-level data fusion models of Poria samples were less than those of low-level data fusion models, indicating that low-level fusion presented stronger prediction ability than mid-level fusion according to cross validation.

As always, The $LC_{242-210}$ fusion model did not build successfully. The fusion of LC_{242} and LC_{210} could not gain satisfactory discrimination and even could not construct the model, and it was likely caused by the similar chemical information provided by both chromatograms. Although they presented different peak shapes, there were many common chromatographic peaks that did not provide complementary and useful information.

2.6.3. PLS-DA of Poriae Cutis

$LC_{242-210}$ model that was not built successfully in low-level fusion finished construction in mid-level fusion. The fact indicated the significance of mid-level data fusion and might be due to the feature extraction. The accuracy rates of FTIR-LC_{210} and FTIR-$LC_{242-210}$ models were equal, but the detail of incorrect identification was different from sensitivity and specificity points of view. Further analysis showed that one sample belonging to Tuodian was judged as the sample from Baliu in FTIR-LC_{210} model and Mengmeng in FTIR-$LC_{242-210}$ model by mistake, respectively. FTIR-LC_{242} and FTIR-$LC_{242-210}$ mid-level fusion models of Poriae Cutis samples presented poorer results than those of Poria samples as well as low-level data fusion models and FTIR model of epidermis samples.

2.6.4. PLS-DA of Combination Data of Two Medicinal Parts

Both FTIR data fusion and LC_{242} data fusion of two medicinal parts had stronger recognition ability when compared to the LC_{210} combination. Both LC_{242} and LC_{210} of two medicinal parts improved performance of single LC_{242} and LC_{210} models. However, the result of FTIR was the opposite. Compared to low-level data fusion, the identification ability of mid-level data fusion did not show any obvious advantage. This might be due to the limitation of our method of feature extraction. In terms of FTIR datasets, only more than 73.29% original information (Table S7) was extracted.

To validate the performance of the PLS-DA model, a 30-iteration permutation test was performed. As shown in Figure S2 that one of permutations plots for Poria $LC_{242-210}$ model, all permutated Q^2 and R^2 values (bottom left) were lower than the corresponding original values (top right). It indicated that the PLS-DA model was considered as an appropriate model without randomness and overfitting. The results showed that all the PLS-DA models were not overfitting.

3. Materials and Methods

3.1. Reagents, Solvents and Standard References

Dehydrotumulosic acid (purity $\geq$ 96%) was supplied by ANPEL Laboratory Technologies Inc. (Shanghai, China). Dehydropachymic acid, pachymic acid, poricoic acid A and dehydrotrametenolic acid (purity $\geq$ 98%) were purchased from Beijing Keliang Technology Co., Ltd. (Beijing, China). HPLC grade acetonitrile and formic acid were purchased from Thermo Fisher Scientific (Fair Lawn, NJ, USA) and Dikma Technologies (Lake Forest, CA, USA), respectively. Purified water was purchased from Guangzhou Watsons Food & Beverage Co., Ltd. (Guangzhou, China). Other chemicals and reagents were analytical grade.

3.2. Samples

Seventy-eight intact cultivated *M. cocos* sclerotia (Figure 6) from eight geographical origins of Yunnan Province, China were collected and identified by Prof. Yuanzhong Wang (Institute of Medicinal Plant, Yunnan Academy of Agricultural Sciences, Kunming, China). Voucher specimens (FL20160217) were deposited in the herbarium of Institute of Medicinal Plant, Yunnan Academy of Agricultural Sciences. After digging sclerotium up, the soil was brushed away. Fresh *M. cocos* sclerotium was air-dried in the shade and then peeled. Both the epidermis and inner part of the dried sclerotium, i.e., Poria and Poriae Cutis, were powdered to a homogeneous size using pulverizer and sieved through No. 60 mesh sieve. The powder was stored in the airproof, dry and dark condition prior to analysis. Detailed information of samples was summarized in Table 2.

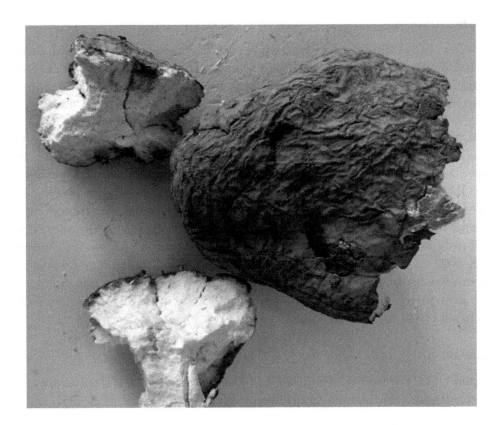

Figure 6. Dried sclerotium of *M. cocos*.

Table 2. The information of *M. cocos* samples.

Class	Location	Abbreviation	Elevation (m)	Latitude (°N)	Longitude (°E)	Parts	Sample Size
1	Beicheng Town, Hongta, Yuxi	BC	1720	24.4319	102.5182	inner part epidermis	10 10
2	Tuodian Town, Shuangbai, Chuxiong	TD	2062	24.6912	101.6493	inner part epidermis	10 10
3	Zhanhe Town, Ninglang, Lijiang	ZH	2560	26.8832	100.9275	inner part epidermis	10 10
4	Dawen Town, Shuangjiang, Lincang	DW	1438	23.3487	100.0047	inner part epidermis	10 10
5	Caodian Town, Yunlong, Dali	CD	2066	25.6360	99.1320	inner part epidermis	10 10
6	Yongping Town, Jinggu, Pu'er	YP	1077	23.4204	100.4044	inner part epidermis	10 10
7	Mengmeng Town, Shuangjiang, Lincang	MM	1052	23.4779	99.8378	inner part epidermis	10 10
8	Baliu Town, Mojiang, Pu'er	BL	1979	23.0676	101.9765	inner part epidermis	8 8

3.3. FTIR Spectra Acquisition

A Fourier transform infrared spectrometer from Perkin Elmer equipped with an attenuated total reflectance (ATR) sampling accessory with a diamond focusing element was employed for FTIR spectroscopy measurement. The sample powder was pressed under a consistent pressure with a pressure tower when collecting spectral. FTIR spectrum of each sample was scanned 16 times successively with a resolution of 4 cm^{-1} in the range of 4000–650 cm^{-1}. After the measurement of one sample was finished, the surface of ATR crystal and the apex of pressure tower were cleaned for the next sample detection. All spectra were background corrected utilizing air spectrum. The laboratory environment was maintained a constant temperature (25 °C) and humidity (30%).

3.4. Chromatographic Analysis

Sample powder was weighed accurately to 0.5 g and extracted with 2.0 mL of methanol by an ultrasound-assisted method for 40 min at ambient temperature. The extract solution was filtered using a 0.22 μm membrane filter. The filtrate was loaded into the auto-sampler vial and stored at 4 °C before injecting into the chromatographic system for analysis.

Analyses of all 156 samples (including Poria and Poriae Cutis) were implemented using a Shimadzu ultra-fast liquid chromatography system equipped with a UV detector, binary gradient pumps, a degasser, an auto sampler and a column oven. The chromatographic separation was achieved using an Inertsil ODS-HL HP column (3.0 × 150 mm, 3 μm particle size) operated at 40 °C. The mobile phase consisted of acetonitrile (A) and 0.05% formic acid (B). Before use, the mobile phase constituents were degassed and filtered through a 0.2 μm filter. The gradient elution sequence was conducted as follows: 0–25 min, 40% A; 25–52 min, 40–69% A; 52–56 min, 69–72% A; 56–58 min, 72–78% A; 58–58.01 min, 78–90% A; and 58.01–60 min, remaining at 90% A (eluting to 65 min for Poriae Cutis samples). Each run was followed by an equilibration period of 3 min with initial conditions (40% A and 60% B). The flow rate was kept at 0.4 mL·min^{-1} and the injection volume was 7 μL. Detective wavelengths were set at 242 nm and 210 nm.

3.5. Method Validation

The developed UFLC method was validated in terms of precision, stability, repeatability, accuracy and linearity under the above chromatographic condition.

A mixed standard solution was determined six times successively within a day and on three consecutive days for evaluating intra- and inter-day precision. For the stability test, the extract of a sample was analyzed at 0, 6, 12, 17, 21 and 24 h, respectively. Six sample solutions prepared individually from the same sample were analyzed in parallel for evaluating the repeatability. The recovery test was performed to evaluate the accuracy by adding reference compounds of three different amounts (low, middle, and high) to the sample with known concentration accurately. The following equation was used to calculate recovery rate: Recovery rate (%) = [(measured amount − original amount)/spiked amount] × 100%.

The standard solutions of five compounds for constructing calibration curves were prepared by diluting the stock solutions with methanol individually. The ranges of concentration in the linearity study were 5.00–999 μg·mL^{-1} (dehydrotumulosic acid), 0.22–6730 μg·mL^{-1} (poricoic acid A), 2.4–480 μg·mL^{-1} (dehydropachymic acid), 10.3–1240 μg·mL^{-1} (pachymic acid) and 0.49–2450 μg·mL^{-1} (dehydrotrametenolic acid). Due to the obvious difference in contents of poricoic acid A of Poria and Poriae Cutis samples, two concentration ranges of 0.22–1121.95 μg·mL^{-1} (Poria) and 0.22–6730 μg·mL^{-1} (Poriae Cutis) were prepared. More than seven levels (in arithmetic progression) of every concentration range were guaranteed. The limit of detection (LOD) and limit of quantification (LOQ) were determined by diluting continuously standard solution until the signal-to-noise ratios (S/N) reached about 3 and 10, respectively.

3.6. Preprocessing of Chromatograms and Spectra

The correlation optimized warping algorithm was applied to correct the retention time shifts of chromatogram using MATLAB software (MathWorks, R2017a, Natick, MA, USA). Then the corrected chromatographic data was reduced by taking one in every three points without affecting the chromatographic features to save computation time, which was inspired by the 'data binning' of Lucio-Gutiérrez et al. [22,23]. The first 11 minutes of data that mainly comprised unseparated peaks and baseline shift were discarded.

Raw FTIR spectra were subjected to advanced ATR correction to reduce the impact of skewing of band intensity using OMNIC 9.7.7 software (Thermo Fisher Scientific). Due to the fact that spectra contained hidden and overlapped absorption peaks, second derivative was used for highlighting slight differences employing SIMCA-P$^+$ 13.0 software (Umetrics, Umeå, Sweden). Derivative spectra were calculated with a Savitzky–Golay filter using a second-order polynomial and a 15-point window. The band of 2600–1750 cm^{-1} was associated to diamond crystal in ATR accessory, of which data were excluded prior to chemometrics analysis. These pre-processed data were used to data fusion and PLS-DA.

3.7. Multiple Chromatograms and Spectra Data Fusion

According to the source of data, there were two kinds of data fusion techniques, including the fusion of multiple complementary pieces of information about a single part and the fusion of information about two parts from one sclerotium. For instance, data matrices of LC-Poria and FTIR-Poria could be fused into a new dataset, and data matrices of FTIR-Poria and FTIR-epidermis could be fused into a dataset. It was important to note that information must correspond in the process of data fusion, namely, the LC and FTIR data of the same Poria sample must correspond, or the FTIR data of inner parts and epidermis from the same sclerotium should correspond.

The data fusion could be classified into three levels in light of the combination of data: low level, mid-level and high level. Low and mid-level fusion has been widely used, and was applied to the identification of geographical origin of *M. cocos*. The scheme of low and mid-level data fusion approaches is shown in Figure 7. In the low-level fusion, pre-processed different datasets were straightforward concatenated into a matrix, and the number of variables was equal to the sum of number of original variables. For the mid-level fusion, the scores obtained independently from different data by PCA were concatenated into a dataset applied for provenance traceability, and the number of variables of the dataset was significantly less than that of original variables. Compared with low level, mid-level data fusion could save more time on the operation. Specific types of the data fusion in this study were shown in Table 1.

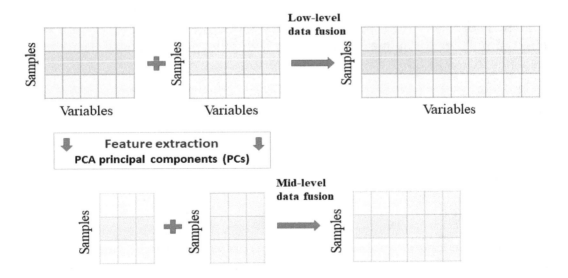

Figure 7. The scheme of the data fusion approaches.

3.8. Evaluation of Model Performance

The calibration and validation sets were selected for assessing the quality of model. The calibration set was used to construct a model that was performed 7-fold cross validation for internal validation, and the validation set was used to externally estimate the practicability of model. To avoid the influence of randomness caused by random sampling, and to obtain a representative calibration set from a pool of samples, the Kennard-Stone algorithm [39] was performed to systematically divide dataset of 78 samples into calibration (52) and validation (26) sets using MATLAB R2017a software (MathWorks).

The performance of discrimination model could be evaluated by sensitivity, specificity and efficiency [40]. The three parameters are dependent on these values: true positive (TP), false positive (FP), true negative (TN) and false negative (FN). TP and TN represent the correctly identified samples in target positive and negative classes, respectively. By analogy, FP and FN represent the incorrectly identified samples in positive and negative classes, respectively.

$$\text{Sensitivity} = \frac{TP}{TP + FN} \tag{1}$$

$$\text{Specificity} = \frac{TN}{TN + FP} \tag{2}$$

$$\text{Efficiency} = \sqrt{\text{sensitivity} \times \text{specificity}} \tag{3}$$

Therein, sensitivity shows the ability to correctly recognize samples belonging to the target class while specificity reflects the model ability to reject samples belonging to all other classes. The measure combining the sensitivity and specificity value is called efficiency.

In addition, the accuracy rate of calibration set, the accuracy rate of validation set, R^2(cum) and Q^2(cum) were also employed for assessing the classification performance. Accuracy was obtained by calculating the proportion of correctly classified samples in the total amount of calibration set (or validation set) samples. R^2 is calculated by following equation: $R^2 = 1 - RSS/SSX$, where RSS is the residual sum of squares of calculated and measured values, and SSX is the total sum of squares after mean centralization [41]. R^2(cum) represents the percentage of explained variance for a defined number of latent variables, indicating how well the model fits the data. Q^2(cum) represents the cross-validated cumulative R2, suggesting how well the model predicts new data. The higher values of these parameters (close to 1 or 100%), the better performance of model.

4. Conclusions

In order to establish an effective method for geographical authentication of *M. cocos*, two data fusion strategies, including low and mid-level fusion, as well as two data combinations, including the fusion of complementary information regarding a single part and the fusion of information about two parts from one sclerotium were compared. FTIR, LC$_{242}$ and LC$_{210}$ were used to characterize the epidermis and inner part of *M. cocos* sclerotium from different places individually and jointly. The results showed that, chromatographic fingerprint was more suitable than content data of five triterpene acids for origin identification. In the fusion of complementary information about single part, good classification performance was achieved obtained by merging LC$_{242}$ chromatograms and FTIR spectra in low-level fusion way. In the fusion of information about two parts from one sclerotium, the predictive ability of the FTIR low-level fusion model of two parts was the most satisfactory, and all analyzed samples were classified correctly.

In most cases, FTIR proved to be more efficient than LC_{242} and LC_{210}, not only in a single data source but in data fusion. Mid-level data fusion was slightly worse than low-level data fusion. The performance of low-level data fusion models was superior to single technique models. Moreover, Poria samples were more suitable for origin identification than Poriae Cutis samples. On the basis of effective and comprehensive fingerprint information, the low-level data fusion strategy could be used for the discrimination of *M. cocos* samples from different origins with the aid of appropriate mathematical algorithms.

Author Contributions: Y.-Z.W. and H.-Y.H. designed the project and revised the manuscript. Q.-Q.W. performed the experiments, analyzed the data and wrote the manuscript.

References

1. Khan, I.; Huang, G.; Li, X.; Leong, W.; Xia, W.; Hsiao, W.L.W. Mushroom polysaccharides from *Ganoderma lucidum* and *Poria cocos* reveal prebiotic functions. *J. Funct. Foods* **2018**, *41*, 191–201. [CrossRef]
2. Miao, H.; Zhao, Y.; Vaziri, N.D.; Tang, D.; Chen, H.; Chen, H.; Khazaeli, M.; Tarbiat-Boldaji, M.; Hatami, L.; Zhao, Y. Lipidomics biomarkers of diet-induced hyperlipidemia and its treatment with *Poria cocos*. *J. Agric. Food Chem.* **2016**, *64*, 969–979. [CrossRef]
3. Lee, S.; Lee, S.; Roh, H.; Song, S.; Ryoo, R.; Pang, C.; Baek, K.; Kim, K. Cytotoxic constituents from the sclerotia of *Poria cocos* against human lung adenocarcinoma cells by inducing mitochondrial apoptosis. *Cells* **2018**, *7*, 116. [CrossRef]
4. Wu, K.; Fan, J.; Huang, X.; Wu, X.; Guo, C. Hepatoprotective effects exerted by *Poria cocos* polysaccharides against acetaminophen-induced liver injury in mice. *Int. J. Biol. Macromol.* **2018**, *114*, 137–142. [CrossRef] [PubMed]
5. Lee, S.; Choi, E.; Yang, S.; Ryoo, R.; Moon, E.; Kim, S.; Kim, K.H. Bioactive compounds from sclerotia extract of *Poria cocos* that control adipocyte and osteoblast differentiation. *Bioorg. Chem.* **2018**, *81*, 27–34. [CrossRef] [PubMed]
6. Zhu, L.; Xu, J.; Zhang, S.; Wang, R.; Huang, Q.; Chen, H.; Dong, X.; Zhao, Z. Qualitatively and quantitatively comparing secondary metabolites in three medicinal parts derived from *Poria cocos* (Schw.) Wolf using UHPLC-QTOF-MS/MS-based chemical profiling. *J. Pharm. Biomed.* **2018**, *150*, 278–286. [CrossRef]
7. Li, Y.; Zhang, J.; Jin, H.; Liu, H.; Wang, Y. Ultraviolet spectroscopy combined with ultra-fast liquid chromatography and multivariate statistical analysis for quality assessment of wild *Wolfiporia extensa* from different geographical origins. *Spectrochim. Acta Part A* **2016**, *165*, 61–68. [CrossRef]
8. Biancolillo, A.; Marini, F. Chapter four—Chemometrics applied to plant spectral analysis. In *Vibrational Spectroscopy for Plant Varieties and Cultivars Characterization, Comprehensive Analytical Chemistry*, 1st ed.; Lopes, J., Sousa, C., Eds.; Elsevier: Amsterdam, The Netherlands, 2018; Volume 80, pp. 69–104.
9. Yuan, T.; Zhao, Y.; Zhang, J.; Wang, Y. Application of variable selection in the origin discrimination of *Wolfiporia cocos* (F.A. Wolf) Ryvarden & Gilb. based on near infrared spectroscopy. *Sci. Rep.* **2018**, *8*, 89.
10. Zhu, L.; Xu, J.; Wang, R.; Li, H.; Tan, Y.; Chen, H.; Dong, X.; Zhao, Z. Correlation between quality and geographical origins of *Poria cocos* revealed by qualitative fingerprint profiling and quantitative determination of triterpenoid acids. *Molecules* **2018**, *23*, 2200. [CrossRef]
11. Chen, J.; Sun, S.; Ma, F.; Zhou, Q. Vibrational microspectroscopic identification of powdered traditional medicines: Chemical micromorphology of *Poria* observed by infrared and Raman microspectroscopy. *Spectrochim. Acta Part A* **2014**, *128*, 629–637. [CrossRef]
12. Orlandi, G.; Calvini, R.; Foca, G.; Pigani, L.; Vasile Simone, G.; Ulrici, A. Data fusion of electronic eye and electronic tongue signals to monitor grape ripening. *Talanta* **2019**, *195*, 181–189. [CrossRef]
13. Wu, X.; Zhang, Q.; Wang, Y. Traceability of wild *Paris polyphylla* Smith var. yunnanensis based on data fusion strategy of FT-MIR and UV-Vis combined with SVM and random forest. *Spectrochim. Acta Part A* **2018**, *205*, 479–488. [CrossRef]
14. Ni, Y.; Li, B.; Kokot, S. Discrimination of *Radix Paeoniae* varieties on the basis of their geographical origin by a novel method combining high-performance liquid chromatography and Fourier transform infrared spectroscopy measurements. *Anal. Methods-UK* **2012**, *4*, 4326. [CrossRef]
15. Borràs, E.; Ferré, J.; Boqué, R.; Mestres, M.; Aceña, L.; Busto, O. Data fusion methodologies for food and beverage authentication and quality assessment—A review. *Anal. Chim. Acta* **2015**, *891*, 1–14. [CrossRef]

16. Casale, M.; Bagnasco, L.; Zotti, M.; Di Piazza, S.; Sitta, N.; Oliveri, P. A NIR spectroscopy-based efficient approach to detect fraudulent additions within mixtures of dried *porcini* mushrooms. *Talanta* **2016**, *160*, 729–734. [CrossRef]

17. Bureau, S.; Cozzolino, D.; Clark, C.J. Contributions of Fourier-transform mid infrared (FT-MIR) spectroscopy to the study of fruit and vegetables: A review. *Postharvest. Biol. Technol.* **2019**, *148*, 1–14. [CrossRef]

18. Li, Y.; Zhang, J.; Wang, Y. FT-MIR and NIR spectral data fusion: A synergetic strategy for the geographical traceability of *Panax notoginseng*. *Anal. Bioanal. Chem.* **2018**, *410*, 91–103. [CrossRef]

19. Wu, Z.; Zhao, Y.; Zhang, J.; Wang, Y. Quality assessment of *Gentiana rigescens* from different geographical origins using FT-IR spectroscopy combined with HPLC. *Molecules* **2017**, *22*, 1238. [CrossRef]

20. Wang, Y.; Shen, T.; Zhang, J.; Huang, H.; Wang, Y. Geographical authentication of *Gentiana rigescens* by high-performance liquid chromatography and infrared spectroscopy. *Anal. Lett.* **2018**, *51*, 2173–2191. [CrossRef]

21. Obisesan, K.A.; Jiménez-Carvelo, A.M.; Cuadros-Rodriguez, L.; Ruisánchez, I.; Callao, M.P. HPLC-UV and HPLC-CAD chromatographic data fusion for the authentication of the geographical origin of palm oil. *Talanta* **2017**, *170*, 413–418. [CrossRef]

22. Lucio-Gutiérrez, J.R.; Garza-Juárez, A.; Coello, J.; Maspoch, S.; Salazar-Cavazos, M.L.; Salazar-Aranda, R.; Waksman De Torres, N. Multi-wavelength high-performance liquid chromatographic fingerprints and chemometrics to predict the antioxidant activity of *Turnera diffusa* as part of its quality control. *J. Chromatogr. A* **2012**, *1235*, 68–76. [CrossRef] [PubMed]

23. Lucio-Gutiérrez, J.R.; Coello, J.; Maspoch, S. Enhanced chromatographic fingerprinting of herb materials by multi-wavelength selection and chemometrics. *Anal. Chim. Acta* **2012**, *710*, 40–49. [CrossRef] [PubMed]

24. Zhang, L.; Liu, Y.; Liu, Z.; Wang, C.; Song, Z.; Liu, Y.; Dong, Y.; Ning, Z.; Lu, A. Comparison of the roots of *Salvia miltiorrhiza* Bunge (Danshen) and its variety *S. miltiorrhiza* Bge f. Alba (Baihua Danshen) based on multi-wavelength HPLC-fingerprinting and contents of nine active components. *Anal. Methods-UK* **2016**, *8*, 3171–3182. [CrossRef]

25. Horn, B.; Esslinger, S.; Pfister, M.; Fauhl-Hassek, C.; Riedl, J. Non-targeted detection of paprika adulteration using mid-infrared spectroscopy and one-class classification–Is it data preprocessing that makes the performance? *Food Chem.* **2018**, *257*, 112–119. [CrossRef]

26. Cael, J.J.; Koenig, J.L.; Blackwell, J. Infrared and Raman spectroscopy of carbohydrates. Part VI: Normal coordinate analysis of V-amylose. *Biopolymers* **1975**, *14*, 1885–1903. [CrossRef]

27. Li, S.; Wang, L.; Song, C.; Hu, X.; Sun, H.; Yang, Y.; Lei, Z.; Zhang, Z. Utilization of soybean curd residue for polysaccharides by *Wolfiporia extensa* (Peck) Ginns and the antioxidant activities in vitro. *J. Taiwan Inst. Chem. E* **2014**, *45*, 6–11. [CrossRef]

28. Akihisa, T.; Uchiyama, E.; Kikuchi, T.; Tokuda, H.; Suzuki, T.; Kimura, Y. Anti-tumor-promoting effects of 25-methoxyporicoic acid A and other triterpene acids from *Poria cocos*. *J. Nat. Prod.* **2009**, *72*, 1786–1792. [CrossRef]

29. Lee, S.; Lee, D.; Lee, S.O.; Ryu, J.; Choi, S.; Kang, K.S.; Kim, K.H. Anti-inflammatory activity of the sclerotia of edible fungus, *Poria cocos* Wolf and their active lanostane triterpenoids. *J. Funct. Foods* **2017**, *32*, 27–36. [CrossRef]

30. Ying, Y.; Shan, W.; Zhang, L.; Zhan, Z. Lanostane triterpenes from *Ceriporia lacerate* HS-ZJUT-C13A, a fungal endophyte of *Huperzia serrata*. *Helv. Chim. Acta* **2013**, *95*, 2092–2097. [CrossRef]

31. Maquelin, K.; Kirschner, C.; Choo-Smith, L.P.; van den Braak, N.; Endtz, H.P.; Naumann, D.; Puppels, G.J. Identification of medically relevant microorganisms by vibrational spectroscopy. *J. Microbiol. Methods* **2002**, *51*, 255–271. [CrossRef]

32. Skov, T.; van den Berg, F.; Tomasi, G.; Bro, R. Automated alignment of chromatographic data. *J. Chemom.* **2006**, *20*, 484–497. [CrossRef]

33. Ballabio, D.; Consonni, V. Classification tools in chemistry. Part 1: Linear models. PLS-DA. *Anal. Methods-UK* **2013**, *5*, 3790. [CrossRef]

34. Ståhle, L.; Wold, S. Partial least squares analysis with cross-validation for the two-class problem: A Monte Carlo study. *J. Chemom.* **1987**, *1*, 185–196. [CrossRef]

35. Barker, M.; Rayens, W. Partial least squares for discrimination. *J. Chemom.* **2003**, *17*, 166–173. [CrossRef]

36. Sjöström, M.; Wold, S.; Söderström, B. PLS discriminant plots. In *Pattern Recognition in Practice*; Gelsema, E.S., Kanal, L.N., Eds.; Elsevier: Amsterdam, The Netherlands, 1986; pp. 461–470.

37. Wold, S.; Johansson, E.; Cocchi, M. PLS: Partial least squares projections to latent structures. In *3D QSAR in Drug Design: Theory, Methods and Applications*; Kubinyi, H., Ed.; KLUWER ESCOM Science Publisher: Leiden, The Netherlands, 1993; pp. 523–550.

38. Wold, S.; Esbensen, K.; Geladi, P. Principal component analysis. *Chemom. Intell. Lab.* **1987**, *2*, 37–52. [CrossRef]

39. Kennard, R.W.; Stone, L.A. Computer aided design of experiments. *Technometrics* **1969**, *11*, 137–148. [CrossRef]

40. Oliveri, P.; Downey, G. Multivariate class modeling for the verification of food-authenticity claims. *TrAC Trends Anal. Chem.* **2012**, *35*, 74–86. [CrossRef]

41. Aa, J.Y. Analysis of metabolomic data: Principal component analysis. *Chin. J. Clin. Pharmacol. Ther.* **2010**, *15*, 481–489.

Screening 89 Pesticides in Fishery Drugs by Ultrahigh Performance Liquid Chromatography Tandem Quadrupole-Orbitrap Mass Spectrometer

Shou-Ying Wang [1,2], Cong Kong [1,3,*], Qing-Ping Chen [1,2] and Hui-Juan Yu [1,3,*]

[1]   Laboratory of Quality & Safety Risk Assessment for Aquatic products (Shanghai),
      Ministry of Agriculture and Rural Affairs, East China Sea Fisheries Research Institute,
      Shanghai 200090, China; magnolia7319@163.com (S.-Y.W.); chenqp128@163.com (Q.-P.C.)
[2]   College of Food Science & Technology, Shanghai Ocean University, Shanghai 201306, China
[3]   Key Laboratory of East China Sea Fishery Resources Exploitation, Ministry of Agriculture and Rural Affairs,
      East China Sea Fisheries Research Institute, Chinese Academy of Fishery Sciences, Shanghai 200090, China
*   Correspondence: kongcong@gmail.com (C.K.); xdyh-7@163.com (H.-J.Y.)

Academic Editors: Marcello Locatelli, Angela Tartaglia, Dora Melucci, Abuzar Kabir, Halil Ibrahim Ulusoy, Victoria Samanidou, Evagelos Gikas and Patrick Chaimbault

Abstract: Multiclass screening of drugs with high resolution mass spectrometry is of great interest due to its high time-efficiency and excellent accuracy. A high-scale, fast screening method for pesticides in fishery drugs was established based on ultrahigh performance liquid chromatography tandem quadrupole-Orbitrap high-resolution mass spectrometer. The target compounds - were diluted in methanol and extracted by ultrasonic treatment, and the extracts were diluted with MeOH-water (1:1, v/v) and centrifuged to remove impurities. The chromatographic separation was performed on an Accucore aQ-MS column (100 mm × 2.1 mm, 2.6 μm) with gradient elution using 0.1% formic acid in water (containing 5 mmol/L ammonium formate) and 0.1% formic acid in methanol (containing 5 mmol/L ammonium formate) in Full Scan/dd-MS2 (TopN) scan mode. A screening database, including mass spectrometric and chromatographic information, was established for identification of compounds. The screening detection limits of methods ranged between 1–500 mg/kg, the recoveries of real samples spiked with the concentration of 10 mg/kg and 100 mg/kg standard mixture ranged from 70% to 110% for more than sixty compounds, and the relative standard deviations (RSDs) were less than 20%. The application of this method showed that target pesticides were screened out in 10 samples out of 21 practical samples, in which the banned pesticide chlorpyrifos were detected in 3 out of the 10 samples.

Keywords: fishery drugs; high-resolution orbitrap mass spectrometry; pesticide; screening

1. Introduction

Aquaculture is estimated to provide half of aquatic products by 2030 from the farming of freshwater or marine areas [1]. There is inevitably going to be a need for intensive aquaculture developed to supply more products from this industry. According to the "Green food—fishery medicine application guideline (NY/T 755-2013)" in the Agricultural Industry Standards of the People's Republic of China [2], fishery medicine refers to the substances that prevent or treat diseases in aquaculture animals or purposefully regulate the physiology of animals, including chemicals, antibiotics, Chinese herbal medicines and biological products. It is also known as chemical inputs or veterinary medicinal products (VMPs) applied in aquaculture in Europe and the United States [3,4]. Chemical inputs from aquaculture include antifoulants, antibiotics, parasiticides, anesthetics and disinfectants [5], while parasiticides in fishery mainly contain avermectins, pyrethroids, hydrogen peroxide, and organophosphates [5,6].

Based on the Guidelines, ten kinds of fishery drugs originated from pesticide have been banned for aquatic animals and plants. However, the illegal or excessive addition of pesticides in the fishery drug, as well as uncontrollable and uncertain administration during culture process can lead to the accumulation and residue of these pesticides in aquatic product. Illegal and unregulated use of pesticide may occur in many aquaculture areas, and further threaten the food safety for human health. To protect the quality and safety of aquatic products, as well as the sustainable ecosystem, surveillance of pesticides components in fishery drugs should be conducted.

Ultrahigh performance liquid chromatography coupled to high-resolution mass spectrometry (UHPLC-HRMS) is a promising strategy for multi-component screening of pesticides [7–9]. HRMS could record full scan of the precursors or fragmented ions with high-resolution, as well as the relative isotopic abundance, and is virtually able to distinguish unlimited number of compounds from one set of analyzed data [10,11]. In the past, the chromatography coupled to Time of Flight Mass Spectrometry (ToF-MS) was used in the development of multiclass components screening methods [9,12,13]. However, comparing to ToF-MS, the orbitrap mass spectrometer can fast scan and simultaneously switch between positive and negative acquisition modes if there's no need to change mobile phase of chromatography unit [14,15]. The combination of quadrupole and Orbitrap for high-resolution mass spectrometry can acquire data with high throughput, excellent accuracy and better sensitivity, which provides an ideal platform for multiclass risk compound screening [16]. Therefore, more methods of screening detection with Orbitrap MS were developed. With this instrument, the data-dependent data acquisition mode scans the full mass distribution of all precursors and then selectively fragments them sequentially for secondary mass scanning according to their abundance. This scan mode allows the quantification of compounds with precursor ion abundancy and identification with corresponding fragment ions [17]. Moreover, due to the stable and high-resolution mass spectrum recorded at standard data provide enough dependency, the identification of targeted compound can be conducted by comparing their database rather than practically acquire data for standards every time [18,19].

In previous studies, the analysis of 139 pesticide residues in fruit and vegetable commodities was established based on the Q-Orbitrap MS, allowing the retrospective analysis of the data feature which cannot be achieved with QqQ [17]. Jia et al. have developed an untargeted screening method for 137 veterinary drugs and their metabolites (16 categories) in tilapia using UHPLC-Orbitrap MS [20]. Turnipseed established a wide-scope screening method for 70 veterinary drugs in fish, shrimp and eel using LC-Orbitrap MS [7]. Recently, a non-target data acquisition for target analysis workflow based on UHPLC/ESI Q-Orbitrap was examined for its performance in screening pesticide residues in fruit and vegetables [21]. However, there is a lack of works on the multi-component screening detection in fishery drugs, especially for pesticide component screening. A fast screening method for a wide range of pesticides detection can be preferred, as much more reagent, time, and labor can be saved to detect more harmful components for safety evaluation.

Our study aims to develop a more generic screening method for a wider scope of pesticides with a self-built database, which can keep the advantages of robustness, simplicity, and time-efficiency. In the current work, we investigated 89 possible pesticides that can be used in fishery-related industry and remained in aquatic products. The chromatographic and high-resolution mass spectra for these compounds were acquired with a UHPLC-quadrupole-Orbitrap HRMS after optimizing parameters. The useful fragment ions with high-resolution were explored and selected. Then, a database including the retention time, isotope pattern, ionization mode and adduct, characteristic fragment ions, was established. Identification rules for data comparison with real samples were also investigated. Finally, a fast pesticide screening method for fishery drug was developed in combination with a rapid pretreatment.

2. Results and Discussion

2.1. Full MS-ddMS^2 Scan for Identification and Qualification

Full MS-ddMS^2 detection mode was applied on UHPLC-ESI-Q-Orbitrap HRMS system, which is a different data acquisition from single (multiple) reaction monitoring on triple quadrupole mass spectrometry. The Orbitrap analyzer collected accurate mass of all precursor ions as the first identification step of compounds. The precursors of high abundance were isolated through quadrupole in the next round scanning. Each of the precursors can be fragmented sequentially in the HCD multipole, re-collected in C-trap, and analyzed through Orbitrap mass spectrometer. It should be noted that the accurate mass of precursors instead of their fragmentation ions was continuously tracked and can be integrated for peak identification. Therefore, the precursor ions can be used for quantification and their corresponding fragmented ions for each peak of precursor ion can be used for identification in combination.

Under the guideline of European SANTE/11813/2017 and Commission Decision 2002/657/EC [22], identification of the concerned analytes with high-resolution mass spectrometry can be performed. The chromatographic information, their mass information should attain given identification points (IPs) to get confirmed results. If the high-resolution mass spectrometric data were collected, 2 IPs are earned if the precursor ion match, and 2.5 IPs for each of their product ions [23,24]. For the identification of all compounds, 4.5 IPs are required. In our work, the m/z of isotope, and its relative abundance for precursors were also identified, which leads to higher IPs for structure identification. Therefore, our identification rule should be stricter and more reliable than current regulations, which can result in less false positive result according to our experiment on fortified samples.

2.2. Mobile Phase

Due to the excellent performance of Accucore aQ-MS column in the analysis of multiclass compounds of different polarities, it was employed for chromatography separation of these target compounds. MeOH-water and MeCN-water binary mobile phase were investigated for the separation of the 89 compounds. In order to improve the efficiency of analyte ionization, 5 mM of ammonia formate and 0.1% formic acid (FA) were added in both phases. The result showed no triggered MS/MS spectrum for fenitrothion, chlorpyrifos, phorate, or dichlorvos since the automatic gain control AGC does not satisfy the setting value 5×10^5, when MeCN was applied as mobile phase at the concentration of 50 ng/mL under the full scan/dd-MS2 acquisition, which was considered as a negative result in our experiment. Moreover, signal intensities of more than 10 compounds decreased by 1–2 orders compared with MeOH as the mobile phase. Compounds with significant difference of signal intensity are shown in Figure 1. There were unremarkable differences for the rest pesticides on either mobile phase. According to Figure 1, MeOH is a better mobile phase, as more compounds showed higher response on mass spectrometer. Therefore, MeOH-water system with buffers and formic acid was selected for eluting these compounds from the column, and which is similar to Raina's research concerning of determination OPs in the air based on LC-MS/MS [25]. Neither MeCN nor MeOH could separate 89 pesticides completely. However, with the mass spectrometer, these compounds are not necessarily to be separated, as the different m/z can be easily acquired and extracted for different co-eluted compounds, with a pure chromatographic signal for individual compound. It should be noted that proper chromatographic elution of these compounds is still important, as it can avoid matrix effect and potentially competitive ionization between each other if high content compounds are present.

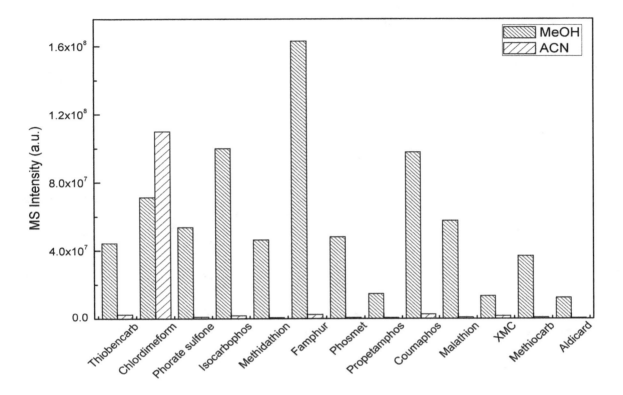

Figure 1. Pesticides with significant changes in sensitivity in MeOH and acetonitrile (ACN) mobile phases with 5 mM ammonium formate and 0.1% formic acid (FA) at the concentration of 50 ng/mL.

2.3. Buffers

The addition of formic acid helped improve the ionization efficiency and further increased the sensitivity of analytes, which has been validated in our optimization work. In our research, different concentration of buffers (ammonium formate, 0 mM, 2 mM, and 5 mM) in mobile phase with 0.1% FA were examined for 50 ng/mL mix standards solutions in the same gradient elution. Results showed better chromatographic peaks for most of the compounds when 2 or 5 mM ammonium formate was added in the mobile phase. As it is shown in Figure 2, signals were enhanced by approximately 10 times for propetamphos, famphur, methidathion, and indoxacarb are obtained when buffers were used in the mobile phase. Furthermore, the retention time of some compounds have been delayed after addition of 5 mM ammonium formate in the mobile phase. Buffers are beneficial to the retention and separation for many compounds, especially for acephate, propetamphos, methomyl, and indoxacard, and they further increase the sensitivity, even though a soft/lower intensity on mass spectrum was shown for phorate, dichlorvos, and chlorpyrifos-methyl when 5 mM ammonium formate added. As a result, 5 mM of ammonium formate was added in both mobile phases.

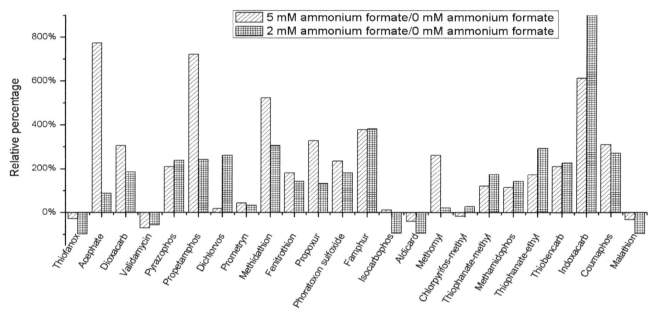

Figure 2. Relative signal enhancement or depression for typical compounds with different concentrations of ammonia formate or without this buffer in the mobile phase in the same gradient.

2.4. Mass Spectrometry

In principle, the higher the resolution of mass spectrum, the identification for target compounds is more accurate. A resolution of 140,000 can be achieved with Orbitrap in our work. However, the analyze time for each scan would be extended significantly and result in a lower data sampling rate. Therefore, enough information for peak integration or critical fragments of precursors will be compromised, as there are only around 15 s of elution time for each compound in chromatography. Similar to our previous work for veterinary drug screening [26,27], full scan/dd-MS2 (TopN) was applied for mass data acquisition, in which an inclusion list of the target compounds was preset. The MS resolution for full scan and fragment acquisition are 70,000 and 17,500, respectively. It could allow the discrimination of low abundant ions undetectable under low resolution [28], and further minimize possibility of false positive [8,29]. For dd-MS2 acquisition, if the high abundant ions were preset in the inclusion list, they were fragmented and scanned sequentially once their precursors were detected. Based on the set parameters, the probability that the instrument fails to trigger MS/MS spectrum acquisition for a detected chemical is greatly reduced. No false negative results were determined for any analyte spiked above its SDL. If there were compounds showing no fragmentation acquisition at the lower concentration, which can be identified by precursor m/z abundance greater than 5×10^5, isotope abundance and retention times with narrower deviation to avoid false negatives. Otherwise the compound is counted as undetected. In this work, the top 2 abundant ions were successively fragmented and transferred into the Orbitrap for data acquisition. Under the electrospray ionization, 76 of these compounds formed precursor ions as $[M + H]^+$, 8 of these compounds ionized as $[M + Na]^+$, and 5 pyrethroids formed additions as $[M + NH_4]^+$. PCP Na and 4 phenylpyrazoles formed negative ions as $[M - H]^-$. Three different normalized collision energy (stepped NCE) allowed the high-efficiency fragmentation of different precursors at their best.

2.5. Sample Preparation

It is critical for high recovery determination to choose the solvent of extraction. In this research, pesticides of interest are of multiclass and of quite different chemical or physical properties. To dissolve or extract different analytes with high or low polarity, MeOH and 10% ethyl acetate in MeOH were used as extract solvents for pure Chinese herb drugs, which contains complex matrices and impurities. Results showed better extracting efficiency when MeOH was used. In terms of the recovery of these target compounds, more than half of targets showed better recovery than 10% ethyl acetate in MeOH. As it is shown in Figure 3, seven compounds including phorate, mevinphos, fenobucarb, chlordimeform, propoxur, XMC, and propamocarb showed more than 35% decrease of recovery. Therefore, MeOH was preferred as a solvent for the analysis of pesticides in these drugs.

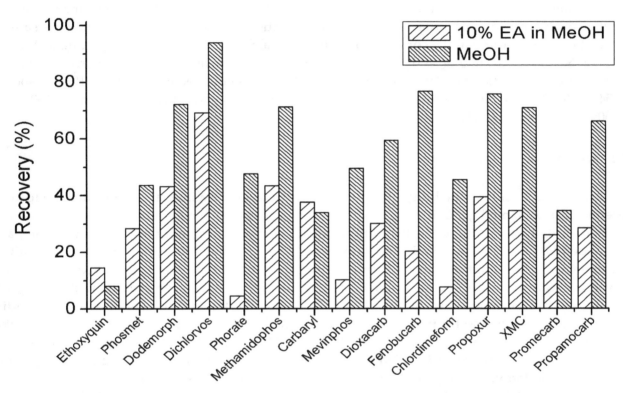

Figure 3. Recoveries of typical compounds with or without the use of 10% ethyl acetate in the methanol.

2.6. Matrix Effect

Matrix effect should be considered in the detection process, which includes intrinsic organic or inorganic compounds after extraction and cleanup, and extrinsic inorganic ions, organic acids, detergent, etc. These interfere material can comprehensively enhance or suppress the response of the target compounds. In our research, the matrix effect ($ME\%$) was calculated based on the following equation [30,31]:

$$ME\% = (\frac{A}{B} - 1) \times 100\%$$

where A is the integration area in matrix-matched standard solution and B is the integration area in a standard solution with identical concentration for each compound. In general, the matrix effect within ±20% can be regarded as acceptable and the calibration can be performed without considering matrix effect. Otherwise, it should be considered during quantification [16,32].

The fishery drugs were dissolved and diluted up to 1000 times, which would significantly decrease the matrix effect. Standards diluted with more than 90% blank matrix solution were used to test for the matrix effect. The result showed less than 20% matrix effect for all the compounds of interest at the concentrations of 100 and 500 ng/mL for compounds with SDL above 100 ng/mL. Because of the acceptable matrix effect, it is feasible to use a methanol–water (1:1, v/v) solution to dilute a series of standard solutions, for quantification of positive compounds.

2.7. Method Validation

2.7.1. Screening Detection Limit

According to SANTE/11813/2017 [33], the screening detection limit (SDL) was examined with similar process, but less replicates, which has been applied in many reported works [34–36]. Fishery drug of Pure Chinese herb was fortified with mixed standard solutions at different concentrations in six duplicates together with their non-spiked counterparts, which were used for the examination of the screening detection limit, and all compounds satisfied 100% detection criterion at their SDL. Simultaneously, an additional criterion, identity confirmation through the 13C/12C-ratio, was satisfied for each target compound at the corresponding theoretical SDL [34–36]. In our experiment, 1 mg/kg, 10 mg/kg, 50 mg/kg, 100 mg/kg, and 500 mg/kg of these mix target samples were prepared respectively. All these fortified samples were pretreated following the aforementioned method (2.4). Results showed that 54, 80, 85, 86, and 89 compounds were screened positive at 1 mg/kg, 10 mg/kg, 50 mg/kg, 100 mg/kg, and 500 mg/kg, respectively.

2.7.2. Accuracy and Repeatability

The accuracy and repeatability of the screening method were investigated under the fortified concentrations of 10 mg/kg and 100 mg/kg in fishery drug of pure Chinese herb. For compounds at the detection limit of 500 ng/mL on the mass spectrometer, fortified samples of 500 mg/kg were prepared independently. Under the fortified concentration of 10 mg/kg and 100 mg/kg, compounds with the instrument detection limit of 10 ng/mL and below can be readily detected. Over sixty compounds showed the recovery of 70%–110% at spiked 10 and 100mg/kg; fifteen compounds with 110%–120% at 10 mg/kg; twelve compounds with 110%–120% at 100 mg/kg; and three compounds including chlorpyrifos, phosmet, and tributylphos-phorotrithioate had recoveries of over 125% at both spiked levels. Over 95% of compounds identified at both fortified levels had RSD of less than 15%. Compounds were not identified at the lower fortified level but detected at 100 mg/kg including amitraz, phorate, fenitrothion, validamycin, and prothiofos, with the recovery of 59.3%–125% and RSD of 6.17%–14.7%. Compounds only detected at the spiked level of 500 mg/kg are bromophos ethyl, cyfluthrin, parathion, with recovery of 85.3%–105% and RSD of less than 20%. All the quantification results were obtained with less than 20% RSD. Because of the soft matrix enhancement, there were some compounds with high recoveries at both fortified levels for quantification with the standard matched solvent, especially for chlorpyrifos, phosmet, and tributylphos-phorotrithioate. The details of recovery and RSD are presented in Table 1. It is noticed that some compounds did not meet the recovery criteria at one or both of the fortified levels, which could be attributed to high volatility and easy converting properties.

2.7.3. Calibration and linearity

As the matrix effect on the response of the fishery drug sample is quite low, and the recovery results satisfied the semiquantification analysis for most of the compounds in positive samples, the standard solution without matrix matched, and internal standards can be amenable for calibration of positive samples from the perspective of economic costs. In our research, different concentrations of mixed pesticide standards were prepared directly with MeOH–water (1:1, v/v). Results on mass spectrometer demonstrated that the R-squared of 81 pesticides were no less than 0.990, and 5 other pesticides, including chlorpyrifos, flumethrin, flucythrinate, tau-fluvalinate, and deltamethrin showed R-square between 0.982 and 0.990. The detailed linear profile for 82 compounds is listed in the electronic Supplementary Material (Table S1). The distribution pie chart of the linear range of these compounds is presented in Figure 4.

Table 1. Recovery and relative standard deviation (RSD) of each drug at different spike levels in feedstuff matrices.

Compound	10 mg·kg⁻¹ Recovery (%)	RSD (%, n = 3)	100 mg·kg⁻¹ Recovery (%)	RSD (%, n = 3)	Compound	10 mg·kg⁻¹ Recovery (%)	RSD (%, n = 3)	100 mg·kg⁻¹ Recovery (%)	RSD (%, n = 3)
Aminocarb	104	5.99	109	4.35	2,3,5-Trimethacarb	80	1.78	103	0.71
Carbaryl	81.8	3.91	99.9	8.49	3,4,5-Trimethylphenol	86.4	4.62	101	2.91
Carbendazim	120	3.77	85.6	4.04	Acephate	102	4.75	116	4.83
Carbofuran	92.1	6.42	94	3.82	Aldicarb sulfone	94.7	10.2	103	4.07
Chlordimeform	81.7	4.96	85	3.17	Aldicarb sulfoxide	86.6	6.19	103	5.97
Coumaphos	105	7.77	109	8.11	Aldicard	89.9	18.2	105	2.97
Dimethoate	110	11.1	107	10.7	Avermectin B1a	91.4	7.87	84.5	4.24
Dodemorph	95.2	7.08	82.1	2.96	Bendiocarb	80.7	3.5	88.9	7.96
Famphur	113	7.1	111	5.44	Bifenthrin	86.5	21	86.7	5.1
Fenobucarb	82.3	3.33	105	2.89	Chlorpyrifos-methyl	94.5	23	124	10.1
Fuberidazole	111	5.44	85	4.01	Deltamethrin	77.2	9.61	79.2	4.54
Imazalil	92.6	5.67	82.2	4.31	Dichlorvos	113	7.61	117	7.23
Indoxacarb	84.2	7.71	72.9	4.11	Dioxacarb	96	5.6	104	5.34
Isocarbophos	96.9	12.3	104	8.42	Doramectin	75.1	8.76	86.6	4.26
Malathion	88	6.52	113	6.66	Ethoxyquin	94.4	5.48	30.2	16.5
Methiocarb	86.1	8.01	104	5.37	Fenvalerate	90.5	21.6	78.8	1.83
Mevinphos	84.8	10.2	115	1.98	Fipronil	80.8	9.54	89.3	2.1
Monocrotophos	113	5.67	121	4.98	Fipronil-desulfinyl	88	8.67	80.8	6.54
Omethoate	123	12.7	117	12.7	Fipronil-sulfide	83.1	8.68	86.6	3.43
PCP Na	83.1	7.62	88	0.989	Fipronil-sulfone	82.9	6.88	80.8	4.43
Phoratoxon sulfoxide	92.2	14	110	4.59	Flucythrinate	97.9	7.69	80	2.09
Phosalone	105	7.34	103	4.11	Flumethrin	65.3	5.66	77.9	0.732
Phoxim	111	10	102	7.12	Isoprocarb	80	1.78	103	0.71
Pirimicarb	85.5	7.96	99.7	2.84	Ivermectin B1a	59.7	4.64	86.3	1.29
Pirimiphos-methyl	113	8.83	119	3.26	Methamidophos	108	6.42	124	6.08
Promecarb	89.1	5.83	102	2.98	Methidathion	98	4.15	113	4.73
Prometryn	88.9	7.63	78.8	0.618	Methomyl	94	6.7	111	4.89
Propamocarb	96.3	8.1	109	5.37	Phorate sulfone	97.3	5.51	107	1.58
Propazine	93.1	8.22	72.5	2.52	Propetamphos	113	5.06	123	4.27
Propiconazole	115	6.92	77.3	0.198	Robenidine	90	12.2	64.5	1.73

Table 1. *Cont.*

Compound	10 mg·kg⁻¹ Recovery (%)	RSD (%, n = 3)	100 mg·kg⁻¹ Recovery (%)	RSD (%, n = 3)
Propoxur	85.2	4.67	91.2	6.94
Pyrazophos	119	12.1	122	6.95
Quinalphos	113	8.76	112	5.45
Simazine	86.7	3.21	65.7	1.7
Simetryne	98.2	4.56	76	3.28
Thiabendazole	119	6.16	80.9	3.43
Thiobencarb	117	6.87	76.8	1.54
Triazophos	112	5.76	116	3.33
trichlorfon	109	4.85	109	6.27
Chlorpyrifos	138	7.17	154	7.63
Phorate sulfoxide	109	9.55	126	3.67
Phosmet	131	10.5	177	17
Thiophanate-ethyl	128	5.49	76.9	1.36
Thiophanate-methyl	128	4.96	77.9	1.86
Tributylphos-phorotrithioate	150	7.98	147	4.56

Compound	10 mg·kg⁻¹ Recovery (%)	RSD (%, n = 3)	100 mg·kg⁻¹ Recovery (%)	RSD (%, n = 3)
Tau-fluvalinate	65.5	13.6	83.6	4.04
Thiofanox	80.6	2.06	95.4	3.46
Thiofanox sulphone	87.7	9.49	92.4	0.0845
Thiofanox -sulphoxide	82.9	3.47	101	2.42
XMC	75.6	9.88	85.8	6.46
Xylazine	90.5	8.82	76.8	1.8
Amitraz	—	—	61.9	14.2
Fenitrothion	—	—	124	13.8
Phorate	—	—	125	12.1
Prothiofos	—	—	100	14.7
Validamycin	—	—	59.3	6.17
Bromophos ethyl*	—	—	85.3	19.4
Cyfluthrin*	—	—	91.5	1.13
Parathion*	—	—	105	3.53

* fortified at 500 mg/kg.

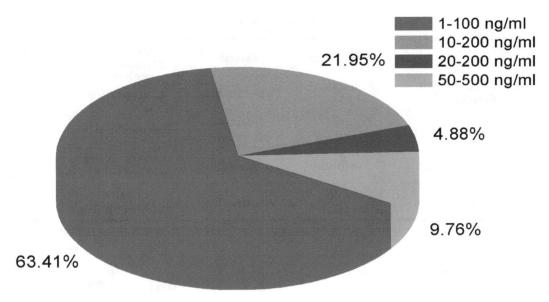

Figure 4. The percentages of different linear range of 82 targeted pesticides.

2.8. Practical Screening

The method was further applied in screening of 21 fishery drug samples (pesticides, water-clean agents and antibacterial agents). Samples were prepared according to sample preparation (4.4) prior to analysis. For the compounds with concentration out of the linear range, samples were re-diluted with the dilution factor of samples adjusted to ensure the concentration to be quantitatively evaluated based on our linear range. The screening was carried out following the home-built database and preset rules. Quantification was conducted through the peak areas of precursor ions in positive samples and was externally calibrated. Based on the database and preset identification rules, 10 out of the 21 fishery drug samples were screened positive with pesticides. 10 samples were detected with unspecified components. As is shown in Table 2, the identified pesticides were chlorpyrifos, ivermectin B1a, phoxim, avermectin B1a, and carbendazim. Three samples contained forbidden drug chlorpyrifos (Figure 5A), and 5 samples contain avermectin and ivermectin (Figure 5B) of more than 3 g/L. Their chromatographic and fragment information was highly identical to the standards, as are shown in Figure 5. Detailed information of the screened positive samples was presented in Table 2.

Table 2. Screening results of practical fishery drugs.

Code	Trade Name	Dosage Form	Listed	Detected Compounds	Contents (mg/kg or mg/L)
4	Insecticide for fish	Aqueous solution	NA	Chlorpyrifos	2.66
5	Insecticide for fish and shrimp	Soluble concentrate	Avermectin	Ivermectin B1a Chlorpyrifos Avermectin B1a	347 1.33 7479
6	Pesticide for water	Soluble concentrate	Bioactive ingredient	Ivermectin B1a Avermectin B1a	207 3482
14	Insecticide for fish	Soluble concentrate	NA	Phoxim	2.20
15	Avermectin solution	Soluble concentrate	Avermectin	Avermectin B1a	5937
16	Benzalkonium Bromide Solution	Aqueous solution	NA	Avermectin B1a	55,587

Table 2. *Cont.*

Code	Trade Name	Dosage Form	Listed	Detected Compounds	Contents (mg/kg or mg/L)
17	Beta-Cypermethrin Solution	Aqueous solution	Cypermethrin	Chlorpyrifos	9.11
20	Insecticide for water	Soluble concentrate	Bioactive ingredient	Ivermectin B1a	8214
21	Pesticide for water	Gel solution	Avermectin	Ivermectin B1a Avermectin B1a	121 3736
22	Insecticide for water	Soluble concentrate	Avermectin	Phoxim Avermectin B1a	19.7 1931

NA: not available. Listed: active compounds were listed in the label of fishery drugs.

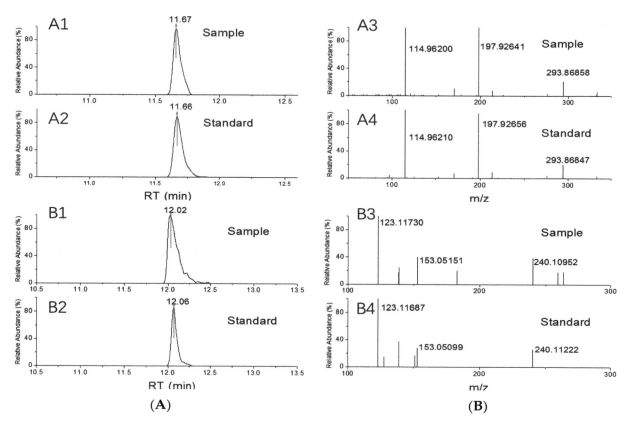

Figure 5. Comparison of the spectra of chlorpyrifos (**A**) and ivermectin B1a (**B**) detected in the positive samples. A1, B1 and A2, B2 are the chromatogram of real samples and standards, respectively. A3, B3 and A4, B4 is the MS/MS spectrum of the positive sample and the standard respectively.

3. Materials and Methods

3.1. Instruments and Reagents

The ultrahigh-performance liquid chromatography (UHPLC) system (Dionex UltiMate 3000, Thermo Fisher Scientific, San-Francisco, USA) coupled to quadrupole Orbitrap mass spectrometer with electrospray ionization (Q-Exactive, Thermo Fisher Scientific) was used for data acquisition.

Eight-nine Pesticides were selected for target screening as listed in Table 3. Carbofuran and dichlorvos were obtained from MANHAGE Biotech. Inc. (Beijing, China), thiofanox-sulfone, thiometon, aldicarb-sulfone, phoratoxon sulfoxide, PCP Na were purchased from Accustandard Inc. (New Haven, CT, USA) The other 85 pesticides standards were supplied by Dr. Ehrenstorfer GmbH (Augsburg,

Germany). Acetonitrile (MeCN) and Methanol (MeOH) of HPLC grade were obtained from J.T. Baker (Phillipsburg, NJ, USA). Formic acid (FA, 98%, LC-MS grade, Fisher Scientific, Spain, or HPLC grade) was obtained from FLUKA. All the other relevant reagents were purchased from common domestic suppliers. Pure water was obtained through Water Milli Q ELEMENT purification unit (Millipore, Bedford, USA).

3.2. Preparation of Standards

Standards stock solution: c.a. 5 mg solid standards was dissolved with MeOH in 10 mL beaker, and then transferred to a 50 mL flask and diluted with MeOH. For compounds will less solubility in MeOH, 0.1 mL formic acid (98%, HPLC grade) was firstly added and the mixture was sonicated until the solids were completely dissolved. Five microliters of liquid standards were pipetted into a 10 mL-beaker and weighed to get accurate mass. After that, they were dissolved in methanol following similarly procedure as the solid standards. All these single standards stock solutions were c.a. 100 μg/mL. The purchased standards solutions were not diluted until further preparation of mixed standards solution. Mixed standards solutions were prepared by mixing standards of the same category, which were finally diluted to 5 μg/mL. The standards were categorized into organophosphorus, carbamate, organochlorine, imidazole, pyrethroid, triazole, phenylpyrazole, avermectin, and miscellaneous. All the standards solutions were stored in refrigerator at −42 °C.

Matrix-matched standards were used to evaluate the matrix effect, where the standards were dissolved into a matrix of Chinese herbal fishery drug, which was negative for pesticides before spiking.

3.3. Methods

3.3.1. Elution Conditions

Accucore aQ-MS column (2.1 × 100 mm, 2.6 μm, Thermo Fisher Scientific, USA), was employed to perform sample separation with a thermostat at 30 °C. The binary mobile phases (MP) were 0.1% FA in water (containing 5 mM Ammonium Formate, A) and 0.1% FA in MeOH (containing 5 mM Ammonium Formate, B). Their gradient elution was started with 2% B, linearly increasing to 20% in 4 min and continuously ramped to 40% within 1.5 min. Subsequentially, B was increased to 98% in the subsequent 5 min, and kept for 2.4 min. Then, B was restored to the initial conditions 2%B in the following 2.1 min, and kept for 5 min to re-equilibrate for the next injection. The whole elution process for one injection analysis took 20 min. The flow rate was kept at 0.3 mL·min^{-1}. The injection volume for analysis was 10 μL for each sample.

3.3.2. Mass Spectrometer Condition

Parameters for electrospray were as following: spray voltage, 3200 V (positive mode), 2800 V (negative mode), sheath gas flow rate at 40 L·min^{-1}, auxiliary gas flow rate at 10 L·min^{-1}, sweep gas flow rate at 1.0 L·min^{-1}, auxiliary gas temperature at 350 °C, capillary temperature at 325 °C and S-lens RF level at 60 V. The scan mode for high-resolution mass spectrometry acquisition was Full MS/dd-MS2 (with inclusion list) mode. Recorded mass range for full mass record was between m/z 100–1000 (positive mode) and 150–1000 (negative mode), at resolution of 70,000. The Full MS/dd-MS2 (with inclusion list) mode can simultaneously record the precursor mass and the MS/MS (fragmentation) spectra for selected precursors. The MS/MS acquisition for fragment scanning of the selected ions was carried out at the isolation window of 2.0 m/z and the resolution of 15,000. For each round of fragmentation acquisition, the top 2 (TopN, 2, loop count 1) abundant precursors above the threshold 5 × 10^5 were sequentially transferred into the C-Trap (AGC, 5 × 10^5, Max IT, 100 ms) for collision at normalized energies (NCE, 20, 50, 80) in HCD multipole and pumped to Orbitrap for MS/MS acquisition.

All the units of UHPLC-ESI-Q-Orbitrap HRMS system were controlled through the Tracefinder software.

Table 3. Chromatographic, mass spectrometric information and limit of detection of 89 targeted pesticides.

Compounds	RT (min)	Extracted Mass (m/z)	Fragment Ions (m/z)	LOD ng/mL	Compounds	RT (min)	Extracted Mass (m/z)	Fragment Ions (m/z)	LOD ng/mL
Bendiocarb	8.95	224.09173	167.07027/109.02841	1	Aldicarb sulfone	5.15	223.0747	86.06004/76.03930	1
Chlorpyrifos	11.75	349.93356	114.9615/197.92730	1	Aminocarb	4.39	209.12845	152.10699/137.08352	1
Coumaphos	10.96	363.02174	226.99263/306.95913	1	Carbaryl	9.27	202.08626	132.04439/124.08827	1
Dimethoate	7.44	230.0069	142.99623/170.96978	1	Carbendazim	6.3	192.07675	160.05054/132.05562	1
Dodemorph	9.78	282.27914	116.1069/98.09643	1	Carbofuran	8.94	222.11247	123.04406/165.09101	1
Famphur	9.55	326.02803	93.00999/142.99263	1	Chlordimeform	6.63	197.084	117.05730/152.02615	1
Fuberidazole	7.4	185.07094	157.07602/156.06820	1	Dioxacarb	7.46	224.09173	123.04406/167.07027	1
Isocarbophos	9.68	312.04299	269.99844/236.00554	1	Ethoxyquin	9.7	218.15394	148.07569/190.12264	1
Malathion	10.26	353.02529	195.06031/227.03238	1	Fenobucarb	10.02	208.13321	208.13321/95.04914	1
Mevinphos	8.12	225.05225	127.01547/67.01784	1	Fipronil	10.62	434.93143	329.95845/183.01646	1
Monocrotophos	6.65	224.06824	127.01547/58.02874	1	Fipronil-desulfinyl	10.54	386.96444	350.98667/281.99146	1
Omethoate	4.17	214.02974	214.02974/182.98754	1	Fipronil-sulfide	10.72	418.93651	57.9746/170.00863	1
Phorate sulfoxide	9.47	277.01502	114.96133/142.93848	1	Fipronil-sulfone	10.85	450.92634	414.94857/243.98839	1
Phoratoxon sulfoxide	7.61	261.03786	114.96150/128.97698	1	Imazalil	9.49	297.0556	158.97628/69.04472	1
Phosalone	11.06	367.99414	114.96150/138.01033	1	Indoxacarb	11.14	528.07799	203.01866/168.02107	1
Phosmet	9.98	318.00181	160.03930/133.02841	1	Methiocarb	10.14	226.08963	169.06816/121.06479	1
Phoxim	11.1	299.06138	129.04472/95.04945	1	Pentachlorophenol	11.74	262.83973	262.83973/262.83973	1
Pirimiphos-methyl	11.02	306.10358	108.05562/67.02907	1	Pirimicarb	8.38	239.15025	182.12879/72.04439	1
Prometryn	10.22	242.14339	200.09644/158.04949	1	Promecarb	10.29	208.13321	208.13321/109.06479	1
Propazine	10.1	230.1167	188.06975/146.02280	1	Propamocarb	4.25	189.15975	102.0496/74.02365	1
Pyrazophos	11.08	374.0934	194.05602/222.08732	1	Propiconazole	10.96	342.07706	158.97628/69.06988	1
Quinalphos	10.84	299.06138	147.05529/163.03245	1	Propoxur	8.89	210.11247	111.04406/95.04914	1
Simazine	9.01	202.0854	132.03230/124.08692	1	Robenidine	10.38	334.06208	138.01050/155.03705	1
Simetryne	9.13	214.11209	96.05562/124.08692	1	Thiobencarb	11.14	258.07139	258.07139/125.01525	1
Thiabendazole	7.26	202.04334	175.03245/131.06037	1	Thiophanate-ethyl	9.68	371.08422	151.03245/282.03654	1
Triazophos	10.42	314.07228	162.06619/114.96133	1	Thiophanate-methyl	8.85	343.05292	151.03245/311.02671	1
Tributyl-phosphorotrithioate	12.11	315.10344	57.06988/168.99052	1	Xylazine	7.26	221.1107	90.03720/164.05285	1
Trichlorfon	7.2	256.92985	127.01547/220.95318	1	2,3,5-Trimethacarb	9.56	194.11756	135.04406/107.04914	2
Doramectin	12.29	921.49708	777.41843/449.25097	2	3,4,5-Trimethylphenol	9.75	194.11756	135.04406/107.04914	2
Isoprocarb	9.56	194.11756	137.09609/95.04914	2	Phorate sulfone	9.55	293.00993	246.96807/114.96133	2
Methidathion	9.86	302.96913	145.00662/71.02399	2	Thiofanox sulphoxide	7.14	235.11109	57.06988/104.01646	2
XMC	9.39	180.10191	123.08044/113.99744	2	Avermectin B1a	12.23	895.48143	123.11683/153.05462	5
Aldicard	8.32	213.06682	95.04914/141.00593	5	Deltamethrin	12.04	521.00699	278.90762/89.05971	5
Dichlorvos	7.18	220.95318	127.01565/78.99452	5	Flucythrinate	11.73	469.19334	114.09134/199.0929	5
Propetamphos	10.34	282.09234	138.01370/156.02395	5	Flumethrin	12.33	532.0853	114.09134/73.02982	5

Table 3. *Cont.*

Compounds	RT (min)	Extracted Mass (*m/z*)	Fragment Ions (*m/z*)	LOD ng/mL	Compounds	RT (min)	Extracted Mass (*m/z*)	Fragment Ions (*m/z*)	LOD ng/mL
Thiofanox	9.56	241.09812	58.06513/184.07906	5	Ivermectin B1a	12.48	897.49708	753.41843/329.20872	5
Thiofanox sulphone	7.3	251.106	251.10600/57.06988	5	Methamidophos	2.57	142.00861	142.00861/112.01598	5
Bifenthrin	12.45	440.15987	181.10118/166.07770	10	Amitraz	11.97	294.19647	163.12298/122.09643	20
Phlorpyrifos-methyl	11.29	321.90226	142.99263/289.87605	10	Tau-fluvalinate	12.13	503.13438	181.06479/114.09134	20
Acephate	3.46	184.01918	113.00250/142.99263	20	Fenitrothion	10.18	278.02466	142.99263/149.02332	50
Aldicarb sulfoxide	4.78	207.07979	89.04195/69.05730	20	Methomyl	5.84	163.05357	88.02155/106.03211	50
Fenvalerate	12.05	437.16265	437.16265/114.09134	20	Phorate	11.11	261.0201	261.02010/75.02630	50
Validamycin	1.26	498.21812	142.08626/124.07569	50	Prothiofos	12.22	344.97009	258.92025/132.96046	100
Bromophos ethyl	12.21	392.8878	336.82520/161.96337	500	Cyfluthrin	12.01	451.0986	191.00250/114.09134	500
Parathion	10.86	292.04031	114.96133/138.00081	500					

3.4. Sample Preparation

A 20 mL centrifuge tube was filled with 100 mg samples and added with 20 mL MeOH. One hundred microliters of liquid sample was pipetted directly into another centrifuge tube. The tube was vortexed for 1 min, and ultrasonicated for 15 min. Then, the solution was vortexed again and silenced for 2 min. Then, MeOH-water (1:1, *v/v*) was used to dilute 0.5 mL of supernatant by 5 fold. After vortex and silence, 1 mL of the solution was transferred to an Eppendorf tube (1.5 mL) and centrifuged at 10,000 rpm/min for 15 min to remove the precipitate. The upper supernatant was transferred into a vial for analysis.

Chinese herbal drugs were used for method validation in the research, which was representative of a complex matrix of fishery products. Fishery drugs of pure Chinese herb products composed of granular herbal extract were purchased from a local fishery store.

3.5. Database for Screening, Qualitative and Quantitative Rules

The names, categories, CAS numbers, formulas, expected mass of the suspected compounds were searched and collected to establish a basic database. Then, the standard solution of each compound (100 ng·mL^{-1}) were analyzed through the UHPLC-ESI-Q-Orbitrap HRMS system using the aforementioned parameters. Therefore, *m/z* of precursor ion, retention time (RT) and fragment ions (FI) were acquired by experiments. In parallel, the isotope pattern for each precursor was automatically calculated by Tracefinder software. All information was organized and built in Tracefinder. It was used to perform screening according to the database with the following screening rules: *m/z* deviation of precursor ion was 3×10^{-6}, allowed RT deviation was ±15 s, at least one fragment ion match with allowed *m/z* deviation at 2×10^{-5}, and the fit threshold for precursor isotope pattern was more than 75% with allowed mass deviation within 10 ppm, and allowed isotope intensity deviation of less than 25%. If the screening rules passed for a compound, it was qualitatively identified as positive. Furthermore, a series of mixed standards solution of 1–500 ng/mL were prepared for quantification of the positive compounds. The integrated peak area of precursor ions for positive compounds was used for external calibration and quantification. The instrument detection limit (LOD), the minimum concentration that the compound could be identified under the qualitative rules, was tested at the optimized parameters. The detailed information for these compounds of interest in the database is shown in Table 3.

4. Conclusions

In summary, a database of 89 pesticides was built, including both chromatographic and HRMS information. The data was acquired after parameter optimization on ultrahigh performance liquid chromatography interfaced quadrupole-Orbitrap mass spectrometer. Based on the database, the screening rule for these compounds was further established by comparing their precursors, fragments, retention time and isotopes. The fast, high-throughput identification and rough quantification of these compounds was achieved. The method was successfully applied for the pesticides risk assessment of fishery drugs. However, as the detection mode established on potential and known pesticides, where their chromatographic and mass spectrometric information were examined and collected, the unknown, non-target risk compounds were ignored. Further work will be focused on non-target screening based on characteristic fragments for recognizing and monitoring risk factors. Overall, our current method can be used as a fast, reliable, efficient and practical tools for the fishery drug risk assessment, which saves more time, and expenses.

Author Contributions: C.K. and H.-J.Y. conceived and designed the experiments; S.-Y.W. and Q.-P.C. performed the experiments; C.K. and S.-Y.W. analyzed the data; C.K. and S.-Y.W. wrote the paper.

Acknowledgments: C.K. thanks Wenlei Zhai for help in paper writing and revision.

References

1.   Broughton, E.I.; Walker, D.G. Policies and practices for aquaculture food safety in china. *Food Pol.* **2010**, *35*, 471–478. [CrossRef]

2.   *Green Food—Fishery Medicine Application Guideline Industry Standard*; Agriculture: Beijing, China, 2013; Volume NY/T 755–2013. Available online: http://foodmate.net/standard/yulan.php?itemid=39306 (accessed on 23 December 2013).

3.   Rico, A.; Geng, Y.; Focks, A.; Van den Brink, P.J. Modeling environmental and human health risks of veterinary medicinal products applied in pond aquaculture. *Environ. Toxicol. Chem.* **2013**, *32*, 1196–1207. [CrossRef] [PubMed]

4.   Rico, A.; Satapornvanit, K.; Haque, M.M.; Min, J.; Nguyen, P.T.; Telfer, T.C.; van den Brink, P.J. Use of chemicals and biological products in asian aquaculture and their potential environmental risks: A critical review. *Rev. Aquacult.* **2012**, *4*, 75–93. [CrossRef]

5.   Burridge, L.; Weis, J.S.; Cabello, F.; Pizarro, J.; Bostick, K. Chemical use in salmon aquaculture: A review of current practices and possible environmental effects. *Aquaculture* **2010**, *306*, 7–23. [CrossRef]

6.   Haya, K.; Burridge, L.E.; Davies, I.M.; Ervik, A. A review and assessment of environmental risk of chemicals used for the treatment of sea lice infestations of cultured salmon. In *Environmental Effects of Marine Finfish Aquaculture. Handbook of Environmental Chemistry*; Springer: Berlin/Heidelberg, Germany, 2005; Volume 5M, pp. 305–340.

7.   Turnipseed, S.B.; Storey, J.M.; Lohne, J.J.; Andersen, W.C.; Burger, R.; Johnson, A.S.; Madson, M.R. Wide-scope screening method for multiclass veterinary drug residues in fish, shrimp, and eel using liquid chromatography–quadrupole high-resolution mass spectrometry. *J. Agric. Food Chem.* **2017**, *65*, 7252–7267. [CrossRef] [PubMed]

8.   Asghar, M.A.; Zhu, Q.; Sun, S.; Peng, Y.; Shuai, Q. Suspect screening and target quantification of human pharmaceutical residues in the surface water of wuhan, china, using UHPLC-Q-ORBITRAP HRMS. *Sci. Total Environ.* **2018**, *635*, 828–837. [CrossRef] [PubMed]

9.   Muehlwald, S.; Buchner, N.; Kroh, L.W. Investigating the causes of low detectability of pesticides in fruits and vegetables analysed by high-performance liquid chromatography-time-of-flight. *J. Chromatogr. A* **2018**, *1542*, 37–49. [CrossRef] [PubMed]

10. Mol, H.G.; Zomer, P.; De Koning, M. Qualitative aspects and validation of a screening method for pesticides in vegetables and fruits based on liquid chromatography coupled to full scan high resolution (orbitrap) mass spectrometry. *Anal. Bioanal. Chem.* **2012**, *403*, 2891–2908. [CrossRef] [PubMed]

11. Zuo, L.; Sun, Z.; Wang, Z. Tissue distribution profiles of multiple major bioactive components in rats after intravenous administration of xuebijing injection by UHPLC-Q-ORBITRAP HRMS. *Biomed. Chromatogr.* **2019**, *33*, e4400. [CrossRef] [PubMed]

12. Núñez, O.; Gallart-Ayala, H.; Martins, C.P.B.; Lucci, P. New trends in fast liquid chromatography for food and environmental analysis. *J. Chromatogr. A* **2012**, *1228*, 298–323. [CrossRef]

13. Wang, J.; Leung, D.; Chow, W.; Chang, J.; Wong, J.W. Development and validation of a multiclass method for analysis of veterinary drug residues in milk using ultrahigh performance liquid chromatography electrospray ionization quadrupole orbitrap mass spectrometry. *J. Agric. Food Chem.* **2015**, *63*, 9175–9187. [CrossRef] [PubMed]

14. Kaufmann, A.; Teale, P. Capabilities and limitations of High-Resolution Mass Spectrometry (HRMS): Time-of-flight and orbitraptm. In *Chemical Analysis of Non-Antimicrobial Veterinary Drug Residues in Food*; Kay, J.F., MacNeil, J.D., Wang, J., Eds.; Wiley: Hoboken, NJ, USA, 2016; Volume 3, pp. 93–139.

15. Kaufmann, A.; Walker, S. Comparison of linear intrascan and interscan dynamic ranges of orbitrap and ion-mobility time-of-flight mass spectrometers. *Rapid Commun. Mass Spectrom.* **2017**, *31*, 1915–1926. [CrossRef] [PubMed]

16. Kong, C.; Wang, Y.; Huang, Y.; Yu, H. Multiclass screening of >200 pharmaceutical and other residues in aquatic foods by ultrahigh-performance liquid chromatography-quadrupole-orbitrap mass spectrometry. *Anal. Bioanal. Chem.* **2018**, *410*, 1–9. [CrossRef] [PubMed]

17. Del Mar Gomez-Ramos, M.; Rajski, L.; Heinzen, H.; Fernandez-Alba, A.R. Liquid chromatography orbitrap mass spectrometry with simultaneous full scan and tandem MS/MS for highly selective pesticide residue analysis. *Anal. Bioanal. Chem.* **2015**, *407*, 6317–6326. [CrossRef] [PubMed]

18. Casado, J.; Santillo, D.; Johnston, P. Multi-residue analysis of pesticides in surface water by liquid chromatography quadrupole-orbitrap high resolution tandem mass spectrometry. *Anal. Chim. Acta* **2018**, *1024*, 1–17. [CrossRef] [PubMed]

19.	Hernandez, F.; Ibanez, M.; Portoles, T.; Cervera, M.I.; Sancho, J.V.; Lopez, F.J. Advancing towards universal screening for organic pollutants in waters. *J. Hazard. Mater.* **2015**, *282*, 86–95. [CrossRef] [PubMed]

20.	Jia, W.; Chu, X.; Chang, J.; Wang, P.G.; Chen, Y.; Zhang, F. High-throughput untargeted screening of veterinary drug residues and metabolites in tilapia using high resolution orbitrap mass spectrometry. *Anal. Chim. Acta* **2017**, *957*, 29–39. [CrossRef]

21.	Wang, J.; Chow, W.; Wong, J.W.; Leung, D.; Chang, J.; Li, M. Non-target data acquisition for target analysis (ndata) of 845 pesticide residues in fruits and vegetables using uhplc/esi q-orbitrap. *Anal. Bioanal. Chem.* **2019**, *411*, 1421–1431. [CrossRef]

22.	Martinello, M.; Borin, A.; Stella, R.; Bovo, D.; Biancotto, G.; Gallina, A.; Mutinelli, F. Development and validation of a quechers method coupled to liquid chromatography and high resolution mass spectrometry to determine pyrrolizidine and tropane alkaloids in honey. *Food Chem.* **2017**, *234*, 295–302. [CrossRef]

23.	Rochat, B. Proposed confidence scale and id score in the identification of known-unknown compounds using high resolution ms data. *J. Am. Soc. Mass Spectrom.* **2017**, *28*, 709–723. [CrossRef]

24.	Schymanski, E.L.; Singer, H.P.; Slobodnik, J.; Ipolyi, I.M.; Oswald, P.; Krauss, M.; Schulze, T.; Haglund, P.; Letzel, T.; Grosse, S.; et al. Non-target screening with high-resolution mass spectrometry: Critical review using a collaborative trial on water analysis. *Anal. Bioanal. Chem.* **2015**, *407*, 6237–6255. [CrossRef]

25.	Raina, R.; Sun, L. Trace level determination of selected organophosphorus pesticides and their degradation products in environmental air samples by liquid chromatography-positive ion electrospray tandem mass spectrometry. *J. Environ. Sci. Health, Part B* **2008**, *43*, 323–332. [CrossRef]

26.	Kong, C.; Zhou, Z.; Wang, Y.; Huang, Y.-F.; Shen, X.-S.; Huang, D.-M.; Cai, Y.-Q.; Yu, H.-J. Screening of chemical drugs in fishery inputs by ultrahigh performance liquid chromatography-orbitrap high resolution mass spectroscopy. *Chin. J. Anal. Chem.* **2017**, *45*, 245–252.

27.	Wang, Y.; Huang, Y.-F.; Han, F.; Kong, C.; Zhou, Z.; Cai, Y.-Q.; Yu, H.-J. Screening 175 veterinary drugs in fishery feed by ultrahigh performance liquid chromatography-orbitrap high resolution mass spectrometry. *Chin. J. Anal. Lab.* **2018**, *37*, 1013–1019.

28.	Rajski, L.; Gomez-Ramos Mdel, M.; Fernandez-Alba, A.R. Large pesticide multiresidue screening method by liquid chromatography-orbitrap mass spectrometry in full scan mode applied to fruit and vegetables. *J. Chromatogr. A* **2014**, *1360*, 119–127. [CrossRef]

29.	Bateman, K.P.; Kellmann, M.; Muenster, H.; Papp, R.; Taylor, L. Quantitative–qualitative data acquisition using a benchtop orbitrap mass spectrometer. *J. Am. Soc. Mass Spectrom.* **2009**, *20*, 1441–1450. [CrossRef]

30.	Li, P.; Zeng, X.-Y.; Cui, J.-T.; Zhao, L.-J.; Yu, Z.-Q. Determination of seven urinary metabolites of organophosphate esters using liquid chromatography-tandem mass spectrometry. *Chin. J. Anal. Chem.* **2017**, *45*, 1648–1654. [CrossRef]

31.	Trufelli, H.; Palma, P.; Famiglini, G.; Cappiello, A. An overview of matrix effects in liquid chromatography-mass spectrometry. *Mass Spectrom. Rev.* **2011**, *30*, 491–509. [CrossRef]

32.	Łozowicka, B.; Rutkowska, E.; Jankowska, M. Influence of quechers modifications on recovery and matrix effect during the multi-residue pesticide analysis in soil by GC/MS/MS and GC/ECD/NPD. *Environ. Sci. Pollut. Res. Int.* **2017**, *24*, 7124–7138. [CrossRef]

33.	European commission. Guidance Document on Analytical Quality Control and Validation Procedures for Pesticide Residues Analysis in Food and Feed. Sante/11813/2017. Available online: https://ec.europa.eu/food/sites/food/files/plant/docs/pesticides_mrl_guidelines_wrkdoc_2017-11813.pdf (accessed on 1 March 2018).

34.	Portolés, T.; Ibáñez, M.; Garlito, B.; Nácher-Mestre, J.; Karalazos, V.; Silva, J.; Alm, M.; Serrano, R.; Pérez-Sánchez, J.; Hernández, F.; et al. Comprehensive strategy for pesticide residue analysis through the production cycle of gilthead sea bream and atlantic salmon. *Chemosphere* **2017**, *179*, 242–253. [CrossRef]

35.	De Paepe, E.; Wauters, J.; Van Der Borght, M.; Claes, J.; Huysman, S.; Croubels, S.; Vanhaecke, L. Ultra-high-performance liquid chromatography coupled to quadrupole orbitrap high-resolution mass spectrometry for multi-residue screening of pesticides, (veterinary) drugs and mycotoxins in edible insects. *Food Chem.* **2019**, *293*, 187–196. [CrossRef]

36.	Huysman, S.; Van Meulebroek, L.; Vanryckeghem, F.; Van Langenhove, H.; Demeestere, K.; Vanhaecke, L. Development and validation of an ultra-high performance liquid chromatographic high resolution q-orbitrap mass spectrometric method for the simultaneous determination of steroidal endocrine disrupting compounds in aquatic matrices. *Anal. Chim. Acta* **2017**, *984*, 140–150. [CrossRef]

6

Chemometrics Approaches in Forced Degradation Studies of Pharmaceutical Drugs

Benedito Roberto de Alvarenga Junior and Renato Lajarim Carneiro *

Department of Chemistry, Federal University of São Carlos, São Carlos 13565-905, Brazil;
benedito.alvarenga@outlook.com
* Correspondence: renato.lajarim@ufscar.br

Academic Editor: Marcello Locatelli

Abstract: Chemometrics is the chemistry field responsible for planning and extracting the maximum of information of experiments from chemical data using mathematical tools (linear algebra, statistics, and so on). Active pharmaceutical ingredients (APIs) can form impurities when exposed to excipients or environmental variables such as light, high temperatures, acidic or basic conditions, humidity, and oxidative environment. By considering that these impurities can affect the safety and efficacy of the drug product, it is necessary to know how these impurities are yielded and to establish the pathway of their formation. In this context, forced degradation studies of pharmaceutical drugs have been used for the characterization of physicochemical stability of APIs. These studies are also essential in the validation of analytical methodologies, in order to prove the selectivity of methods for the API and its impurities and to create strategies to avoid the formation of degradation products. This review aims to demonstrate how forced degradation studies have been actually performed and the applications of chemometric tools in related studies. Some papers are going to be discussed to exemplify the chemometric applications in forced degradation studies.

Keywords: forced degradation; degradation products; stress test; chemometrics

1. Chemometrics

The Swedish word "kemometri" appeared for the first time in 1971 by a combination between the terms chemistry and -metri. In 1972, the English homologous term chemometrics (chemo + metrics) was referred by Prof. Svante Wold that named his group as Forskningsgruppen för Kemometri (Research Group for Chemometrics) or Kemometrigruppen (Chemometrics Group), and in the next year, it was published the first article with the term kemometri [1,2]. The International Chemometrics Society explained the term "chemometrics" for the first time in 1974. International journals, in the 1980s, had special issues on chemometrics. In 1986–1987, the publishers Wiley and Elsevier created the chemometrics journals "The Journal of Chemometrics" and "Chemometrics and Intelligent Laboratory Systems," respectively [3].

The definition of chemometrics is intimately linked to what it is expected to gain from using it. This definition has presented some inconsistencies between authors over the years, once each one belongs to fields with different aims [4].

According to Pure and Applied Chemistry (IUPAC), the full definition of chemometrics, considering no preference of area, is the science of relating measurements performed on a chemical system or process to the state of the system through application of mathematical or statistical methods. IUPAC also highlights that, in chemometrics, the data are treated commonly in a multivariate approach, and although there are cases in theoretical chemistry that use the same mathematical and statistical techniques in some application, it should aim primarily to extract useful chemical information of measured data [5].

This definition evidences clearly the utilization of chemometrics in all stages of the chemical measurement process, from definition of optimal experimental conditions, data collection, and processing of data. Chemometrics has its roots in analytical chemistry [6], but it is totally interdisciplinary and has been applied in many different areas [7], such as food sciences [8–12], assessment of adulteration, geographical origin [13–15], metabolomics [16–18], engineering [19,20], forensics [21–25], pharmaceutical studies [26–30], cultural studies [31–33], environmental chemistry [34], etc. Chemometric tools are fundamental to solve real life problems [35].

In fact, when chemometric is applied appropriately with suitable interpretations, it enables to obtain a better data visualization even from experimental of poor quality (low resolution and high level of noise), making the relations between analytical signals and experimental parameters clearer [36]. The development of methods for analysis of degradation products is a hard work, time consuming, and an expensive task. In this context, chemometric tools are an alternative approach to carry out studies related to impurities in pharmaceutical drugs, contributing for acquiring relevant information from the system or turning the analytical method greener.

2. Degradation Products

The efficacy and safety of drugs are determined by toxicological and pharmacological profiles and adverse side effects due to the dosage and impurities [37–39]. According to the International Council for Harmonization and Technical Requirements for Pharmaceuticals for Humans Use (ICH), a drug impurity is any component that is not a chemical entity defined as an active pharmaceutical ingredient or excipient [40]. The impurities can be classified regarding their origin: inorganic impurities (reagents, ligands and catalysts, heavy metals or other residual metals, and inorganic salts), organic impurities (starting materials, by-products, intermediates, degradation products, reactants, ligands, and catalysts), and solvents (organic and inorganic liquids used in preparation of solutions or in the synthesis of a new drug substance). Therefore, any extra material present in the drug, even if it does not have pharmacological activity, is considered an impurity [39]. Although the term "impurity" is commonly assigned as synonymous of degradation products, it is worth highlighting that these compounds belong to a subgroup inside the impurity definition [41]. The United States Pharmacopoeia adopts the term "Related Compounds" for the main degradation products and impurities from synthesis.

The yielding of degradation products depends of several variables, chemical stability being the most important one. The degradation of APIs involves the formation or breaking of covalent bonds in chemical processes such as oxidation, reduction, thermolysis, and hydrolysis reactions. These processes can usually be accelerated when the drug is exposed to light, high temperatures, acidic or basic conditions, humidity, oxidative environment, incompatible excipients, and even due to its contact with packaging during its shelf-life [41].

2.1. The Generation of Degradation Products

Stability of API is a critical parameter in the development of a drug product, which should be considered in the formulation, analytical methods, package, storage, shelf life determination, safety, and toxicological studies [42,43].

The degradation of an API can result in the loss of effectiveness and can also lead to adverse effects due to degradation products [44]. Therefore, understating the processes that contribute to generation of degradation products is extremely important to create strategies aimed at the prevention and/or minimization of the API's degradation.

The oxidative degradation is one of the leading causes of drugs degradation, once it involves the removal of an electropositive atom, radical, electron, or the addiction of an electronegative atom or radical. The major part of API's oxidation occurs slowly due to the action of molecular oxygen, and some procedures used during manufacturing and storage are employed to stabilize the API in the product. For that, it is necessary to know the variables that increase the extension of oxidation. One form of preventing the oxidation process is to substitute oxygen inside pharmaceutical recipients by nitrogen or dioxide carbon. The contact of drug with metal ions, which can catalyze the oxidation, should be also avoided, as well as high storage temperatures [45].

Temperature is another variable that has significant influence on degradation and is often used in forced degradation studies. The same product can present different shelf lives depending on how and where it is stored. For example, countries in which equatorial climate predominates have higher average temperature than the ones with tropical climate, and this difference promotes different degradation conditions and, consequently, different shelf lives [46].

Several pharmaceutical drugs have low stability in aqueous medium and must be evaluated under hydrolysis conditions. First, to evaluate the hydrolysis of an API, it is necessary to perform tests in a wide range of pH (solution or suspension) once the hydrogen and hydroxyl ions are able to influence the degradation ratio [47–49]. Then, hydrolytic forced degradation studies are performed by submitting the API to acid, basic, and neutral conditions, in a fashion that the experimental variables have to be adapted if it is observed high degradation of API, in order to avoid the formation of secondary degradation products [48].

Photostability studies should also be performed to demonstrate the extension of reactions when the APIs are exposed to light. The photolytic reactions are caused when the drug absorbs the ultraviolet/visible (UV-Vis) light (wavelength 300 to 800 nm), which promote the molecule to an excited state and can increase its reactivity in some sites of the molecule. The UV-Vis radiation also can lead to cleavage of chemical bonds, yielding new molecules. The extension of photodegradation is dependent of the wavelength of the incident radiation and the absorptivity of the molecule. In other words, this process depends of the presence of specific functional groups [50].

Nonetheless, it is worth mentioning that even when an API is shown to be chemically stable in stress tests, the stress conditions can degrade this API when excipients are present.

2.2. Forced Degradation Studies

Since the release of the first guidelines, massive changes to the definition of quality in pharmaceutical drugs have taken place, and several countries are extending the requirements of regulatory agencies to generic drugs and already commercialized products [51]. Forced degradation studies, also called "stress tests," have been used in the pharmaceutical industry for a long time [50], but the International Conference on Harmonization (ICH) only issued the formal request Q1A with a guideline "Stability Testing of New Drug Substance and Products" in 1993 [52]. In general terms, forced degradation studies are processes that involve the degradation of drugs under extreme conditions to accelerate the yielding of degradation products. The information obtained from these studies are usually used to determine the chemical stability, pathways of degradation, to identify the degradation products, conditions of storage, self-life, excipient compatibility, and also allow the development of selective analytical methods [52–54].

Today, the control of impurities has been established by ICH Q3A and Q3B guidelines, which are addressed for registration applications about the content and qualification of impurities classified as degradation products, which are observed during manufacturing or stability studies of the new drug product. Furthermore, the registration application should present a validated analytical procedure suitable for the detection and quantification of degradation products, which should include or evidence the method's specificity for specified and unspecified degradation products according to ICH Q2A and Q2B guidelines for analytical validation. When the impurities are available in the validation method phase, the discriminatory capacity of drug and impurities is validated through spiking drug substance with levels of impurities. On the other hand, if impurity or degradation product standards are unavailable, the drug substance should be submitted to stress conditions (light, heat, humidity, acid/base hydrolysis, and oxidation). Therefore, in general, the forced degradation studies are performed in the developing stability-indicating method, and the method validation should take into account the chromatographic separation of the degradation products.

Several works in the literature deal with studies of forced degradation and stability as synonymous, but it is worth highlighting that there are some differences between them. Stability studies consist of submitting the pharmaceutical drug in milder conditions over a long period (months or years) and, besides determining some degradation products, allow the establishment of the product's shelf life. Forced degradation studies are often performed by exposing the API or the product in drastic conditions for some hours or days. These extreme conditions are able to provide, as a general rule, substantial degradation of the API, usually from 10 to 30%. The set of whole degradation products found in every degradation condition composes a "potential" degradation profile. If just few degradation products are found, the degradation profile is then denominated as "real degradation profile." The method to evaluate the degradation products should be selective and developed considering the occurrence of every degradation product [55].

The forced degradation studies are critical in the development of drug products and aims the following points:

- To obtain the potential degradation potential of an API or drug product;

- To discover the degradation mechanism, such as hydrolysis, thermolysis, oxidation, photolysis, etc.;

- To elucidate the molecular structure of degradation product;

- To solve problems regarded to the API stability;

- To identify the conditions where the API or the drug product are more susceptible to degradation in order to ensure the quality of the final product, bringing to pharmaceutical industry enough knowledge for development, packaging, manufacture, manipulation, and storage;

- To obtain more stable formulations;

- To develop analytical methods that can be used to quantify the API without interference of its degradation products and to quantify these degradation products [48,56,57].

The degradation products are commonly analyzed by high-performance liquid chromatography (HPLC) coupled with ultraviolet/visible (UV-Vis) and/or mass spectrometric (MS) detectors. UV-Vis detectors are able to provide only information related to chromophores groups, but they are excellent for quantification. MS detectors are not robust as UV-Vis detectors for quantification, but MS presents high sensitivity (traces level) and gives important data to characterize the degradation products through fragmentation profile, accurate mass (for detectors of High Resolution such as Q-ToF, Orbitrap, and Fourier-transform ion cyclotron resonance (FT-ICR)), as well as information about the origin of fragments using multiple stage (MS^n) and neutral loss scan. When more information is necessary to elucidate a chemical structure, the nuclear magnetic resonance (NMR) technique is required. NMR presents low sensitivity, but it is able to resolve conformational, structural, and optical isomers. All these techniques generate a great amount of data, and the manual data mining is very time and money consuming. In this context, chemometric tools can present a way to organize and pre-process data, optimize parameters of HPLC, MS, and NMR techniques, obtain the maximum knowledge about them, and clarify a lot of useful information [51,58,59].

2.3. Strategies to Select the Degradation Conditions

Forced degradation studies are performed in batches with solutions at different pHs, in the presence of hydrogen peroxide, UV-Vis radiation, metallic cations (Fe^{3+} and Cu^{2+}), and high temperatures [48].

Usually, the influence of pH is evaluated using 0.1 mol L^{-1} of HCl or NaOH [48]. The degradation by radiation is performed under UV-Vis light, which should not be lesser than 1.2 million of lux per hour and a power of 200 Wh m^{-2} [60]. For oxidant condition, the literature recommends using hydrogen peroxide (H_2O_2) in concentration from 0.1% to 3.0% at room temperature (25 °C). The evaluation of temperature is usually performed between 40 to 80 °C, but it could be higher for recalcitrant APIs. Other additional variables can be taken into consideration in the global stability studies of an API or the final product, such as humidity and microbiological stability [22,57,61,62].

According to ICH, in "Expert Committee on Specifications for Pharmaceutical Preparations" document, the recommended degradation should be between 10 to 30% of the API. This degradation range commonly allows for the evaluation of the main degradation products, avoiding the yielding of secondary degradation products [63]. In Brazil, the regulatory agency ANVISA recommends not less than 10% of degradation of API, and a technical justification is needed in the case where such degradation is not obtained [64].

It is worth highlighting that the cited conditions for forced degradation studies are just initial attempts, and the ideal condition could be more extreme or mild, depending of the chemical recalcitrance of the API. Table 1 summarizes degradation conditions of some papers that performed forced degradation studies.

Table 1. Degradation conditions for pharmaceutical drugs in forced degradation studies.

API: Year	Acid	Base	Neutral	Thermolysis	Oxidation	Photolysis
Zidovudine: 2017 [65]	2 M HCl	2 M NaOH	-	Acid/base at 80 °C for 72 h	10% H_2O_2 at room temperature for 10 h	1.2×10^6 lx × h of fluorescent light and 200 W h/m² UV light
Toloxatone: 2018 [66]	1 M HCl	0.01 M NaOH	H_2O	All hydrolysis at 80 °C for 2 h	0.01% H_2O_2 at room temperature for 2 h	2700 kJ/m²/h of UV-VIS and UVC 7.5 W/m²
Amlodipine: 2015 [67]	1 M HCl at 80 °C for 30 min	1 M NaOH at 80 °C for 1 h	H_2O at 80 °C for 2 h	50 °C for 48 h	15% H_2O_2 at room temperature for 48 h	1.2×10^6 lx × h of fluorescent light and 200 Wh/m² UV-A light for 14 days
Acebutolol: 2018 [68]	1 M HCl	2 M HCl	H_2O	All hydrolysis at 80 °C	3% H_2O_2 at 80 °C	Not less than 1.2×10^6 lx × h and ultraviolet energy of not less than 200 W h/m²
Stevioside: 2018 [69]	0.1 M HCl/0.1 M H_3PO_4	0.1 M NaOH	H_2O	All hydrolysis at 80 °C for 8 h	10% H_2O_2 at 25 °C for 72 h	UV$_{254nm}$ lamp for 48 h
Pentoxifylline: 2013 [70]	2 M HCl at 70 °C for 4 h	2 M NaOH at 70 °C for 4 h	H_2O at 70 °C for 4 h	Dry heat under at 105 °C for 4 h	30% H_2O_2 at 70 °C for 4 h	Sunlight for 8 h
Leflunomide: 2015 [71]	0.1–5 M at 85 °C for 8 h	0.1 M NaOH at 85 °C for 8 h	H_2O at 85 °C for 8 h	50 °C for 30 days	30% H_2O_2 at room temperature for 24 h	UV and white light for 14 days
Actarit: 2014 [72]	0.1 M HCl at 70 °C for 24 h	0.1 M NaOH at 70 °C for 24 h	H_2O at 70 °C for 14 days	Dry heat at 70 °C for 14 days	3% H_2O_2 for 14 days	UV light
Nicardipine: 2014 [73]	1 M HCl at 60 °C for 1 h	0.1–0.5 M NaOH at 50–80 °C for 1 h	-	-	5% H_2O_2 at 30–50 °C for 1 h	UV$_{254–365nm}$ light at room temperature
Clopidogrel bisulfate: 2010 [74]	1 M HCl	1 M NaOH	-	All hydrolysis at 80 °C for 1 h	5% H_2O_2	-
Biapenem: 2009 [75]	pH from 2.5 to 7.5 at 80 °C for 40 min			From room temperature to 100 °C in pH 3.5	-	-
Irbesartan: 2010 [76]	1 M HCl at 80 °C for 24 h	2 M NaOH at 80 °C for 48 h	H_2O at 80 °C for 48 h	50 °C	30% H_2O_2 at room temperature for 2 days	8500 lx fluorescent and 0.05 W/m² UV light

2.4. Acceptable Limits of Impurities

After obtaining the degradation profile, a critical analysis should be performed to verify the purity of the chromatographic band of the API and to evaluate the variables that can promote degradation of the API. The degradation products are analyzed according to their amount in relation to the API in the final product, after the regular stability time (without any stress condition). The evaluation considers the maximum amount of API administered per day, and the limit of degradation products are expressed as a percentage (or mass) relative to the API. The amount of degradation products defines if it is necessary to perform notification, identification, or qualification [40,57,77]. Table 2 shows the acceptance criterion used by ICH, FDA, and ANVISA for the amount of impurities found in relation of a daily administrated API. The acceptance criteria have the following meaning:

- Reporting threshold: A limit of impurity that is not necessary to be reported.
- Identification threshold: A limit of impurity does not need to be structurally identified.
- Qualification threshold: The maximum amount of impurity that is not necessary to be qualified. Being "qualified" is the process of acquisition and evaluation of data that establishes biological security of an impurity or a degradation profile at the specified levels [40].

Table 2. Thresholds for degradation products.

	Maximum Daily Dose	**Threshold**
Reporting Threshold	≤1 g >1 g	0.1% 0.05%
Identification Threshold	<1 mg 1 mg–10 mg >10 mg–2 g >2 g	1.0% or 5 μg TDI, whichever is lower 0.5% or 20 μg TDI, whichever is lower 0.2% or 2 mg TDI, whichever is lower 0.10%
Qualification Threshold	<10 mg 10 mg–100 mg >100 mg–2 g >2 g	1.0% or 50 μg TDI, whichever is lower 0.5% or 200 μg TDI, whichever is lower 0.2% or 3 mg TDI, whichever is lower 0.15%

3. Applications of Chemometric Tools in Forced Degradation Studies

3.1. Design of Experiment (DoE)

In every area is important to know how variables act on the system. In general, processes aim to enhance the quality of the final product, taking into account the minimization of cost and time. To achieve these goals, it is necessary to perform the optimization of variables of the system to gain knowledge about the behavior of variables in order to determine the influence of each variable [78,79]. The optimization of variables in a system is more commonly performed using one-variable-at-a-time approach (OVAT), where one variable, or also called factor, is changed at a time, causing a change in the monitored response. However, this univariate approach does not consider the interactions between variables, and therefore, it does not ensure the discovery of the optimum point in an optimization process [80]. The design of experiments arises as an alternative multivariate approach for studying the behavior of a system [81]. In this approach, the factors are simultaneously evaluated, and the experiments are performed in an organized way in order to acquire information about all the system performing a minimum number of experiments [82,83].

Some terms in DoE must to be clear for better understanding, as variables, levels, and responses. Variables or factors are independent experimental inputs capable of changing the responses of the system. Such factors are temperature, pH, irradiation time, reaction time, concentration of reactants, and so on. It is worth reiterating that variables can be changed independently of each other, but the response is dependent of synergism between them [84].

Levels are different values that a variable can assume within experimental domain. The variable temperature in an optimization process, for example, can be studied at three levels: at 30, 50 and 70 °C.

Responses or independent variables are the monitored parameters. Typical responses are cost, time of analysis, resolution between chromatographic peaks, percentage of API degradation, etc.

The values studied for each variable are coded in levels as high (+1), central (0), low (−1), and other levels, which depend on the design. This codification normalizes the independent variables, avoiding any wrong interpretation of data. The processes involved in DoE allow it to fit the empirical data to a function, creating a linear or quadratic model and considering the interactions between variables of the system [85]. Figure 1 shows the experimental domain of the most common experimental designs for screening and optimization steps.

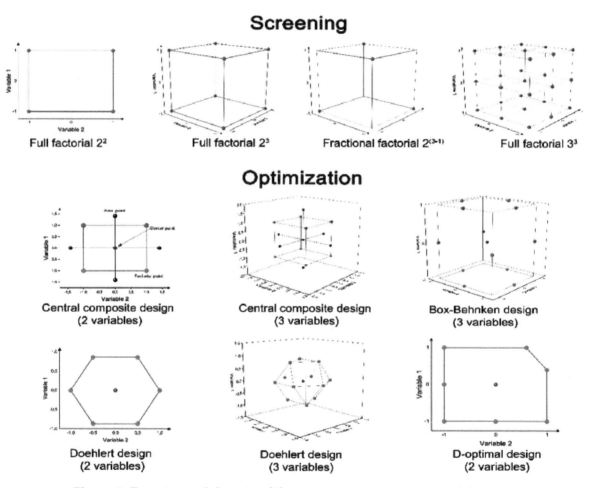

Figure 1. Experimental domain of the most common experimental designs.

In sum, the DoE presents the following advantages:

- Determining how many experiments are necessary to achieve the goal;
- Reducing the number of experiments;
- Observing the synergic and antagonist interactions between variables;
- Allowing for the possibility to create mathematical models and surface response to describe the behavior of the variables and to predict the system's response within an experimental domain;
- Decreasing the time, costs, and generation of lesser amounts of chemical waste, which contributes for the green chemistry principles [79].

In the context of forced degradation studies, the DoE has been mainly used for the development and optimization of chromatographic methods and for multivariate evaluation of stress conditions.

The use of DoE in the development and optimization of chromatographic conditions is not exclusive for forced degradation studies; instead, its application has spread to several fields that use chromatography as a tool [86–88]. Krishna et al. [89] performed forced degradation studies of eberconazole nitrate (EBZ) submitting it to hydrolytic (acid, basic, and neutral), thermal, oxidative, and photolytic degradation. In this work, a full factorial 3^3 design was used to identify the best conditions of the mobile phase for drug analysis. As is already well known in chromatography, the organic modifier in the mobile phase (methanol in this case), pH (10 mM potassium dihydrogen orthophosphate), and ion pair agent (tetra butyl ammonium hydroxide, TBAH) are important variables and alter the capacity factor (k) of the mobile phase. These variables were evaluated in three levels (−1, 0, and +1) following a full factorial design with 27 experiments (3^3 Full Factorial). Table 3 presents the real value of variables, and Table 4 shows the 27 different experiments.

Table 3. Real and coded values of variables considered in design of experiment.

Variable	Level (−1)	Level (0)	Level (+1)
TBHAH (mM)	5	7.5	10
pH	2.6	2.9	3.2
Organic phase (v/v)	20	25	30

Table 4. Conditions of experiments performed in full factorial 3^3 design.

Experiment	x_1	x_2	x_3	Experiment	x_1	x_2	x_3	Experiment	x_1	x_2	x_3
1	−1	−1	−1	10	−1	−1	0	19	−1	−1	1
2	0	−1	−1	11	0	−1	0	20	0	−1	1
3	1	−1	−1	12	1	−1	0	21	1	−1	1
4	−1	0	−1	13	−1	0	0	22	−1	0	1
5	0	0	−1	14	0	0	0	23	0	0	1
6	1	0	−1	15	1	0	0	24	1	0	1
7	−1	1	−1		−1	1	0	25	−1	1	1
8	0	1	−1	17	0	1	0	26	0	1	1
9	1	1	−1	18	1	1	0	27	1	1	1

The ranges studied in design were selected according to previous studies and considered the physicochemical properties of EZB. Other chromatographic parameters such as column dimensions, flow rate, injection volume, wavelength for detection, as well as the procedure performed in each degradation condition, can be found in reference [89].

As a result, a Pareto chart of standardized effects showed the quantification of each variable on the capacity factor, where organic phase and TBAH presented the higher influence on the response. Both linear and quadratic regressions showed no significance for pH inside its range of variation. The results of experimental design also allowed the authors to create contour plots, and they emphasized the usefulness of studying the interaction effects of variables on capacity factor. It was observed through contour plots that, by increasing concentration of TBAH, the capacity factor of EBZ was increased, and the same behavior occurred when the organic modifier decreased. Furthermore, pH did not affect the capacity factor in the investigated experimental domain. At the end, the optimum conditions (pH 2.8, 10 mM TBAH, and methanol 25% (v/v)) made it possible to find a capacity factor equal to 2.06.

Table 5 shows some papers that used the experiment design to optimize the chromatographic conditions to analyze the degradation products yielded in forced degradation studies.

Table 5. Design of experiments used in some papers to optimize chromatographic conditions for analyses of degradation products.

API	Design	Ref
Teriflunomide	Full factorial 3^3	[90]
Simvastatin	Plackett Burman/Box-Behnken	[91]
Linagliptin	Full factorial	[92]
Ticagrelor	Fractional Factorial Resolution V/Central composite	[93]
Imatinib mesylate	Box Behnken	[94]
Fusidic acid	Taguchi/Central Composite	[95]
Cloxacillin	Plackett Burman	[96]
Vilazodone hydrochloride	Central composite experimental	[97]
Darifenacin hydrobromide	Central composite	[98]
Edaravone	Placket Burman/Box Behnken	[99]
Sofosbuvir and Ledipasvir	Box Behnken	[100]

In the papers presented in Table 5, the DoEs were used to evaluate the chromatographic parameters in order to obtain the best chromatographic method. The meaning of the best chromatographic method depends of the intention of the analyst—better resolution for the API, higher number of peaks in order to detect all degradation compounds, cost-and-time saving methods, etc.

Another purpose for forced degradation studies found by Sonawane and Gide [101] was the application of experimental design for the optimization of forced degradation of luliconazole (LCZ), 4-(2,4-dichlorophenyl)-1,3-dithiolan-2-ylidene-1-imidazolylacetonitrile), which is recommended for the treatment of fungal infections. The LCZ was submitted to acidic (HCl), alkaline (NaOH), oxidative (H_2O_2), thermolytic (under reflux), and photolytic (direct sunlight) stress conditions, and a full factorial design was chosen to identify the conditions to obtain a degradation of this API between 10 and 20%. The 2^3 factorial design for acid and alkaline conditions took into account the variables concentration of the degradant agent (x_1), temperature (x_2), and time of exposure (x_3) to achieve the desired degradation. The variable temperature was not included in oxidative degradation, and the design became a 2^2 factorial design. The same design was performed to dry heat and wet heat degradation, but including the variable temperature and discarding the variable concentration. For photolytic degradation, LCZ powder was exposed to direct sunlight for 48 h and compared with control in dark, but DoE was not applied. The level of the variables for each stress condition is presented at Table 6. The 2^3 factorial design was performed in a total of eight experiments, and the 2^2 factorial in a total of four experiments for each degradation (oxidative, dry heat, and wet heat) by design. Table 7 shows the experiments and the obtained results by liquid chromatography.

Table 6. Real values of the variables used in the design of experiments.

Variable	High Level (+1)					Low Level (−1)				
	Acid	Basic	Oxid.	Dry Heat	Wet Heat	Acid	Basic	Oxid.	Dry Heat	Wet Heat
Conc. (x_1)/mol×L^{-1}	1	0.1	30%	-	-	0.1	0.01	3%	-	-
Time (x_2)/min	75	30	24 h	360	120	15	10	2h	30	30
Temperature (x_3)/°C	100	100	-	200	100	60	60	-	50	60

Table 7. Design of experiments with coded values and % of degradation of active pharmaceutical ingredient (API) for acid, basic, and oxidative conditions.

2^3 Full Factorial Design						2^2 Full Factorial Design			
Exp.	X_1	X_2	X_3	Acid Condition	Basic Condition	Exp.	X_1	X_2	Oxidative Condition
1	−1	−1	−1	0%	0%	1	−1	−1	0%
2	+1	−1	−1	4%	3%	2	−1	+1	48%
3	−1	+1	−1	10%	8%	3	+1	−1	51%
4	+1	+1	−1	23%	11%	4	+1	+1	100%
5	−1	−1	+1	8%	19%				
6	+1	−1	+1	32%	26%				
7	−1	+1	+1	21%	38%				
8	+1	+1	+1	41%	43%				

The dry and wet heat degradation did not present any degradation of luliconazole, but photolytic degradation obtained 8%. Concerning acid, alkali and oxidative conditions, the degradation ranges were 0–41%, 0–43%, and 0–100%, respectively. Multivariate regressions were performed on the results for each degradation (acid, alkali, and oxidative) in order to obtain the regression models (equations) for the studied experimental domain. These regression models are used to predict suitable conditions to achieve the desired percentage of degradation. These conditions provided degradation of 11%, therefore, a relative error equal to 9%. More details about the equations in each degradation condition as well as surface response created to better visualization of the results can be found in the reference [101]. The DoE in this work allowed the authors to gain knowledge about stability of LCZ, presenting the degradation condition where LCZ is more susceptible to undergo degradation and indicating the variables that present higher influence on the degradation of LCZ. Finally, the chemometrics tools aid to predict the values of variables to obtain the desired degradation.

Another example was presented by Kurmi et al. [102]. that used DoE to develop the stability-indicating method and also found the stress conditions for forced degradation of furosemide in the range of 20–30%.

Despite the fact that DoE is a very interesting tool to find the most suitable conditions in the degradation studies and avoiding the generation of secondary degradation products, there are few papers presenting such approach.

3.2. About Fusion QbD®

As mentioned previously, forced degradation studies are performed in the development stability-indicating method phase. DoE is extremely useful to build a set of screening, optimization and robustness experiments. In this context, some HPLC method development software platforms are commercially available to automatically perform the experimental design. This software, such as Fusion QbD, uses concepts of experimental design and creates a sequence of experiments considering all relevant chromatographic parameters. It is possible to build, for example, a set of screening experiments considering more than one type of chromatography columns, multi-solvents, and other chromatographic variables. After the creation of a set of methods, guided by the DoE principles, and after running the sequence of experiments, the software generates mathematical models and makes predictions to find the better chromatographic method. As Fusion QbD is integrated with the chromatography system, all functions of HPLC are explored, and it allows users to reach maximum efficiency and speed in the method developing process [103]. Others specialized software is also used to create basic designs, such as Origin [104], Matlab [105], Minitab [106], Design-Expert [107], and Statistica [108].

3.3. Principal Component Analysis (PCA)

Principal component analysis (PCA) is one of the most used chemometric tools for data exploration through the reduction of a system's dimensionality [23,109,110]. This technique allows the user to establish the numerical adjustment of a linear model for describing the central relationships among process variables [111]. The PCA aims mainly to extract the most useful information from data. Besides, this chemometric tool helps simplify the description of the data for the analysis of variables [112].

The use of PCA enables the user to represent objects with new variables that are linear combinations of the original variables. These linear combinations, denominated principal components (PCs), are calculated considering directions of maximum variance, in a fashion that they may also be perpendicular to each other [23]. The first PC describes the maximum variance of the sample. The second PC describes the most considerable variability that the first one was not able to describe. The directions of the most dispersed samples are generally described in the first PC, since it corresponds to the vector with more information about the linear combinations of the original variables [113]. Figure 2 presents a graphical representation of PCA, where the axes are changed in order to maximize the explained variance using a smaller number of dimensions.

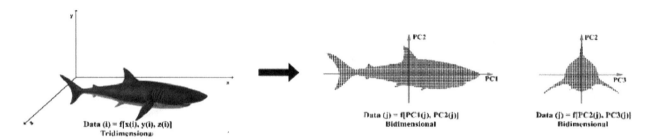

Figure 2. Representation of principal component analysis (PCA). Original data at left side, PC1 × PC2 in the middle and PC2 × PC3 at right side.

In the literature, three papers were found involving PCA associated with degradation products of pharmaceutical drugs. Two of them will be discussed in the next paragraphs, and the other one will be discussed later, in the MCR-ALS context.

Tôrres et al. [114] performed accelerated degradation studies of captopril and applied Multivariate Statistical Process Control (MSPC) for monitoring and identifying any changes in samples in order to guarantee the product quality. The details of all procedure data treatment can be found in reference [114]. The captopril stability was evaluated leaving 24 blisters of tablets of the same batch in a climatic chamber at $40 \pm 2\,°C$ and $75 \pm 5\%$ of relative humidity. One blister per week was analyzed by liquid chromatography, for six months, totalizing 24 chromatograms. In order to build the process control chart, a sample set of Captopril was used under normal operation conditions in the calibration (training stage), and in the validation stage, samples were used under normal operation conditions, as were samples presenting expired shelf life. Hotelling's T^2 statistic and Square Prediction Error (SPE) were used for sample monitoring. PCA is a useful tool in the Hotelling's T^2 statistic, since it reduces the number of variables to be monitored, changing the original variables by the scores in the PCA, without significant information loss from dataset. The PCA along with the multivariate control charts contributes to identify possible failures and changes early in the process, making this method useful to ensure the quality control of product [114]. The same authors also performed a similar work using the mid (MIR) and near (NIR) infrared techniques [115].

Skibinski et al. [66] performed forced degradation of toloxatone, which is a pharmaceutical drug used as an antidepressant. These studies were carried out in basic (0.01 M NaOH), acidic (1 M HCl), neutral (water), photo UV-Vis, photo UVC, and oxidative (0.01% H_2O_2) degradation conditions. The samples (including the control solution) were evaluated in a LCMS (ToF) totalizing 21 chromatographic profiles. The stress conditions provided eight unique degradation products of toloxatone [66].

After aligning of chromatographic profiles, PCA analysis showed a visible grouping of the stressed samples. The author noticed that stressed basic samples gave rise to a separated cluster from other stressed samples in the scores analysis obtained from PCA, while neutral and acidic samples were close to the control samples. On the other hand, it was possible to separate in groups the samples carried out under photo UV-VIS, photo UVC, and oxidation conditions. The first three components of PCA model were able to explain almost 71% of the total variance. This work shows that PCA analysis can be used as a tool to characterize the chromatographic profiles.

3.4. Partial Least Squares (PLS)

Partial least squares (PLS) regression is a multivariate regression technique, the most important one in the chemometrics. It is used to stablish quantitative relationships between a vector of information (UV-Vis, Raman, NIR, MID-IR, NMR spectra or chromatogram, diffractogram, etc.) and properties to be quantified (concentration of an analyte, crystalline phase of API, etc.) [116–119].

As example, the concentrations of an analyte in calibration samples are organized in a vector y, and the chemical data (spectra) are organized in a matrix X. In the classic multivariate regression, the regression coefficient $\mathbf{b}$ is found by $\mathbf{b} = \mathbf{y} \times X^+$, where X^+ is the pseudoinverse of X. The regression equation (model) can be written in the matrix form as $\mathbf{y} = \mathbf{b} \times X$. However, there is some issues related to the use of classical multivariate regression, such as the need of high number of samples and the problem of the correlation among the variables in the matrix X. Then, in a similar way as PCA, PLS calculations simultaneously decomposes X and $\mathbf{y}$ in order to maximize the correlation among the scores of X and $\mathbf{y}$. After defining coefficients $\mathbf{b}$, it can be applied to determine the concentration in external samples [120].

Some algorithms have been proposed to perform PLS, and the most common are PLS1 and PLS2, for one response and for multiple responses, respectively. Although PLS2 is used for multiple responses, it is recommended only in the cases where there is high correlation among the responses [121].

Recently, Sayed et al. [122] developed a stability-indicating method using PLS to determine mometasone furoate (MF) pure or in pharmaceutical formulation in the presence of its degradation products. The forced degradation was performed only in basic conditions once other previous works have demonstrated its susceptibility in undergoing alkaline hydrolysis. The multilevel multifactor experimental design was applied to prepare mixtures of calibration set constituted by 14 samples, which were scanned over the range of 220–350 nm. The UV spectra of 11 different mixtures of MF and its degradation products were used to predict the concentration of MF. The PLS model applied in the determination of MF presented good results, obtaining in calibration set mean recovery of 100.2% and RMSEC 0.002% meanwhile validation set presented mean recovery of 97.24% and RMSEP 0.04%. The recoveries in pharmaceutical samples were also satisfactory (98.47–102.66%), demonstrating no interference from excipients or alkaline degradation products in the quantification and the power of PLS method for quantification of MF [122]. Besides, in this same work, a new TLC densitometric method and the chemometric tools CLS and PCR were found, which were applied to develop quantification models for the MF in pharmaceutical samples.

Attia et al. [123] also developed spectrometric methods for determination of cefoxitin-sodium in the presence of its alkaline degradation product using different chemometric tools. PLS was applied to quantify cefoxitin-sodium in pharmaceutical sample. To obtain degradation product, the basic forced degradation was performed using NaOH 0.1 M for 10 min, which was neutralized with HCl 0.1 M. More details about the procedure to prepare the working solution are in reference [123]. The PLS model was built considering 13 mixtures denominated calibration set and 12 mixtures as a validation set obtained through experimental design. The number of factors was optimized through cross-validation method, as performed in reference [122]. The genetic algorithm (GA) was coupled with PLS to improve the prediction capability of models eliminating variables without information. In fact, the efficiency of the calibration of GA-PLS was better than only PLS, given lower RMSEC and RMSEP values for GA-PLS. The analysis of cefoxitin-sodium in presence of degradation products and in the pharmaceutical sample presented mean recovery of 100.54% and 99.86 ± 1.347%, respectively, using GA-PLS. The proposed method presented no significant difference compared to the standard method. Different chemometric tools were proposed and all of them showed a solvent reduction and sample consumption, making the methods greener. Table 8 present papers found in the literature that use in some moment the PLS tool in forced degradation studies of pharmaceutical products.

Table 8. Works involving forced degradation studies and the partial least squares (PLS) tool.

Author	API	Forced Degradation Condition	Chemometric Tool	Year	Ref.
Attia et al.	Cefprozil	Basic hydrolysis	PLS; SRACLS	2016	[124]
Alamein et al.	Pimozide	Acid and basic hydrolysis	CLS; PCR; PLS	2015	[125]
Hegazy et al.	Linezolid	Acid and basic hydrolysis; oxidative	PLS; PCR; Parafac; N-PLS	2014	[126]
Hegazy et al.	Imidapril hydrochloride	Basic hydrolysis; oxidative	PCR; PLS	2014	[127]
Souza et al.	Captopril	Thermolysis	PLS	2012	[128]
Abou Al Alamein	Zafirlukast	Basic hydrolysis	PLS	2012	[129]
Naguib	Bisacodyl	Acid hydrolysis	PLSR; SRACLS	2011	[130]
Abdelwahab	Atenolol; Chlorthalidone	Acid and basic hydrolysis	PCR; PLS	2010	[131]
Wagieh et al.	Oxybutynin hydrochloride	Basic hydrolysis	PCR; PLS	2010	[132]
Moneeb	Rabeprazole sodium	Acid hydrolysis	CLS; PCR; PLS	2008	[133]
S Fayed et al.	Cilostazol	Acid hydrolysis	PLS; CRACLS	2007	[134]
Ragno et al.	Lacidipine	Photodegradation	PLS; PCR; MLRA	2006	[135]
Shehata et al.	Rofecoxib	Basic hydrolysis; photodegradation	PLS; CRACLS	2004	[136]

3.5. Multivariate Curve Resolution (MCR)

Multivariate curve resolution (MCR) has been widely used to analyze several types of data in different application fields [137–139]. MCR constitutes a bilinear model based on the classical least squares (CLS) that decomposes data matrix into two submatrices, which have chemical information of the compounds involved in the system [137,139–141].

This approach is also known to be spectral unmixing tool once it allows mathematically solving analyte signals of a complex mixture where they are overlapped in one or more dimensions of data, as chromatograms and spectra of analyte in the presence of interferents in analysis without resolution. MCR aims to differentiate the individual contributions of components of a mixture providing the pure signals (spectra) and the proportions of analytes through concentration profile [138,139,142]. MCR comes from the Beer's law, where concentration is proportional to the absorbance. In this way, a spectral data set can be deconvoluted in the pure spectra from the analytes and their relative concentration. The general equation for MCR is $X = C \times S^t$, where the spectral matrix X is deconvoluted in the concentration matrix and the pure spectra matrix.

Most papers related to forced degradation studies and MCR-ALS aimed for the evaluation of photodegradation. Except for basic hydrolysis condition, other degradation conditions were not found in the literature.

Marín-García et al. [143] investigated photodegradation of tamoxifen in aqueous medium using Multivariate Curve Resolution-Alternating Least Squares (MCR-ALS). The photodegradation experiments were conducted at 35 °C in a cabinet equipped with light at two different irradiation power conditions (400 and 765 W/m^2) according to ICH requirements. To monitor the photodegradation of tamoxifen, the UV-VIS spectra were collected from 0 to 160 min for irradiation power 400 W/m^2, and from 0 to 120 min for 765 W/m^2. The UV spectra allowed to obtain the evolution of the photodegradation process. MCR-ALS analysis of the UV data allowed to observe the estimation of the kinect profiles for the possible presence of at least four species, three of them being degradation products. Besides, it was possible to obtain the relative concentration of each specie along time.

During photodegradation some molecules cannot be detected by UV-Vis due to the loss of chromophore groups. The authors overcame this situation using a LC-DAD-MS technique to obtain deeper knowledge about species formed in photodegradation. In this case, MCR-ALS analysis provides the C and S matrixes that contain, respectively, the elution profile and pure UV-VIS or MS spectra for each substance. These matrixes showed a new component, which represents a fourth degradation product. This new specie was not observed in the UV-VIS monitoring, it rises during photodegradation but disappears at the end of the process. Furthermore, the authors elucidated the degradation product structures. This work shows MCR-ALS's ability to monitor and solve mixtures of degradation products formed during photodegradation process [143].

Another work reported in the literature was conducted by Feng et. al. [144], which investigated the basic degradation for paracetamol using two-way dimensional UV-Vis associated to MCR-ALS. Forced degradation was performed using a quartz cell where paracetamol and NaOH solutions were added, and the UV-VIS spectra were collected from 1 s to 24 h. Initially, a PCA was applied on UV-VIS data, and it suggested the existence of four components. Later, the concentration profiles were obtained from evolving factor analysis (EFA), and it confirmed the number of chemical components involved in degradation reaction. In the MCR-ALS deconvolution, it was applied to the constraints non-negativity for spectral and concentration profiles and unimodality for the concentration profile. Through the concentration profile and spectra profile plots, it was possible to perform a critical analysis of the formation and consumption of the species during alkaline degradation. It was possible to observe that there were a reactant, a degradation product, and two intermediates. The authors compared the results with HPLC analysis, which proved the existence of two intermediates, and the concentration profile were in agreement with the one recovered by MCR-ALS using UV-Vis. Besides, the authors also proposed a degradation pathway in alkaline media. The use of MCR-ALS in forced degradation studies allowed to verify the drug stability and kinect of degradation of paracetamol [144]. Other papers regarding forced degradation studies and MCR-ALS are presented in Table 9.

Table 9. Works involving forced degradation studies and the Multivariate Curve Resolution-Alternating Least Squares (MCR-ALS) tool.

Author	API	Forced Degradation Condition	Chemometric Tool	Year	Ref.
Gómez-Canela	5-Fluorouracil	Photodegradation	MCR-ALS	2017	[145]
Bērziņš et al.	Furazidin	Basic hydrolysis	HS-MCR-ALS	2016	[146]
Luca et al.	Amiloride	Photodegradation	MCR-ALS	2012	[147]
Sílvia Mas et al.	ketoprofen	Photodegradation	MCR-ALS; HSMCR	2011	[148]
Luca et al.	Nitrofurazone	Photodegradation	HS-MCR-ALS	2010	[149]
Javidnia et al.	Nitrendipine and felodipine	Photodegradation	MCR	2008	[150]
Shamsipur et al.	Nifedipine	Photodegradation	MCR	2003	[151]

3.6. Artificial Neural Network (ANN)

Artificial neural networks (ANNs) are powerful chemometric tools based on artificial intelligence. They can model nonlinear data through learning processes in a similar way to the human brain [36,152]. ANN models are able to map the input data in a set of appropriate outputs following a "learning by examples." In other words, the structure of data is learned through training algorithms [153].

To the best of our knowledge, two works regarding to forced degradation studies and artificial neural network are reported in the literature, and only one of them uses ANNs as the main tool [123,154].

Golubović et al. [154] used ANNs to develop quantitative structure-retention relationships (QSRRs) model to optimize isocratic RP-HPLC method of candesartan cilexetil in the presence of seven degradation products obtained from acid, alkaline, neutral hydrolysis, photolysis, and oxidation conditions. QSRRs is able to relate chromatographic retention parameters and molecular structure, and it becomes a valuable tool to the prediction of chromatographic behavior and separation of complex mixtures.

Initially, to investigate the variables that could influence the chromatographic behavior, a 2^{5-1} fractional factorial design was performed. The following variables were included in the design: percentage of acetonitrile in the mobile phase, buffer pH and ionic strength, temperature of the column, and flow rate of the mobile phase. All variables showed to be significant and, therefore, were considered as inputs in the ANN modeling, except flow rate, which was maintained as a constant.

The molecular structure is an essential variable in QSRR model and is encoded by descriptors. Roughly, molecular descriptors are obtained by logic and mathematical procedures that transform chemical information in a useful number of some standardized experiments. The selection of molecular descriptors was based on intermolecular interactions suggested by theory of liquid chromatography. In the ANN modeling it were included the descriptors which present low correlation between them, such as polarizability, H-donor sites, H-acceptor sites, and octanol/water distribution coefficient.

It was used a multi-layer feedforward, the most common ANNs, constituted by one input layer (descriptors and significant chromatographic variables), number of hidden neurons connected to both input and output neurons (retention factor). In the network training stage, the overall agreement between computed and target output for a set training is maximized. In order to avoid overfitting, the predictive power of network was evaluated using a validation set. Both training and validation sets were defined through a Box-Behnken design, varying from −1 to +1 level. A total of 344 cases for ANN optimization were obtained, which were divided into 280 cases for the training set, 32 for external validation, and 32 to validation set. For training, validation, and external validation data sets, coefficients of determination (R^2) were obtained between experimental and predicted retention factor (K_{exp} and K_{ANN} respectively) equal to 0.9993, 0.9969 and 0.9956, respectively. Therefore, high R^2 and low RSME values demonstrate an excellent predictive ability of model and non-occurrence of overfitting during the training process.

This kind of mathematical model is an important tool in forced degradation studies since degradation products derive from the API and, therefore, are chemically similar. The creation of models able to predict the behavior of active substance and all degradation products contribute to defining the optimal chromatographic conditions during the optimization process [154].

4. Conclusions

Chemometric tools can bring considerable gains in forced degradation studies. DoE is the most used chemometric tool in such studies, especially in the development of suitable chromatographic methods to monitor the API. However, the application of DoE directly in stress experiments is also promising, as it is possible to quantify the individual effect of stress variables as well as the synergy between them, simulating what may occur in real life. The other widely used tool is PLS, since its use allows the quantification of the API directly in UV-Vis spectrophotometry analyzes, since it performs multivariate quantification, which makes possible quantification of species without resolution.

The PCA technique is not applied in these studies since it is an exploratory method, and its application is more related to process monitoring and classification methods for raw material identification.

The other tools, despite being very useful in such studies, are more complex, and their application is limited for non-chemometricians.

Author Contributions: Both authors contribute equally to produce this review.

References

1. Kiralj, R.; Ferreira, M.M.C. The past, present, and future of chemometrics worldwide: Some etymological, linguistic, and bibliometric investigations. *J. Chemom.* **2006**, *20*, 247–272. [CrossRef]
2. Swarbrick, B.; Westad, F. An Overview of Chemometrics for the Engineering and Measurement Sciences. In *Handbook of Measurement in Science and Engineering*; John Wiley & Sons, Inc.: Hoboken, NJ, USA, 2016; p. 2309.
3. Kumar, R.; Sharma, V. Chemometrics in forensic science. *Trac Trends Anal. Chem.* **2018**, *105*, 191–201. [CrossRef]
4. Brown, S. The chemometrics revolution re-examined. *J. Chemom.* **2017**, *31*, e2864. [CrossRef]
5. Hibbert David, B. Vocabulary of concepts and terms in chemometrics (IUPAC Recommendations 2016). *Pure Appl. Chem.* **2016**, *88*, 407. [CrossRef]
6. Caballero, B.; Finglas, P.; Toldrá, F. *Encyclopedia of Food and Health*; Academic Press: Cambridge, MA, USA, 2015.
7. Pomerantsev, A.L.; Rodionova, O.Y. Chemometric view on "comprehensive chemometrics". *Chemom. Intell. Lab. Syst.* **2010**, *103*, 19–24. [CrossRef]
8. Ferreira, S.L.C.; Silva Junior, M.M.; Felix, C.S.A.; da Silva, D.L.F.; Santos, A.S.; Santos Neto, J.H.; de Souza, C.T.; Cruz Junior, R.A.; Souza, A.S. Multivariate optimization techniques in food analysis—A review. *Food Chem.* **2019**, *273*, 3–8. [CrossRef]
9. De Luca, S.; De Filippis, M.; Bucci, R.; Magrì, A.D.; Magrì, A.L.; Marini, F. Characterization of the effects of different roasting conditions on coffee samples of different geographical origins by HPLC-DAD, NIR and chemometrics. *Microchem. J.* **2016**, *129*, 348–361. [CrossRef]
10. Briandet, R.; Kemsley, E.K.; Wilson, R.H. Discrimination of Arabica and Robusta in Instant Coffee by Fourier Transform Infrared Spectroscopy and Chemometrics. *J. Agric. Food Chem.* **1996**, *44*, 170–174. [CrossRef]
11. Santos, P.M.; Pereira-Filho, E.R.; Rodriguez-Saona, L.E. Rapid detection and quantification of milk adulteration using infrared microspectroscopy and chemometrics analysis. *Food Chem.* **2013**, *138*, 19–24. [CrossRef]
12. Amorello, D.; Orecchio, S.; Pace, A.; Barreca, S. Discrimination of almonds (Prunus dulcis) geographical origin by minerals and fatty acids profiling. *Nat. Prod. Res.* **2016**, *30*, 2107–2110. [CrossRef]
13. Wu, X.M.; Zuo, Z.T.; Zhang, Q.Z.; Wang, Y.Z. Classification of Paris species according to botanical and geographical origins based on spectroscopic, chromatographic, conventional chemometric analysis and data fusion strategy. *Microchem. J.* **2018**, *143*, 367–378. [CrossRef]
14. Chen, H.; Lin, Z.; Tan, C. Fast discrimination of the geographical origins of notoginseng by near-infrared spectroscopy and chemometrics. *J. Pharm. Biomed. Anal.* **2018**, *161*, 239–245. [CrossRef] [PubMed]
15. Uríčková, V.; Sádecká, J. Determination of geographical origin of alcoholic beverages using ultraviolet, visible and infrared spectroscopy: A review. *Spectrochim. Acta Part A* **2015**, *148*, 131–137. [CrossRef] [PubMed]
16. Kanginejad, A.; Mani-Varnosfaderani, A. Chemometrics advances on the challenges of the gas chromatography– mass spectrometry metabolomics data: A review. *J. Iran. Chem. Soc.* **2018**, *15*, 2733–2745. [CrossRef]
17. Liu, S.; Liang, Y.Z.; Liu, H.T. Chemometrics applied to quality control and metabolomics for traditional Chinese medicines. *J. Chromatogr. B* **2016**, *1015–1016*, 82–91. [CrossRef]
18. Savorani, F.; Rasmussen, M.A.; Mikkelsen, M.S.; Engelsen, S.B. A primer to nutritional metabolomics by NMR spectroscopy and chemometrics. *Food Res. Int.* **2013**, *54*, 1131–1145. [CrossRef]

19. Bhushan, N.; Hadpe, S.; Rathore, A.S. Chemometrics applications in biotech processes: Assessing process comparability. *Biotechnol. Prog.* **2012**, *28*, 121–128. [CrossRef] [PubMed]

20. Xu, Q.S.; Xu, Y.D.; Li, L.; Fang, K.T. Uniform experimental design in chemometrics. *J. Chemom.* **2018**, *32*, e3020. [CrossRef]

21. Gadžurić, S.B.; Podunavac Kuzmanović, S.O.; Vraneš, M.B.; Petrin, M.; Bugarski, T.; Kovačević, S.Z. Multivariate Chemometrics with Regression and Classification Analyses in Heroin Profiling Based on the Chromatographic Data. *Iran. J. Pharm. Res. IJPR* **2016**, *15*, 725–734.

22. Materazzi, S.; Gregori, A.; Ripani, L.; Apriceno, A.; Risoluti, R. Cocaine profiling: Implementation of a predictive model by ATR-FTIR coupled with chemometrics in forensic chemistry. *Talanta* **2017**, *166*, 328–335. [CrossRef]

23. Muehlethaler, C.; Massonnet, G.; Esseiva, P. The application of chemometrics on Infrared and Raman spectra as a tool for the forensic analysis of paints. *Forensic Sci. Int.* **2011**, *209*, 173–182. [CrossRef] [PubMed]

24. Thanasoulias, N.C.; Parisis, N.A.; Evmiridis, N.P. Multivariate chemometrics for the forensic discrimination of blue ball-point pen inks based on their Vis spectra. *Forensic Sci. Int.* **2003**, *138*, 75–84. [CrossRef] [PubMed]

25. Brereton, R.G. Pattern recognition in chemometrics. *Chemom. Intell. Lab. Syst.* **2015**, *149*, 90–96. [CrossRef]

26. Roggo, Y.; Degardin, K.; Margot, P. Identification of pharmaceutical tablets by Raman spectroscopy and chemometrics. *Talanta* **2010**, *81*, 988–995. [CrossRef]

27. da Silva, V.H.; Soares-Sobrinho, J.L.; Pereira, C.F.; Rinnan, Å. Evaluation of chemometric approaches for polymorphs quantification in tablets using near-infrared hyperspectral images. *Eur. J. Pharm. Biopharm.* **2019**, *134*, 20–28. [CrossRef]

28. Dinç, E.; Büker, E. Spectrochromatographic determination of dorzolamide hydrochloride and timolol maleate in an ophthalmic solution using three-way analysis methods. *Talanta* **2019**, *191*, 248–256. [CrossRef]

29. Sakr, M.; Hanafi, R.; Fouad, M.; Al-Easa, H.; El-Moghazy, S. Design and optimization of a luminescent Samarium complex of isoprenaline: A chemometric approach based on Factorial design and Box-Behnken response surface methodology. *Spectrochim. Acta Part A Mol. Biomol. Spectrosc.* **2019**, *208*, 114–123. [CrossRef]

30. Rodionova, O.Y.; Titova, A.V.; Demkin, N.A.; Balyklova, K.S.; Pomerantsev, A.L. Qualitative and quantitative analysis of counterfeit fluconazole capsules: A non-invasive approach using NIR spectroscopy and chemometrics. *Talanta* **2019**, *195*, 662–667. [CrossRef]

31. Visco, G.; Avino, P. Employ of multivariate analysis and chemometrics in cultural heritage and environment fields. *Environ. Sci. Pollut. Res.* **2017**, *24*, 13863–13865. [CrossRef]

32. Musumarra, G.; Fichera, M. Chemometrics and cultural heritage. *Chemom. Intell. Lab. Syst.* **1998**, *44*, 363–372. [CrossRef]

33. Madariaga, J.M. Analytical chemistry in the field of cultural heritage. *Anal. Methods* **2015**, *7*, 4848–4876. [CrossRef]

34. Barreca, S.; Mazzola, A.; Orecchio, S.; Tuzzolino, N. Polychlorinated Biphenyls in Sediments from Sicilian Coastal Area (Scoglitti) using Automated Soxhlet, GC-MS, and Principal Component Analysis. *Polycycl. Aromat. Compd.* **2014**, *34*, 237–262. [CrossRef]

35. Granato, D.; Santos, J.S.; Escher, G.B.; Ferreira, B.L.; Maggio, R.M. Use of principal component analysis (PCA) and hierarchical cluster analysis (HCA) for multivariate association between bioactive compounds and functional properties in foods: A critical perspective. *Trends Food Sci. Technol.* **2018**, *72*, 83–90. [CrossRef]

36. Panchuk, V.; Yaroshenko, I.; Legin, A.; Semenov, V.; Kirsanov, D. Application of chemometric methods to XRF-data—A tutorial review. *Anal. Chim. Acta* **2018**, *1040*, 19–32. [CrossRef] [PubMed]

37. Jain, D.; Basniwal, P.K. Forced degradation and impurity profiling: Recent trends in analytical perspectives. *J. Pharm. Biomed. Anal.* **2013**, *86*, 11–35. [CrossRef]

38. Rao, R.N.; Nagaraju, V. An overview of the recent trends in development of HPLC methods for determination of impurities in drugs. *J. Pharm. Biomed. Anal.* **2003**, *33*, 335–377.

39. Holm, R.; Elder, D.P. Analytical advances in pharmaceutical impurity profiling. *Eur. J. Pharm. Sci.* **2016**, *87*, 118–135. [CrossRef]

40. ICH. *Impurities in New Drug Substances Q3a(R2)*; Published by food and Drug Administration: Silver Spring, MD, USA, 2008. Available online: https://www.fda.gov/media/71272/download (accessed on 10 August 2019).

41. Melo, S.R.d.O.; Mello, M.H.d.; Silveira, D.; Simeoni, L.A. Advice on Degradation Products in Pharmaceuticals: A. *PDA J. Pharm. Sci. Technol.* **2014**, *68*, 221–238. [CrossRef]

42. Pan, C.; Liu, F.; Motto, M. Identification of pharmaceutical impurities in formulated dosage forms. *J. Pharm. Sci.* **2011**, *100*, 1228–1259. [CrossRef]

43. Sastry, R.P.; Venkatesan, C.; Sastry, B.; Mahesh, K. Identification and characterization of forced degradation products of pralatrexate injection by LC-PDA and LC–MS. *J. Pharm. Biomed. Anal.* **2016**, *131*, 400–409. [CrossRef]

44. Skibiński, R.; Trawiński, J.; Komsta, Ł.; Bajda, K. Characterization of forced degradation products of clozapine by LC-DAD/ESI-Q-TOF. *J. Pharm. Biomed. Anal.* **2016**, *131*, 272–280. [CrossRef] [PubMed]

45. Attwood, D.; Florence, A.T.; Rothschild, Z. *Princípios Físico-Químicos em Farmácia Volume 4*; Edusp: São Paulo, Brazil, 2003.

46. Gallardo, C.; Rojas, J.J.; Flórez, O.A. La temperatura cinética media en los estudios de estabilidad a largo plazo y almacenamiento de los medicamentos. *Vitae* **2004**, *11*, 67–72.

47. Allen Jr, L.V.; Popovich, N.G.; Ansel, H.C. *Formas Farmacêuticas e Sistemas de Liberação de Fármacos-9*; Artmed Editora: Porto Alegre, Brazil, 2013.

48. Blessy, M.; Patel, R.D.; Prajapati, P.N.; Agrawal, Y.K. Development of forced degradation and stability indicating studies of drugs—A review. *J. Pharm. Anal.* **2014**, *4*, 159–165. [CrossRef]

49. Qiu, F.; Norwood, D.L. Identification of pharmaceutical impurities. *J. Liq. Chromatogr. Relat. Technol.* **2007**, *30*, 877–935. [CrossRef]

50. Ahmad, I.; Ahmed, S.; Anwar, Z.; Sheraz, M.A.; Sikorski, M. Photostability and photostabilization of drugs and drug products. *Int. J. Photoenergy* **2016**, *2016*. [CrossRef]

51. Singh, S.; Handa, T.; Narayanam, M.; Sahu, A.; Junwal, M.; Shah, R.P. A critical review on the use of modern sophisticated hyphenated tools in the characterization of impurities and degradation products. *J. Pharm. Biomed. Anal.* **2012**, *69*, 148–173. [CrossRef]

52. ICH. *Stability Testing of New Drug Substances and Products Q1A (R2)*; Published by Food and Drug Administration: Silver Spring, MD, USA, 2003. Available online: https://www.fda.gov/media/71707/download (accessed on 11 August 2019).

53. Singh, S.; Junwal, M.; Modhe, G.; Tiwari, H.; Kurmi, M.; Parashar, N.; Sidduri, P. Forced degradation studies to assess the stability of drugs and products. *Trac Trends Anal. Chem.* **2013**, *49*, 71–88. [CrossRef]

54. Chen, W.-H.; Lin, Y.-Y.; Chang, Y.; Chang, K.-W.; Hsia, Y.-C. Forced degradation behavior of epidepride and development of a stability-indicating method based on liquid chromatography–mass spectrometry. *J. Food Drug Anal.* **2014**, *22*, 248–256. [CrossRef]

55. ANVISA Perguntas & Respostas. Assunto: RDC 53/2015 e Guia 4/2015. Available online: http://portal.anvisa.gov.br/documents/33836/418522/Perguntas+e+Respostas+-+RDC+53+2015+e+Guia+04+2015/6b3dec42-546c-4953-943f-4047b8b50f87 (accessed on 10 August 2019).

56. Canavesi, R.; Aprile, S.; Varese, E.; Grosa, G. Development and validation of a stability-indicating LC-UV method for the determination of pantethine and its degradation product based on a forced degradation study. *J. Pharm. Biomed. Anal.* **2014**, *97*, 141–150. [CrossRef]

57. Bhardwaj, S.P.; Singh, S. Study of forced degradation behavior of enalapril maleate by LC and LC-MS and development of a validated stability-indicating assay method. *J. Pharm. Biomed. Anal.* **2008**, *46*, 113–120. [CrossRef]

58. Palaric, C.; Molinié, R.; Cailleu, D.; Fontaine, J.-X.; Mathiron, D.; Mesnard, F.; Gut, Y.; Renaud, T.; Petit, A.; Pilard, S. A Deeper Investigation of Drug Degradation Mixtures Using a Combination of MS and NMR Data: Application to Indapamide. *Molecules* **2019**, *24*, 1764. [CrossRef] [PubMed]

59. Fatima, S.; Beg, S.; Samim, M.; Ahmad, F.J. *Application of Chemometric Approach for Development and Validation of High Performance Liquid Chromatography Method for Estimation of Ropinirole Hydrochloride*; John Wiley & Sons, Inc.: Hoboken, NJ, USA, 2019.

60. ICH. *Photostability Testing of New Drug Substances and Products Q1B*; Published by Food and Drug Administration: Silver Spring, MD, USA, 1996. Available online: https://www.fda.gov/media/71713/download (accessed on 15 August 2019).

61. Bakshi, M.; Singh, S. Development of validated stability-indicating assay methods—Critical review. *J. Pharm. Biomed. Anal.* **2002**, *28*, 1011–1040. [CrossRef]

62. Bansal, G.; Singh, M.; Jindal, K.C.; Singh, S. Ultraviolet-photodiode array and high-performance liquid chromatographic/mass spectrometric studies on forced degradation behavior of glibenclamide and development of a validated stability-indicating method. *J. Aoac Int.* **2008**, *91*, 709–719. [PubMed]

63. World Health Organization. *WHO Expert Committee on Specifications for Pharmaceutical Preparations: Thirty-Ninth Report*; World Health Organization: Geneva, Switzerland, 2005; Volume 39.

64. Sanitária, A.N.d.V. *Resolução De Diretoria Colegiada—RDC N° 53*; Diário Oficial da União: Brasília, Brazil, 2015.

65. Devrukhakar, P.S.; Shankar, M.S.; Shankar, G.; Srinivas, R. A stability-indicating LC–MS/MS method for zidovudine: Identification, characterization and toxicity prediction of two major acid degradation products. *J. Pharm. Anal.* **2017**, *7*, 231–236. [CrossRef] [PubMed]

66. Skibiński, R.; Trawiński, J.; Komsta, Ł.; Murzec, D. Characterization of forced degradation products of toloxatone by LC-ESI-MS/MS. *Saudi Pharm. J.* **2018**, *26*, 467–480. [CrossRef] [PubMed]

67. Tiwari, R.N.; Shah, N.; Bhalani, V.; Mahajan, A. LC, MSn and LC–MS/MS studies for the characterization of degradation products of amlodipine. *J. Pharm. Anal.* **2015**, *5*, 33–42. [CrossRef]

68. Rakibe, U.; Tiwari, R.; Mahajan, A.; Rane, V.; Wakte, P. LC and LC–MS/MS studies for the identification and characterization of degradation products of acebutolol. *J. Pharm. Anal.* **2018**, *8*, 357–365. [CrossRef]

69. Martono, Y.; Rohman, A.; Martono, S.; Riyanto, S. Degradation study of stevioside using RP-HPLC and ESI-MS/MS. *Malays. J. Fundam. Appl. Sci.* **2018**, *14*, 138–141. [CrossRef]

70. Korany, M.A.; Haggag, R.S.; Ragab, M.A.A.; Elmallah, O.A. A validated stability indicating DAD–HPLC method for determination of pentoxifylline in presence of its pharmacopeial related substances. *Bull. Fac. Pharm.* **2013**, *51*, 211–219. [CrossRef]

71. Saini, B.; Bansal, G. Isolation and characterization of a degradation product in leflunomide and a validated selective stability-indicating HPLC–UV method for their quantification. *J. Pharm. Anal.* **2015**, *5*, 207–212. [CrossRef]

72. Abiramasundari, A.; Joshi, R.P.; Jalani, H.B.; Sharma, J.A.; Pandya, D.H.; Pandya, A.N.; Sudarsanam, V.; Vasu, K.K. Stability-indicating assay method for determination of actarit, its process related impurities and degradation products: Insight into stability profile and degradation pathways. *J. Pharm. Anal.* **2014**, *4*, 374–383. [CrossRef] [PubMed]

73. Al-Ghannam, S.M.; Al-Olayan, A.M. Stability-indicating HPLC method for the determination of nicardipine in capsules and spiked human plasma. Identification of degradation products using HPLC/MS. *Arab. J. Chem.* **2014**, in press. [CrossRef]

74. Alarfaj, N.A. Stability-indicating liquid chromatography for determination of clopidogrel bisulfate in tablets: Application to content uniformity testing. *J. Saudi Chem. Soc.* **2012**, *16*, 23–30. [CrossRef]

75. Xia, M.; Hang, T.J.; Zhang, F.; Li, X.M.; Xu, X.Y. The stability of biapenem and structural identification of impurities in aqueous solution. *J. Pharm. Biomed. Anal.* **2009**, *49*, 937–944. [CrossRef] [PubMed]

76. Shah, R.P.; Sahu, A.; Singh, S. Identification and characterization of degradation products of irbesartan using LC–MS/TOF, MSn, on-line H/D exchange and LC–NMR. *J. Pharm. Biomed. Anal.* **2010**, *51*, 1037–1046. [CrossRef]

77. US Department of Health and Human Services. *Guidance for Industry ANDAs: Impurities in Drug Substances*; US Department of Health and Human Services, Food and Drug Administration: Washington, DC, USA, 1999.

78. Robinson, T.J.; Borror, C.M.; Myers, R.H. Robust parameter design: A review. *Qual. Reliab. Eng. Int.* **2004**, *20*, 81–101. [CrossRef]

79. Breitkreitz, M.C.; Souza, A.M.d.; Poppi, R.J. Experimento didático de quimiometria para planejamento de experimentos: Avaliação das condições experimentais na determinação espectrofotométrica de ferro II com o-fenantrolina. *Química Nova* **2014**, *37*, 564–573.

80. Dejaegher, B.; Vander Heyden, Y. Experimental designs and their recent advances in set-up, data interpretation, and analytical applications. *J. Pharm. Biomed. Anal.* **2011**, *56*, 141–158. [CrossRef]

81. Mäkelä, M. Experimental design and response surface methodology in energy applications: A tutorial review. *Energy Convers. Manag.* **2017**, *151*, 630–640. [CrossRef]

82. Tye, H. Application of statistical 'design of experiments' methods in drug discovery. *Drug Discov. Today* **2004**, *9*, 485–491. [CrossRef]

83. Altekar, M.; Homon, C.A.; Kashem, M.A.; Mason, S.W.; Nelson, R.M.; Patnaude, L.A.; Yingling, J.; Taylor, P.B. Assay Optimization: A Statistical Design of Experiments Approach. *Clin. Lab. Med.* **2007**, *27*, 139–154. [CrossRef] [PubMed]

84. Bezerra, M.A.; Santelli, R.E.; Oliveira, E.P.; Villar, L.S.; Escaleira, L.A. Response surface methodology (RSM) as a tool for optimization in analytical chemistry. *Talanta* **2008**, *76*, 965–977. [CrossRef] [PubMed]

85. Sahu, P.K.; Ramisetti, N.R.; Cecchi, T.; Swain, S.; Patro, C.S.; Panda, J. An overview of experimental designs in HPLC method development and validation. *J. Pharm. Biomed. Anal.* **2018**, *147*, 590–611. [CrossRef] [PubMed]

86. Lafossas, C.; Benoit-Marquié, F.; Garrigues, J.C. Analysis of the retention of tetracyclines on reversed-phase columns: Chemometrics, design of experiments and quantitative structure-property relationship (QSPR) study for interpretation and optimization. *Talanta* **2019**, *198*, 550–559. [CrossRef] [PubMed]

87. Valente, J.F.A.; Sousa, A.; Queiroz, J.A.; Sousa, F. DoE to improve supercoiled p53-pDNA purification by O-phospho-l-tyrosine chromatography. *J. Chromatogr. B* **2019**, *1105*, 184–192. [CrossRef]

88. Mahrouse, M.A.; Lamie, N.T. Experimental design methodology for optimization and robustness determination in ion pair RP-HPLC method development: Application for the simultaneous determination of metformin hydrochloride, alogliptin benzoate and repaglinide in tablets. *Microchem. J.* **2019**, *147*, 691–706. [CrossRef]

89. Krishna, M.V.; Dash, R.N.; Jalachandra Reddy, B.; Venugopal, P.; Sandeep, P.; Madhavi, G. Quality by Design (QbD) approach to develop HPLC method for eberconazole nitrate: Application oxidative and photolytic degradation kinetics. *J. Saudi Chem. Soc.* **2016**, *20*, S313–S322. [CrossRef]

90. Nadella, N.P.; Ratnakaram, V.N.; Srinivasu, N.; Technologies, R. Quality-by-design-based development and validation of a stability-indicating UPLC method for quantification of teriflunomide in the presence of degradation products and its application to in-vitro dissolution. *J. Liquid Chromatogr. Relat. Technol.* **2017**, *40*, 517–527. [CrossRef]

91. Hadzieva Gigovska, M.; Petkovska, A.; Acevska, J.; Nakov, N.; Antovska, P.; Ugarkovic, S.; Dimitrovska, A. Comprehensive Assessment of Degradation Behavior of Simvastatin by UHPLC/MS Method, Employing Experimental Design Methodology. *J. Int. J. Anal. Chem.* **2018**, *2018*, 17. [CrossRef]

92. Jadhav, S.B.; Reddy, P.S.; Narayanan, K.L.; Bhosale, P.N. Development of RP-HPLC, Stability Indicating Method for Degradation Products of Linagliptin in Presence of Metformin HCl by Applying 2 Level Factorial Design; and Identification of Impurity-VII, VIII and IX and Synthesis of Impurity-VII. *Sci. Pharm.* **2017**, *85*, 25. [CrossRef]

93. Wingert, N.R.; Ellwanger, J.B.; Bueno, L.M.; Gobetti, C.; Garcia, C.V.; Steppe, M.; Schapoval, E.E.S. Application of Quality by Design to optimize a stability-indicating LC method for the determination of ticagrelor and its impurities. *Eur. J. Pharm. Sci.* **2018**, *118*, 208–215. [CrossRef] [PubMed]

94. Ren, Z.; Zhang, X.; Wang, H.; Jin, X. Using an innovative quality-by-design approach for the development of a stability-indicating UPLC/Q-TOF-ESI-MS/MS method for stressed degradation products of imatinib mesylate. *RSC Adv.* **2016**, *6*, 13050–13062. [CrossRef]

95. Sharma, G.; Thakur, K.; Raza, K.; Katare, O.P. Stability kinetics of fusidic acid: Development and validation of stability indicating analytical method by employing Analytical Quality by Design approach in medicinal product(s). *J. Chromatogr. B* **2019**, *1120*, 113–124. [CrossRef] [PubMed]

96. Zhang, X.; Hu, C. Application of quality by design concept to develop a dual gradient elution stability-indicating method for cloxacillin forced degradation studies using combined mixture-process variable models. *J. Chromatogr. A* **2017**, *1514*, 44–53. [CrossRef]

97. Kalariya, P.D. Experimental Design Approach for Selective Separation of Vilazodone HCl and Its Degradants by LC-PDA and Characterization of Major Degradants by LC/QTOF–MS/MS. *Chromatographia* **2014**, *77*, 1299–1313. [CrossRef]

98. Murthy, M.V.; Krishnaiah, C.; Srinivas, K.; Rao, K.S.; Kumar, N.R.; Mukkanti, K. Development and validation of RP-UPLC method for the determination of darifenacin hydrobromide, its related compounds and its degradation products using design of experiments. *J. Pharm. Biomed. Anal.* **2013**, *72*, 40–50. [CrossRef]

99. Baghel, M.; Rajput, S.J. Stress degradation of edaravone: Separation, isolation and characterization of major degradation products. *Biomed. Chromatogr.* **2018**, *32*, e4146. [CrossRef]

100. Yeram, P.; Hamrapurkar, P.; Mukhedkar, P. Implementation of Quality by Design approach to develop and validate stability indicating assay method for simultaneous estimation of sofosbuvir and ledipasvir in bulk drugs and tablet formulation. *Int. J. Pharm. Sci.* **2019**, *10*, 180–188.

101. Sonawane, S.; Gide, P. Application of experimental design for the optimization of forced degradation and development of a validated stability-indicating LC method for luliconazole in bulk and cream formulation. *Arab. J. Chem.* **2016**, *9*, S1428–S1434. [CrossRef]

102. Kurmi, M.; Kumar, S.; Singh, B.; Singh, S. Implementation of design of experiments for optimization of forced degradation conditions and development of a stability-indicating method for furosemide. *J. Pharm. Biomed. Anal.* **2014**, *96*, 135–143. [CrossRef]

103. Fusion QbD Quality by Design Software Solutions. Available online: http://www.smatrix.com/ (accessed on 10 March 2019).

104. Originlab. Available online: https://www.originlab.com/ (accessed on 10 March 2019).

105. MATLAB for Artificial Intelligence. Available online: https://au.mathworks.com/ (accessed on 10 March 2019).

106. Minitab 19. Available online: https://www.minitab.com/ (accessed on 10 March 2019).

107. StatEase Statistics Made Easy. Available online: https://www.statease.com (accessed on 10 March 2019).

108. Accelerate Innovation with Data Science. Available online: https://www.tibco.com/products/data-science (accessed on 10 March 2019).

109. Dégardin, K.; Guillemain, A.; Guerreiro, N.V.; Roggo, Y. Near infrared spectroscopy for counterfeit detection using a large database of pharmaceutical tablets. *J. Pharm. Biomed. Anal.* **2016**, *128*, 89–97. [CrossRef] [PubMed]

110. The basic building block of chemometrics. *Analytical Chemistry.* Available online: https://www.intechopen .com/books/analytical-chemistry/pca-the-basic-building-block-of-chemometrics (accessed on 15 August 2019).

111. Godoy, J.L.; Vega, J.R.; Marchetti, J.L. Relationships between PCA and PLS-regression. *Chemom. Intell. Lab. Syst.* **2014**, *130*, 182–191. [CrossRef]

112. Abdi, H.; Williams, L.J. Principal component analysis. *Wiley Interdiscip. Rev. Comput. Stat.* **2010**, *2*, 433–459. [CrossRef]

113. Rutledge, D.N.; Jouan-Rimbaud Bouveresse, D. Independent Components Analysis with the JADE algorithm. *Trac Trends Anal. Chem.* **2013**, *50*, 22–32. [CrossRef]

114. Tôrres, A.R.; Grangeiro, S.; Fragoso, W.D. Multivariate control charts for monitoring captopril stability. *Microchem. J.* **2015**, *118*, 259–265. [CrossRef]

115. Tôrres, A.R.; Grangeiro, S.; Fragoso, W.D. Vibrational spectroscopy and multivariate control charts: A new strategy for monitoring the stability of captopril in the pharmaceutical industry. *Microchem. J.* **2017**, *133*, 279–285. [CrossRef]

116. Bro, R. Multiway calibration. *Multilinear PLS J. Chemom.* **1996**, *10*, 47–61.

117. Trygg, J.; Wold, S. Orthogonal projections to latent structures (O-PLS). *J. Chemom. A J. Chemom. Soc.* **2002**, *16*, 119–128. [CrossRef]

118. Krishnan, A.; Williams, L.J.; McIntosh, A.R.; Abdi, H. Partial Least Squares (PLS) methods for neuroimaging: A tutorial and review. *NeuroImage* **2011**, *56*, 455–475. [CrossRef]

119. Medina, S.; Perestrelo, R.; Silva, P.; Pereira, J.A.M.; Câmara, J.S. Current trends and recent advances on food authenticity technologies and chemometric approaches. *Trends Food Sci. Technol.* **2019**, *85*, 163–176. [CrossRef]

120. Rosipal, R.; Krämer, N. *In Overview and Recent Advances in Partial Least Squares, International Statistical and Optimization Perspectives Workshop Subspace, Latent Structure and Feature Selection*; Springer: New York, NY, USA, 2005; pp. 34–51.

121. Biancolillo, A.; Marini, F. Chemometric Methods for Spectroscopy-Based Pharmaceutical Analysis. *Front. Chem.* **2018**, *6*, 576. [CrossRef] [PubMed]

122. Sayed, R.A.; El-Masri, M.M.; Hassan, W.S.; El-Mammli, M.Y.; Shalaby, A. Validated Stability-Indicating Methods for Determination of Mometasone Furoate in Presence of its Alkaline Degradation Product. *J. Chromatogr. Sci.* **2017**, *56*, 254–261. [CrossRef] [PubMed]

123. Attia, K.A.-S.M.; Abdel-Aziz, O.; Magdy, N.; Mohamed, G.F. Development and validation of different chemometric-assisted spectrophotometric methods for determination of cefoxitin-sodium in presence of its alkali-induced degradation product. *Future J. Pharm. Sci.* **2018**, *4*, 241–247. [CrossRef]

124. Attia, K.A.M.; Nassar, M.W.I.; El-Zeiny, M.B.; Serag, A. Stability indicating methods for the analysis of cefprozil in the presence of its alkaline induced degradation product. *Spectrochim. Acta Part A Mol. Biomol. Spectrosc.* **2016**, *159*, 1–6. [CrossRef] [PubMed]

125. Alamein, A.M.A.A.; Hussien, L.A.E.A.; Mohamed, E.H. Univariate spectrophotometry and multivariate calibration: Stability-indicating analytical tools for the quantification of pimozide in bulk and pharmaceutical dosage form. *Bull. Fac. Pharm.* **2015**, *53*, 173–183. [CrossRef]

126. Hegazy, M.A.E.-M.; Eissa, M.S.; Abd El-Sattar, O.I.; Abd El-Kawy, M. Two and three way spectrophotometric-assisted multivariate determination of linezolid in the presence of its alkaline and oxidative degradation products and application to pharmaceutical formulation. *Spectrochim. Acta Part A Mol. Biomol. Spectrosc.* **2014**, *128*, 231–242. [CrossRef] [PubMed]

127. Hegazy, M.A.-M.; Eissa, M.S.; Abd El-Sattar, O.I.; Abd El-Kawy, M.M. Determination of a novel ACE inhibitor in the presence of alkaline and oxidative degradation products using smart spectrophotometric and chemometric methods. *J. Pharm. Anal.* **2014**, *4*, 132–143. [CrossRef]

128. Souza, J.A.L.; Albuquerque, M.M.; Grangeiro, S.; Pimentel, M.F.; de Santana, D.P.; Simões, S.S. Quantification of captopril disulphide as a degradation product in captopril tablets using near infrared spectroscopy and chemometrics. *Vib. Spectrosc.* **2012**, *62*, 35–41. [CrossRef]

129. Abou Al Alamein, A.M. Validated stability-indicating methods for the determination of zafirlukast in the presence of its alkaline hydrolysis degradation product. *Bull. Fac. Pharm.* **2012**, *50*, 111–119. [CrossRef]

130. Naguib, I.A. Stability indicating analysis of bisacodyl by partial least squares regression, spectral residual augmented classical least squares and support vector regression chemometric models: A comparative study. *Bull. Fac. Pharm.* **2011**, *49*, 91–100. [CrossRef]

131. Abdelwahab, N.S.J.A.M. Determination of atenolol, chlorthalidone and their degradation products by TLC-densitometric and chemometric methods with application of model updating. *Anal. Methods* **2010**, *2*, 1994–2001. [CrossRef]

132. Wagieh, N.E.; Hegazy, M.A.; Abdelkawy, M.; Abdelaleem, E.A. Quantitative determination of oxybutynin hydrochloride by spectrophotometry, chemometry and HPTLC in presence of its degradation product and additives in different pharmaceutical dosage forms. *Talanta* **2010**, *80*, 2007–2015. [CrossRef] [PubMed]

133. Moneeb, M.S. Chemometric determination of rabeprazole sodium in presence of its acid induced degradation products using spectrophotometry, polarography and anodic voltammetry at a glassy carbon electrode. *Pak. J. Pharm. Sci.* **2008**, *21*, 214–224. [PubMed]

134. Fayed, A.S.; Shehata, M.; Ibrahim, A.; Hassan, N.; Weshahy, S.A. Validated stability-indicating methods for determination of cilostazol in the presence of its degradation products according to the ICH guidelines. *J. Pharm. Biomed. Anal.* **2007**, *45*, 407–416. [CrossRef] [PubMed]

135. Ragno, G.; Ioele, G.; De Luca, M.; Garofalo, A.; Grande, F.; Risoli, A. A critical study on the application of the zero-crossing derivative spectrophotometry to the photodegradation monitoring of lacidipine. *J. Pharm. Biomed. Anal.* **2006**, *42*, 39–45. [CrossRef] [PubMed]

136. Shehata, M.A.; Ashour, A.; Hassan, N.Y.; Fayed, A.S.; El-Zeany, B.A. Liquid chromatography and chemometric methods for determination of rofecoxib in presence of its photodegradate and alkaline degradation products. *Anal. Chim. Acta* **2004**, *519*, 23–30. [CrossRef]

137. Jaumot, J.; de Juan, A.; Tauler, R. MCR-ALS GUI 2.0: New features and applications. *Chemom. Intell. Lab. Syst.* **2015**, *140*, 1–12. [CrossRef]

138. Ruckebusch, C.; Blanchet, L. Multivariate curve resolution: A review of advanced and tailored applications and challenges. *Anal. Chim. Acta* **2013**, *765*, 28–36. [CrossRef]

139. Firmani, P.; Hugelier, S.; Marini, F.; Ruckebusch, C. MCR-ALS of hyperspectral images with spatio-spectral fuzzy clustering constraint. *Chemom. Intell. Lab. Syst.* **2018**, *179*, 85–91. [CrossRef]

140. Devos, O.; Schröder, H.; Sliwa, M.; Placial, J.P.; Neymeyr, K.; Métivier, R.; Ruckebusch, C. Photochemical multivariate curve resolution models for the investigation of photochromic systems under continuous irradiation. *Anal. Chim. Acta* **2019**, *1053*, 32–42. [CrossRef]

141. Alcaraz, M.R.; Aguirre, A.; Goicoechea, H.C.; Culzoni, M.J.; Collins, S.E. Resolution of intermediate surface species by combining modulated infrared spectroscopy and chemometrics. *Anal. Chim. Acta* **2019**, *1049*, 38–46. [CrossRef] [PubMed]

142. Cook, D.W.; Oram, K.G.; Rutan, S.C.; Stoll, D.R. Rational Design of Mixtures for Chromatographic Peak Tracking Applications via Multivariate Selectivity. *Anal. Chim. Acta* **2019**, *2*, 100010. [CrossRef]

143. Marín-García, M.; Ioele, G.; Franquet-Griell, H.; Lacorte, S.; Ragno, G.; Tauler, R. Investigation of the photodegradation profile of tamoxifen using spectroscopic and chromatographic analysis and multivariate curve resolution. *Chemom. Intell. Lab. Syst.* **2018**, *174*, 128–141. [CrossRef]

144. Feng, X.; Zhang, Q.; Cong, P.; Zhu, Z. Determination of the paracetamol degradation process with online UV spectroscopic and multivariate curve resolution-alternating least squares methods: Comparative validation by HPLC. *Anal. Methods* **2013**, *5*, 5286–5293. [CrossRef]

145. Gómez-Canela, C.; Bolivar-Subirats, G.; Tauler, R.; Lacorte, S. Powerful combination of analytical and chemometric methods for the photodegradation of 5-Fluorouracil. *J. Pharm. Biomed. Anal.* **2017**, *137*, 33–41. [CrossRef]

146. Bērziņš, K.; Kons, A.; Grante, I.; Dzabijeva, D.; Nakurte, I.; Actiņš, A. Multi-technique approach for qualitative and quantitative characterization of furazidin degradation kinetics under alkaline conditions. *J. Pharm. Biomed. Anal.* **2016**, *129*, 433–440. [CrossRef]

147. De Luca, M.; Ioele, G.; Mas, S.; Tauler, R.; Ragno, G. A study of pH-dependent photodegradation of amiloride by a multivariate curve resolution approach to combined kinetic and acid–base titration UV data. *Analyst* **2012**, *137*, 5428–5435. [CrossRef]

148. Mas, S.; Tauler, R.; de Juan, A. Chromatographic and spectroscopic data fusion analysis for interpretation of photodegradation processes. *J. Chromatogr. A* **2011**, *1218*, 9260–9268. [CrossRef]

149. De Luca, M.; Mas, S.; Ioele, G.; Oliverio, F.; Ragno, G.; Tauler, R. Kinetic studies of nitrofurazone photodegradation by multivariate curve resolution applied to UV-spectral data. *Int. J. Pharm.* **2010**, *386*, 99–107. [CrossRef]

150. Javidnia, K.; Hemmateenejad, B.; Miri, R.; Saeidi-Boroujeni, M. Application of a self-modeling curve resolution method for studying the photodegradation kinetics of nitrendipine and felodipine. *J. Pharm. Biomed. Anal.* **2008**, *46*, 597–602. [CrossRef]

151. Shamsipur, M.; Hemmateenejad, B.; Akhond, M.; Javidnia, K.; Miri, R. A study of the photo-degradation kinetics of nifedipine by multivariate curve resolution analysis. *J. Pharm. Biomed. Anal.* **2003**, *31*, 1013–1019. [CrossRef]

152. Arabzadeh, V.; Sohrabi, M.R.; Goudarzi, N.; Davallo, M. Using artificial neural network and multivariate calibration methods for simultaneous spectrophotometric analysis of Emtricitabine and Tenofovir alafenamide fumarate in pharmaceutical formulation of HIV drug. *Spectrochim. Acta Part A Mol. Biomol. Spectrosc.* **2019**, *215*, 266–275. [CrossRef] [PubMed]

153. Marini, F.; Bucci, R.; Magrì, A.L.; Magrì, A.D. Artificial neural networks in chemometrics: History, examples and perspectives. *Microchem. J.* **2008**, *88*, 178–185. [CrossRef]

154. Golubović, J.B.; Protić, A.D.; Zečević, M.L.; Otašević, B.M. Quantitative structure retention relationship modeling in liquid chromatography method for separation of candesartan cilexetil and its degradation products. *Chemom. Intell. Lab. Syst.* **2015**, *140*, 92–101. [CrossRef]

7

Fast Detection of 10 Cannabinoids by RP-HPLC-UV Method in *Cannabis sativa* L.

Mara Mandrioli [1], Matilde Tura [1], Stefano Scotti [2] and Tullia Gallina Toschi [1,*]

[1] Department of Agricultural and Food Sciences, Alma Mater Studiorum-University of Bologna, Viale Fanin 40, 40127 Bologna, Italy; mara.mandrioli@unibo.it (M.M.); matilde.tura2@unibo.it (M.T.)

[2] Shimadzu Italia, Via G. B. Cassinis 7, 20139 Milano, Italy; sscotti@shimadzu.it

[*] Correspondence: tullia.gallinatoschi@unibo.it

Academic Editor: Marcello Locatelli

Abstract: Cannabis has regained much attention as a result of updated legislation authorizing many different uses and can be classified on the basis of the content of tetrahydrocannabinol (THC), a psychotropic substance for which there are legal limitations in many countries. For this purpose, accurate qualitative and quantitative determination is essential. The relationship between THC and cannabidiol (CBD) is also significant as the latter substance is endowed with many specific and non-psychoactive proprieties. For these reasons, it becomes increasingly important and urgent to utilize fast, easy, validated, and harmonized procedures for determination of cannabinoids. The procedure described herein allows rapid determination of 10 cannabinoids from the inflorescences of *Cannabis sativa* L. by extraction with organic solvents. Separation and subsequent detection are by RP-HPLC-UV. Quantification is performed by an external standard method through the construction of calibration curves using pure standard chromatographic reference compounds. The main cannabinoids dosed (g/100 g) in actual samples were cannabidiolic acid (CBDA), CBD, and Δ9-THC (Sample L11 CBDA 0.88 ± 0.04, CBD 0.48 ± 0.02, Δ9-THC 0.06 ± 0.00; Sample L5 CBDA 0.93 ± 0.06, CBD 0.45 ± 0.03, Δ9-THC 0.06 ± 0.00). The present validated RP-HPLC-UV method allows determination of the main cannabinoids in *Cannabis sativa* L. inflorescences and appropriate legal classification as hemp or drug-type.

Keywords: cannabinoids; *Cannabis sativa* L.; HPLC; validation

1. Introduction

Cannabis is classified into the family of Cannabaceae and initially encompassed three main species: *Cannabis sativa*, *Cannabis indica*, and *Cannabis ruderalis* [1]. Nowadays, Cannabis has only one species due to continuous crossbreeding of the three species to generate hybrids. In fact, all plants are categorized as belonging to *Cannabis sativa* and classified into chemotypes based on the concentration of the main cannabinoids. Depending on the THCA/CBDA ratio, some chemotypes have been distinguished. In particular, chemotype I or "drug-plants" have a TCHA/CBDA ratio >1.0, plants that exhibit an intermediate ratio are classified as chemotype II, chemotype III or "fiber-plants" have a THCA/CBDA ratio <1.0, plants that contain cannabigerolic acid (CBGA) as the main cannabinoid are classified as chemotype IV, and plants that contain almost no cannabinoids are classified as chemotype V [2–5].

Recently, in Italy the interest in *Cannabis sativa* L. has increased mainly due to the latest legislation (Legge n. 242 del 2 dicembre 2016) [6]. As a consequence, there is a request to develop cost-effective and easy-to-use quantitative and qualitative methods for analysis of cannabinoids.

The Italian regulatory framework has classified two types of *Cannabis sativa* L. depending on the content of Δ9-THC. In particular, fiber-type plants of *Cannabis sativa* L., also called "hemp", are characterized by a low content of Δ9-THC (<0.2% *w/w*). If the content of Δ9-THC is >0.6% *w/w*, it is considered as drug-type, also called "therapeutic" or "marijuana".

Industrial hemp is used in several sectors, such as in the pharmaceutical, cosmetic, food, and textile industries, as well as in energy production and building. In general, fiber-type plants are less used in the pharmaceutical field, where drug-type plants are more often employed [5]. However, there is also an increased interest in hemp varieties containing non-psychoactive compounds. In fact, the European Union has approved 69 varieties of *Cannabis sativa* L. for commercial use [7].

Hemp has a complex chemical composition that includes terpenoids, sugars, alkaloids, stilbenoids, quinones, and the characteristic compounds of this plant, namely cannabinoids. *Cannabis sativa* L. has several chemotypes, each of which is characterized by a different qualitative and quantitative chemical profile [5]. The cannabinoids, terpenes, and phenolic compounds in hemp are formed through secondary metabolism [3,8]. The term "cannabinoid" indicates terpenophenols derived from *Cannabis*. More than 90 cannabinoids are known, and some are derived from breakdown reactions [8]. Gaoni and Mechoulam [9] were the first to define cannabinoids "as a group of C_{21} compounds typical of and present in *Cannabis sativa*, their carboxylic acids, analogs, and transformation products". Currently, cannabinoids have been classified according to their chemical structure, mainly seven types of cannabigerol (CBG); five types of cannabichromene (CBC); seven types of cannabidiol (CBD); the main psychoactive cannabinoid Δ9-tetrahydrocannabinol (Δ9-THC) in nine different forms including its acid precursor (Δ9-tetrahydrocannabinolic acid, Δ9-THCA); Δ8-tetrahydrocannabinol (Δ8-THC), which is a more stable isomer of Δ9-THC but 20% less active; three types of cannabicyclol (CBL); five different forms of cannabielsoin (CBE); seven types of Cannabinol (CBN), which is the oxidation artifact of Δ9-THC; cannabitriol (CBT); cannabivarin (CBDV); and tetrahydrocannabivarin (THCV) [10,11]. THC, CBD, CBG, CBN, and CBC are not biosynthesized in *Cannabis sativa*, and the plant produces the carboxylic acid forms of these cannabinoids (THCA, CBDA, CBGA, CBNA, and CBCA). Cannabinoid acids undergo a chemical decarboxylation reaction triggered by different factors, mainly temperature. This decarboxylation reaction leads to the formation of the respective neutral cannabinoids (THC, CBD, CBG, CBN, and CBC) [12,13].

There are several methods to quantify cannabinoids [14–21], some of which require expensive mass spectrometry detectors [22–25]. Furthermore, there is a great deal of uncertainty around the use of gas chromatography (GC) for the titration of cannabinoids due to the high temperature of the injector and detector that can lead to the decarboxylation of cannabinoid acids if not derivatized correctly [26]. Moreover, recent studies have reported that cannabinoid acid decarboxylation is only partial, and as result the actual value is underestimated. An HPLC system allows for determination of the actual cannabinoid composition, both neutral and acid forms, without the necessity of the derivatization step [13].

It is necessary, in addition to honed methods, to develop new procedures with a view to discriminate different *Cannabis* varieties in order to identify and titrate cannabinoids in a simple way. These methods should ideally be fast, easy, robust, and cost-efficient as they can be used not only by research laboratories but also by small companies with a view on quality control.

This study focuses on the development, validation, and step-by-step explanation of a rapid and simple HPLC-UV method for identification and quantification of the main cannabinoids in hemp inflorescences that can be easily reproduced and applied. The method described is focused on the quantification of CBD but can also be applied to check the levels of THC.

2. Results and Discussion

2.1. Method Development

The aim of this work was to develop a new analytical method for determination of the main cannabinoids in hemp samples. In fact, the method described below can be used as a routine quality control procedure and can be applied by the pharmaceutical industry, small laboratories, or even small pharmacies.

A crucial aspect for accurate identification and quantification of analytes is optimization of separation conditions, and therefore various preliminary tests were carried out (e.g., mobile phase, detection wavelength). Different mobile phases were tested, and trials were performed with different compositions and gradient elution to optimize the separation of all 10 target compounds considered (File S2). The greatest difficulty was that of separating CBD and THCV, which in many cases co-eluted. It was also difficult to separate the isomers Δ9-THC and Δ8-THC. The best resolution of cannabinoids was obtained using a chromatographic column and, as an eluent mixture, water with 0.085% phosphoric acid and acetonitrile with 0.085% phosphoric acid.

The quantification of cannabinoids was made at 220 nm after testing different wavelengths (File S2). This wavelength represents the best compromise for all the cannabinoids considered and was selected to detect and integrate all compounds of interest within the dedicated concentration range. As far as chromatographic analysis is concerned, before using the instrument, the system was conditioned for 20 min by fluxing the eluent mixture in the instrument under the same conditions as the method, and then a chromatographic run was performed by injecting 5 μL of acetonitrile to verify that the chromatographic system was adequately cleaned. Simultaneously with the analysis of the sample, standard solutions were injected at different concentrations for the construction of calibration curves and to evaluate the separation and identification of each compound. The identification of cannabinoids was performed by comparing their retention times with those obtained by the injection of pure standards and by an enhancing procedure. Figure 1 shows a chromatogram of a standard mixture of cannabinoids and Figure 2 shows a chromatogram of a sample of hemp.

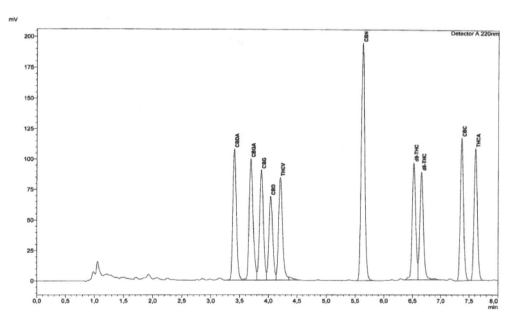

Figure 1. Chromatographic trace of a standard cannabinoid mixture analyzed by RP-HPLC-UV equipped with reverse phase C18 column.

Cannabinoids in different varieties of *Cannabis sativa* L. can be present in very different concentrations. In order to obtain good chromatographic separation and correct quantification, it may be necessary to dilute or concentrate the extract, performing two different injections. For example, in the case of high levels of CBDA or CBD it will be necessary to dilute the extract. For THC, it is often found at low concentration in hemp inflorescences, so it may be necessary to concentrate the extract before injection. In our case, 2 mL of filtered extract was dried using a weak nitrogen flow, and the dry extract was recovered in 500 μL of acetonitrile.

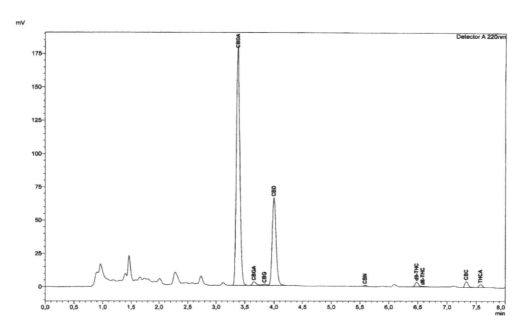

Figure 2. Chromatographic trace of *Cannabis sativa* L. inflorescence extract analyzed by RP-HPLC-UV equipped with a reverse phase C18 column.

2.2. Validation

2.2.1. Precision

The precision of the method was measured by the expression of repeatability (*r*) and reproducibility (*R*). Precision was expressed through coefficient of variation (CV%).

2.2.2. Repeatability, R

Table 1 shows data on the intraday and interday repeatability, evaluated as reported in Section 3.6, which demonstrates very high repeatability. In fact, the relative standard deviation (RSD) varied from 2.59 to 5.65 for intraday repeatability and from 2.83 to 5.05 for interday repeatability. In both cases, the highest RSD was found for CBDA, which is probably due its higher concentration compared to the other cannabinoids.

2.2.3. Reproducibility, R

The RSDs obtained in the reproducibility studies are shown in Table 1. The maximum RSD value was 2.13 for CBGA. The other cannabinoids show RSD values lower than 1.91, and the lowest of the RSDs was 0.09 for CBDA, which is probably due to the higher concentration of this cannabinoid.

2.2.4. Recovery

The tests were performed by using three different concentrations to test the recovery values in the linearity range of the method.

Quantities of CBD (4, 8, and 24 µg/mL) were added, thus assessing concentrations similar to, higher, and lower than those found in samples.

Recovery was determined according to this modality for CBD and was 84.92%.

An evaluation of recovery on all the compounds present in the sample was carried out by proceeding with a further extraction with 10 mL of methanol-chloroform on the sample residue after the usual extraction; in this extract, some cannabinoids were present, and indirectly the percentage of recovery was determined.

Table 1. Validation parameters of RP-HPLC-UV method.

Compound	R^2	[1] LOD (µg/mL)	[2] LOQ (µg/mL)	[3] LOD (µg/mL)	[4] LOQ (µg/mL)	Intraday (Repeatability) RSD	Interday (Repeatability) RSD	Reproducibility RSD	Recovery (%)
CBDA	0.9999	0.34	1.05	0.11	0.37	5.65	5.05	0.09	96.06
CBGA	0.9999	0.32	0.98	0.12	0.40	4.71	4.34	2.13	93.90
CBG	0.9995	0.62	1.87	0.13	0.45	3.34	2.83	0.91	94.60
CBD	0.9995	0.63	1.91	0.17	0.58	4.89	4.44	0.70	84.92
THCV	0.9989	0.95	2.87	0.15	0.49	-	-	N.d. *	N.d. *
CBN	0.9999	0.28	0.84	0.06	0.21	2.59	2.95	0.81	97.08
Δ9-THC	0.9981	1.25	3.79	0.15	0.50	3.05	3.22	0.13	99.69
Δ8-THC	0.9987	1.02	3.10	0.17	0.56	3.81	3.64	0.74	100
CBC	0.9999	0.29	0.88	0.11	0.36	5.3	4.78	0.89	98.68
THCA	0.9998	0.43	1.29	0.11	0.37	5.55	5.01	1.91	95.27

[1] Limit of detection (LOD) determined by the calibration curves (Instrumental LOD = $(3.3 \times \sigma)/m$). [2] Limit of quantification (LOQ) determined by the calibration curves (Instrumental LOQ = $(10 \times \sigma)/m$). [3] LOD determined by the signal-to-noise ratio (Instrumental LOD: S/N = 3). [4] LOQ determined by the signal-to-noise ratio (Instrumental LOQ: S/N = 10). * Not detectable.

The percentage of recovery values, as shown in Table 1, were higher than 84.92% and can be considered very satisfactory. In fact, considering CBD, the percentages are higher than those previously reported in the literature [5].

2.2.5. Detection Limit, LOD

The instrumental limit of detection was determined by the calibration curve, according to the formulas expressed in Section 3.6. The instrumental limit of detection (LOD) values obtained for CBDA and CBGA (Table 1) were lower, while those of CBG and CBD were comparable with similar methods described in literature [5,27]. Low LOD values were found also for the other cannabinoids (THCV, CBN, Δ-9 THC, Δ-8 THC, CBC, THCA), indicating that the method is sensitive.

2.2.6. Quantification Limit, LOQ

The instrumental limit of quantification was determined by a calibration curve, according to the formulas expressed in Section 3.6, considering that the signal-to-noise method is particularly useful to quantify the cannabinoids present at lower concentrations, such as THC. As reported for the LODs, the instrumental limit of quantification (LOQ) values obtained for CBDA and CBGA (Table 1) were also lower than those reported in the literature, while those for CBG and CBD were comparable with those of other methods described for similar procedures [5,27]. In addition, the other cannabinoids (THCV, CBN, Δ-9 THC, Δ-8 THC, CBC, THCA) showed low LOQs. The instrumental noise was registered in μV, by performing 3 blank injections with the ASTM method [28] given by the instrument, and a maximum CV% of 3.49% was calculated for all individual compounds to determine the single LOD and LOQ, which was considered acceptable.

2.2.7. Linearity

In order to evaluate the linearity of the method, eight different points of standard mixture solutions were analyzed in triplicate by HPLC-UV.

The following equations are related to the calibration curves in a concentration range between 0.01–100 μg/mL: CBDA, $y = 18955x - 1612.6$ ($r^2 = 0.9999$); CBGA, $y = 19796x - 3475.7$ ($r^2 = 0.9999$); CBG, $y = 18094x - 9195.3$ ($r^2 = 0.9995$); CBD, $y = 13703x - 6009.5$ ($r^2 = 0.9995$); THCV, $y = 18534x - 15213$ ($r^2 = 0.9989$); CBN, $y = 34148x - 7943.1$ ($r^2 = 0.9999$); Δ9 − THC, $y = 19893x - 31896$ ($r^2 = 0.9981$); Δ8-THC, $y = 17526x - 18267$ ($r^2 = 0.9987$); CBC, $y = 18590x - 4777.1$ ($r^2 = 0.9999$); THCA, $y = 18239x - 8969.3$ ($r^2 = 0.9998$) (Table 1).

With the aid of the equation obtained from the calibration curve, the quantity of each cannabinoid was calculated.

To express the data relative to the content of the individual cannabinoid as a percentage (%, p/p) referred to the dried material, it is necessary to refer to the weight of the sample considering the dilution factor. The linearity in the concentration range analyzed was good for cannabinoid standards, being $r^2 > 0.998$, as reported before.

2.3. Cannabinoids in Hemp Samples

The method developed in this study was applied to quali-quantitative analysis of main cannabinoids in two samples of hemp inflorescences. The samples analyzed, belonging to the same variety of *Cannabis sativa* L., did not show a significant difference in the concentration of the target compounds. As shown in Table 2, CBDA is the only cannabinoid for which a different concentration was determined. The other cannabinoids had a similar or the same concentration (e.g., CBGA, CBG, CBN, Δ-9-THC, and Δ-8-THC) in both samples. THCV was not found in the hemp inflorescence samples analyzed, as shown in Figure 2 and Table 2. Δ-9-THC and Δ-8-THC were found at a low concentration, below the legal limit. Under the current legislation regarding *Cannabis sativa* L. cultivation [6,29], in fact, the total content of THC must not be higher than 0.2% and in any case within 0.6%. Indeed, only the hemp varieties reported in the *Common catalogue of varieties of agricultural plant species* can be cultivated without authorization [6,7]. These kinds of results confirmed that the analyzed samples were correctly classified as hemp, since the quantity of Δ8-THC and Δ9-THC was found to be lower than the limits established by the legislation. According to what is indicated in literature [30], in the hemp variety considered (Futura 75), the most present compound was CBDA, followed by CBD; all the other compounds were in very low amounts ranging from 0.01 to 0.06%. CBGA is the compound from which all other cannabinoids are biosynthesized [5], which is probably why it was found at a low concentration in both samples examined.

The number of cannabinoids in hemp samples is reported in Table 2.

Table 2. Number of cannabinoids in hemp samples.

Sample	Cannabinoids									
	CBDA (%)	CBGA (%)	CBG (%)	CBD (%)	THCV (%)	CBN (%)	Δ9-THC (%)	Δ8-THC (%)	CBC (%)	THCA (%)
L11	0.88 ± 0.04	0.02 ± 0.00	0.02 ± 0.00	0.48 ± 0.02	N.d. *	0.01 ± 0.00	0.06 ± 0.00	0.03 ± 0.00	0.03 ± 0.00	0.03 ± 0.00
CV%	5.05	4.34	2.83	4.44		2.95	3.22	3.64	4.78	5.10
L5	0.93 ± 0.06	0.02 ± 0.00	0.02 ± 0.00	0.45 ± 0.03	N.d. *	0.01 ± 0.00	0.06 ± 0.00	0.03 ± 0.00	0.02 ± 0.00	0.04 ± 0.00
CV%	6.48	1.28	1.73	6.28		1.49	0.21	2.20	2.98	7.17

* Not detectable.

3. Materials and Methods

3.1. Chemicals, Standards and Apparatus

All chemicals used were of analytical grade. Methanol p.a CAS 67-56-1, chloroform p.a CAS 67-66-3, acetonitrile CAS 75-05-8, water CAS 7732-18-5, and orthophosphoric acid CAS 7664-38-2 were purchased from Sigma-Aldrich (St. Louis, MO, USA). Nitrogen, pure gas for analysis CAS 7727-37-9 was purchased from SIAD Spa (Bergamo, Italy). Standard mixture of phytocannabinoids 0.1% in acetonitrile: Cannabidiolic acid (0.01%) CAS 1244-58-2, cannabigerolic acid (0.01%) CAS 25555-57-1, cannabigerol (0.01%) CAS 25654-31-3, cannabidiol (0.01%) CAS 13956-29-1, tetrahydrocannabivarin (0.01%) CAS 31262-37-0, cannabinol (0.01%) CAS 521-35-7, tetrahydrocannabinolic acid (0.01%) CAS 23978-85-0, Δ-9-tetrahydrocannabinol (0.01%) CAS 1972-08-3, Δ-8-tetrahydrocannabinol (0.01%) CAS 5957-75-5, cannabichromene (0.01%) CAS Number 20675-51-8, were purchased from Cayman Chemical Company, (Ann Arbor, MI, USA). Cannabidiol 1.0 mg/mL in methanol CAS 13956-29-1: LGC Standards S.r.l., (Milan, Italy).

Analytical mill, IKA A11 Basic (IKA® Werke GMBH & Co. KG, Germany). Analytical balance with precision of 0.1 mg, mod. E42, (Gibertini, Italy). Vortex vibrating shaker, mod. ST5, (Janke & Kunkel, Germania). Centrifuge mod. ALC, PK 120 (Thermo Electron Corporation, Massachusetts, USA). Termoblock heating block, mod. A120, (Falc, Italy). Natural ventilation stove. Sieve with 1 mm meshes. Tilting shaker. Ultrasound bath Branson 2150, (Danbury-CT, USA). Volumetric flasks of 1, 2, 10 and 25 mL. SOVIREL-type tubes with screw cap. Glass syringes with luer lock attachment, 0.45 μm nylon membrane filters. Microsyringes from 1 to 1000 μL. HPLC Cannabis Analyzer for Potency Prominence-i LC-2030C equipped with a reverse phase C18 column, Nex-Leaf CBX Potency 150 × 4.6 mm, 2.7 μm with a guard column Nex-Leaf CBX 5 × 4.6 mm, 2.7, UV detector and acquisition software LabSolutions version 5.84 (Shimazu, Kyoto, Japan).

3.2. Sampling

The samples were supplied by a company that produces industrial hemp. In particular, two samples (L11 and L5) of inflorescences of *Cannabis sativa* L. Futura 75 were analyzed, having come from the same land and harvested in August 2017, and supplied by Enecta Srl. Sampling of material was carried out on a population of hemp plants, according to a systematic path, so that the sample taken was representative of the particle, excluding the edges, taking the upper third of the selected plant as indicated in Reg. (EU) No 1155/2017 [31]. The sample was dried in an oven at 35 °C ± 1 to constant weight, and gross wood parts and seeds with a length of more than 2 mm were removed. The samples were then subjected to grinding and subsequent sieving through a sieve with 1 mm meshes. The sieved material was transferred into polypropylene containers and stored under nitrogen atmosphere, protected from light at a temperature of −20 °C until extraction. Three independent replicates were performed for each sample, and three HPLC injections were performed for each replication.

3.3. Cannabinoid Extraction

To extract cannabinoids, an aliquot of powder sample, about 25 mg, was weighed using an analytical balance; 10 mL of methanol-chloroform extraction solvent 9:1 (*v/v*) was added as reported by De Backer et al. (2009) [32], Jin et al. (2017) [33], and was placed first for 10 min on an oscillating oscillator set at 350 oscillations per minute and then for 10 min in an ultrasonic bath. The sample was centrifuged for 10 min at 1125 g, and the supernatant was removed. The extraction was performed twice. The two fractions containing cannabinoids were collected in a 25 mL volumetric flask and were brought to volume with methanol/chloroform (9:1, *v/v*). The samples were filtered with a 45 μm nylon filter. Two mL of the filtered extract was transferred to a glass tube. The solvent was removed, leading to dryness with the help of a weak nitrogen flow, and recovered with 500 μL acetonitrile. The solution was injected into an HPLC-UV.

3.4. Preparation of Standard Solution

Appropriate aliquots of a standard mixture of cannabinoids are diluted with acetonitrile to obtain solutions of known concentration, in particular eight points in a concentration range between 0.05 and 100 μg/mL (0.05, 0.50, 4.17, 8.33, 16.70, 25.00, 50.00, 100.00 μg/mL). The standard solutions were prepared to construct calibration curves for the 10 cannabinoids considered: CBDA, CBGA, CBG, CBD, THCV, CBN, Δ9-THC, Δ8-THC, CBC, and THCA. The standard solutions were stored away from light at a temperature of −20 °C. The stability of standard solutions stored at −20 °C was evaluated every week for 3 months with the HPLC-UV system, and no degradation of cannabinoids was found.

3.5. HPLC Conditions

For the RP-HPLC analysis, the column was thermostated at 35 °C, and the autosampler was thermostated to 4 °C. Sample concentration was 4 mg/mL, and injection volume was 5.0 μL. UV detection was used at 220 nm, and gradient elution was used at flow rate of 1.6 mL/min according to the following procedure. Eluent mixture: Water + 0.085% phosphoric acid (A), acetonitrile + 0.085% phosphoric acid (B). Gradient elution: 70% of B up to 3 min, 85% of B to 7 min, 95% of B to 7.01 up to 8.00 min, and 70% of B up to 10 min. The eluent mixture was previously filtered with a Millipore system equipped with a 0.2 μm nylon filter.

3.6. Validation Parameters

3.6.1. Precision

Precision is the closeness of agreement among independent test results, obtained with stipulated conditions and usually in terms of standard deviation or relative standard deviation [34].

Precision was calculated with the following formula: CV% = [(SD/$\bar{x}$) × 100], where SD is the estimate of the standard deviation and $\bar{x}$ is the average of the replications made.

3.6.2. Repeatability, R

The repeatability (intraday) of the method was evaluated by analyzing three replicates of the same sample, injected three times on the same day, performed by the same operator with the same method and instrument. The result corresponds to the arithmetic mean of the three determinations made considering the estimate of the standard deviation (SD) calculated on the three replicates performed.

The repeatability (interday) of the method was evaluated by performing three replicates of the same sample, injected three times on three different days, performed by the same operator with the same method and instrument. The result corresponds to the arithmetic mean of the three determinations made considering the estimate of the standard deviation (SD) calculated on the three replicates performed.

3.6.3. Reproducibility, R

Reproducibility was evaluated by the agreement between the results obtained on the same sample with the same procedure carried out by different operators in the laboratory and was measured with the coefficient of variation.

3.6.4. Recovery

Recovery is the fraction of analyte that was added to the sample being tested. Recovery was expressed as a percentage (R (%)) according to the following formula: R (%) = [(Cf − C)/Cc] × 100, where Cf is the endogenous amount of the cannabinoid in the sample plus the amount of standard added to the analyte under examination. C is the endogenous amount present in the sample not added with the standard. Cc is the amount of the standard analyte added to the sample.

3.6.5. Detection Limit, LOD

The detection limit is the smallest amount or concentration of analyte in the sample that can be reliably distinguished from zero [34]. It can be calculated using the following formula: LOD = (3.3 × σ)/m, where: σ represents the residual standard deviation of the calibration curve and m represents the slope of the calibration curve.

Furthermore, the LOD of the method from the signal (S)/noise (N) ratio can be determined as LOD: S/N = 3.

3.6.6. Quantification limit, LOQ

The quantification limit is the concentration of analyte below which it is determinable with a level of precision that is too low with inaccurate results. The LOQ can be determined according to the following formula: LOQ = (10 × σ)/m, where σ represents the residual standard deviation of the calibration curve and m represents the slope of the calibration curve.

The LOQ of the method can also be determined by the signal-to-noise ratio (S/N): LOQ: S/N = 10.

3.6.7. Linearity

Linearity can be tested by examination of a plot of residuals produced by linear regression of the responses on the concentrations in an appropriate calibration set [34].

In order to quantify the analytes of interest, the equation of the calibration curve obtained for each standard is used. The equation is: y = ax + b, where y = area of the analyte obtained by HPLC/UV analysis, a = slope of the calibration curve, x = unknown concentration (μg/mL) of analyte in the sample, b = intercept of the calibration curve.

4. Conclusions

One of the most relevant problems in analytical determinations for quality control, especially when there are legal problems related with quantitation, such as for cannabis, relates to the proficiency of laboratories. Therefore, detailed and validated procedures that are freely available are essential for the full understanding of any analytical step and its careful application. This is also true for "daily" methods that can be easily applied for quality control, carried out using traditional RP-HPLC and UV-Vis detectors, with less efficient performance than diode-array detectors but with lower costs, rendering them affordable even for small laboratories.

The validated method described herein allows the quantitative determination of the 10 most relevant cannabinoids using a single wavelength (220 nm) in 8 min. A full separation is obtained, even in the elution sequence of a difficult resolution, of the group of peaks related to CBGA, CBG, CBD, and THCV (from 3.5 to 4.5 min).

The method is applied to cannabis inflorescences and involves extraction in methanol/ chloroform, drying of the extract, taking it up in acetonitrile and injection into an HPLC. The method has sensitivity and accuracy to discriminate samples with amounts of Δ-9- and Δ-8-THC (total THC content) that are below the limit of 0.2% from those that are subjected to legal restrictions in many EU countries, with a total THC content above 0.6%, which cannot be classified as hemp. Due to its simplicity and rapidity, it can be used to check raw material or crops during the harvesting period.

A detailed standard operating procedure (SOP), as a supplementary information file, is also available, so that any operator with basic knowledge of HPLC can easily apply it and make all the elution and calibration control checks using commercially available mixtures of standards, which are more affordable and sustainable than single cannabinoid standards in terms of costs and solvents used for calibration.

Author Contributions: Conceptualization, T.G.T., M.M. and M.T.; Methodology, M.M.; Software, M.M. and S.S.; Validation, M.M.; Formal analysis, M.M.; Investigation, T.G.T. and M.M.; Resources, T.G.T. and M.M.; Data curation, M.M., T.G.T. and M.T.; Writing—original draft preparation, M.M., M.T. and T.G.T.; writing—review and editing, T.G.T. and S.S.; Visualization, T.G.T.; Supervision, T.G.T.; project administration, T.G.T.; funding acquisition, T.G.T.

Acknowledgments: The authors gratefully acknowledge Enecta Srl for providing samples. The experimentation was conducted in the context of a PhD project entitled *"Harmonized procedures of analysis of medical, herbal, food and industrial cannabis: development and validation of cannabinoids' quality control methods, of extraction and preparation of derivatives from the plant raw material, according to the product destination"* and funded by ENECTA Srl.

References

1.   Montserrat-de la Paz, S.; Marín-Aguilar, F.; García-Gimenez, M.D.; Fernández-Arche, M.A. Hemp (*Cannabis sativa* L.) seed oil: Analytical and phytochemical characterization of the unsaponifiable fraction. *J. Agric. Food Chem.* **2014**, *62*, 1105–1110. [CrossRef] [PubMed]

2.   Appendino, G.; Chianese, G.; Taglialatela-Scafati, O. Cannabinoids: Occurrence and medicinal chemistry. *Curr. Med. Chem.* **2011**, *18*, 1085–1099. [CrossRef]

3.   Andre, C.M.; Hausman, J.F.; Guerriero, G. *Cannabis sativa*: The plant of the thousand and one molecules. *Front. Plant. Sci.* **2016**, *7*, 19. [CrossRef] [PubMed]

4.   Aizpurua-Olaizola, O.; Soydaner, U.; Öztürk, E.; Schibano, D.; Simsir, Y.; Navarro, P.; Usobiaga, A. Evolution of the cannabinoid and terpene content during the growth of *Cannabis sativa* plants from different chemotypes. *J. Nat. Prod.* **2016**, *79*, 324–331. [CrossRef]

5.   Brighenti, V.; Pellati, F.; Steinbach, M.; Maran, D.; Benvenuti, S. Development of a new extraction technique and HPLC method for the analysis of non-psychoactive cannabinoids in fibre-type *Cannabis sativa* L.(hemp). *J. Pharm. Biomed. Anal.* **2017**, *143*, 228–236. [CrossRef] [PubMed]

6.   Legge 2 Dicembre 2016, n.242. Disposizioni per la Promozione della Coltivazione e della Filiera Agroindustriale della Canapa (16G00258), GU Serie Generale n. 304 del 30-12-2016. Available online: https://www.gazzettaufficiale.it/eli/id/2016/12/30/16G00258/sg (accessed on 1 June 2019).

7.   European Commission (EC). European Union Common Catalogue of Varieties of Agricultural Plant Species, Plant Variety Database. Available online: http://ec.europa.eu/food/plant/plant_propagation_material/plant_variety_catalogues_databases/search//public/index.cfm (accessed on 1 June 2019).

8.   Pisanti, S.; Malfitano, A.M.; Ciaglia, E.; Lamberti, A.; Ranieri, R.; Cuomo, G.; Laezza, C. Cannabidiol: State of the art and new challenges for therapeutic applications. *Pharmacol. Ther.* **2017**, *175*, 133–150. [CrossRef] [PubMed]

9.   Mechoulam, R.; Gaoni, Y. Recent advances in the chemistry of hashish. In *The Chemistry of Organic Natural Products/Progrès dans la Chimie des Substances Organiques Naturelles*; Springer International Publishing: Vienna, Austria, 1967; pp. 175–213.

10. Radwan, M.M.; Wanas, A.S.; Chandra, S.; ElSohly, M.A. Natural Cannabinoids of Cannabis and Methods of Analysis. In Cannabis sativa L.-*Botany and Biotechnology*, 1st ed.; Springer International Publishing AG: Cham, Switzerland, 2017; pp. 161–182.

11. Leghissa, A.; Hildenbrand, Z.L.; Schug, K.A. A review of methods for the chemical characterization of cannabis natural products. *J. Sep. Sci* **2018**, *41*, 398–415. [CrossRef] [PubMed]

12. Citti, C.; Pacchetti, B.; Vandelli, M.A.; Forni, F.; Cannazza, G. Analysis of cannabinoids in commercial hemp seed oil and decarboxylation kinetics studies of cannabidiolic acid (CBDA). *J. Pharm Biomed. Anal.* **2018**, *149*, 532–540. [CrossRef] [PubMed]

13. Citti, C.; Braghiroli, D.; Vandelli, M.A.; Cannazza, G. Pharmaceutical and biomedical analysis of cannabinoids: A critical review. *J. Pharm Biomed. Anal.* **2018**, *147*, 565–579. [CrossRef] [PubMed]

14. Rodrigues, A.; Yegles, M.; Van Elsué, N.; Schneider, S. Determination of cannabinoids in hair of CBD rich extracts consumers using gas chromatography with tandem mass spectrometry (GC/MS–MS). *Forensic. Sci. Int.* **2018**, *292*, 163–166. [CrossRef]

15. Cardenia, V.; Gallina Toschi, T.; Scappini, S.; Rubino, R.C.; Rodriguez Estrada, M.T. Development and validation of a Fast gas chromatography/mass spectrometry method for the determination of cannabinoids in *Cannabis sativa* L. *J. Food Drug Anal.* **2018**, *26*, 1283–1292. [CrossRef] [PubMed]

16. Leghissa, A.; Hildenbrand, Z.L.; Foss, F.W.; Schug, K.A. Determination of cannabinoids from a surrogate hops matrix using multiple reaction monitoring gas chromatography with triple quadrupole mass spectrometry. *J. Sep. Sci* **2018**, *41*, 459–468. [CrossRef] [PubMed]

17. Patel, B.; Wene, D.; Fan, Z.T. Qualitative and quantitative measurement of cannabinoids in cannabis using modified HPLC/DAD method. *J. Pharm Biomed. Anal.* **2017**, *146*, 15–23. [CrossRef] [PubMed]

18. Burnier, C.; Esseiva, P.; Roussel, C. Quantification of THC in Cannabis plants by fast-HPLC-DAD: A promising method for routine analyses. *Talanta* **2019**, *192*, 135–141. [CrossRef] [PubMed]

19. Pellati, F.; Brighenti, V.; Sperlea, J.; Marchetti, L.; Bertelli, D.; Benvenuti, S. New Methods for the Comprehensive Analysis of Bioactive Compounds in *Cannabis sativa* L.(hemp). *Molecules* **2018**, *23*, 2639. [CrossRef]

20. Ciolino, L.A.; Ranieri, T.L.; Taylor, A.M. Commercial cannabis consumer products part 2: HPLC-DAD quantitative analysis of cannabis cannabinoids. *Forensic. Sci. Int.* **2018**, *289*, 438–447. [CrossRef]

21. Fekete, S.; Sadat-Noorbakhsh, V.; Schelling, C.; Molnár, I.; Guillarme, D.; Rudaz, S.; Veuthey, J.L. Implementation of a generic liquid chromatographic method development workflow: Application to the analysis of phytocannabinoids and *Cannabis sativa* extracts. *J. Pharm. Biomed. Anal.* **2018**, *155*, 116–124. [CrossRef]

22. Purschke, K.; Heinl, S.; Lerch, O.; Erdmann, F.; Veit, F. Development and validation of an automated liquid-liquid extraction GC/MS method for the determination of THC, 11-OH-THC, and free THC-carboxylic acid (THC-COOH) from blood serum. *Anal. Bioanal. Chem.* **2016**, *408*, 4379–4388. [CrossRef]

23. Pacifici, R.; Marchei, E.; Salvatore, F.; Guandalini, L.; Busardò, F.P.; Pichini, S. Evaluation of cannabinoids concentration and stability in standardized preparations of cannabis tea and cannabis oil by ultra-high performance liquid chromatography tandem mass spectrometry. *Clin. Chem. Lab. Med.* **2017**, *55*, 1555–1563. [CrossRef]

24. Casiraghi, A.; Roda, G.; Casagni, E.; Cristina, C.; Musazzi, U.M.; Franzè, S.; Gambaro, V. Extraction Method and Analysis of Cannabinoids in Cannabis Olive Oil Preparations. *Planta Med.* **2018**, *84*, 242–249. [CrossRef]

25. Lin, S.Y.; Lee, H.H.; Lee, J.F.; Chen, B.H. Urine specimen validity test for drug abuse testing in workplace and court settings. *J. Food Drug Anal.* **2018**, *26*, 380–384. [CrossRef] [PubMed]

26. Mudge, E.M.; Murch, S.J.; Brown, P.N. Leaner and greener analysis of cannabinoids. *Anal. Bioanal. Chem.* **2017**, *409*, 3153–3163. [CrossRef] [PubMed]

27. Gul, W.; Gul, S.W.; Radwan, M.M.; Wanas, A.S.; Mehmedic, Z.; Khan, I.I.; ElSohly, M.A. Determination of 11 cannabinoids in biomass and extracts of different varieties of Cannabis using high-performance liquid chromatography. *J. AOAC Int.* **2015**, *98*, 1523–1528. [CrossRef] [PubMed]

28. ASTM. *ASTM E685-93(2013), Standard Practice for Testing Fixed-Wavelength Photometric Detectors Used in Liquid Chromatography*; ASTM International: West Conshohocken, PA, USA, 2013. Available online: www.astm.org (accessed on 1 June 2019).

29. Regulation (EU) No 1307/2013 of the European Parliament and of the Council of 17 December 2013 Establishing Rules for Direct Payments to Farmers under the Support Schemes within the Framework of the Common Agricultural Policy and Repealing Council Regulation (EC) No 637/2008 and Council Regulation (EC) No 73/2009. Available online: https://eur-lex.europa.eu/legal-content/EN/TXT/?uri=celex:32013R1307 (accessed on 1 June 2019).

30. del M. Contreras, M.; Jurado-Campos, N.; Sánchez-Carnerero Callado, C.; Arroyo-Manzanares, N.; Fernández, L.; Casano, S.; Ferreiro-Vera, C. Thermal desorption-ion mobility spectrometry: A rapid sensor for the detection of cannabinoids and discrimination of *Cannabis sativa* L. chemotypes. *Sens. Actuators B Chem.* **2018**, *273*, 1413–1424. [CrossRef]

31. Commission Delegated Regulation (EU) 2017/1155 of 15 February 2017 Amending Delegated Regulation (EU) No 639/2014 as Regards the Control Measures Relating to the Cultivation of Hemp, Certain Provisions on the Greening Payment, the Payment for Young Farmers in Control of a Legal Person, the Calculation of the per Unit Amount in the Framework of Voluntary Coupled Support, the Fractions of Payment Entitlements and Certain Notification Requirements Relating to the Single Area Payment Scheme and the Voluntary Coupled Support, and Amending Annex X to Regulation (EU) No 1307/2013 of the European Parliament and of the Council. Available online: https://eur-lex.europa.eu/eli/reg_del/2017/1155/oj (accessed on 1 June 2019).

32. De Backer, D.B.; Debrus, B.; Lebrun, P. Innovative development and validation of an HPLC/DAD method for the qualitative and quantitative determination of major cannabinoids in cannabis plant material. *J. Chromatogr. B Anal. Technol. Biomed. Life Sci.* **2009**, *877*, 4115–4124. [CrossRef] [PubMed]

33. Jin, D.; Jin, S.; Yu, Y.; Lee, C.; Chen, J. Analytical & Bioanalytical Techniques Classification of Cannabis Cultivars Marketed in Canada for Medical Purposes by Quantification of Cannabinoids and Terpenes Using HPLC-DAD and GC-MS. *J. Anal. Bioanal. Tech.* **2017**, *8*, 1–9.

34. Thompson, M.; Ellison, S.L.; Wood, R. Harmonized guidelines for single-laboratory validation of methods of analysis (IUPAC Technical Report). *Pure Appl. Chem.* **2002**, *74*, 835–855. [CrossRef]

8

Protein-Based Fingerprint Analysis for the Identification of *Ranae Oviductus* using RP-HPLC

Yuanshuai Gan [1][ID], Yao Xiao [1], Shihan Wang [2], Hongye Guo [1], Min Liu [1], Zhihan Wang [3] and Yongsheng Wang [1],*

[1] College of Pharmacy, Jilin University, Changchun 130021, China; ganys18@mails.jlu.edu.cn (Y.G.); xiaoyao17@mails.jlu.edu.cn (Y.X.); guohy18@mails.jlu.edu.cn (H.G.); liumin17@mails.jlu.edu.cn (M.L.)

[2] College of Chinese Herbal Medicine, Jilin Agricultural University, Changchun 130118, China; wsh8805@163.com

[3] Department of Physical Sciences, Eastern New Mexico University, Portales, NM 88130, USA; zhihan.wang@enmu.edu

* Correspondence: mikewangwys@outlook.com or wys@jlu.edu.cn

Abstract: This work demonstrated a method combining reversed-phase high-performance liquid chromatography (RP-HPLC) with chemometrics analysis to identify the authenticity of *Ranae Oviductus*. The fingerprint chromatograms of the *Ranae Oviductus* protein were established through an Agilent Zorbax 300SB-C8 column and diode array detection at 215 nm, using 0.085% TFA (v/v) in acetonitrile (A) and 0.1% TFA in ultrapure water (B) as mobile phase. The similarity was in the range of 0.779–0.980. The fingerprint chromatogram of *Ranae Oviductus* showed a significant difference with counterfeit products. Hierarchical clustering analysis (HCA) and principal component analysis (PCA) successfully identified *Ranae Oviductus* from the samples. These results indicated that the method established in this work was reliable.

Keywords: *Ranae Oviductus*; identification; protein; RP-HPLC; fingerprint

1. Introduction

Rana chensinensis is mainly distributed in the Changbai Mountain area, China. *Ranae Oviductus* is the dried oviduct of female *Rana temporaria chensinensis* David. The *Ranae Oviductus* is a potent traditional Chinese medicine that has been used in clinical studies for thousands of years. Today it is widely used as a nutrient food. It has been reported that *Ranae Oviductus* has significant effects in enhancing immunity, anti-fatigue, anti-aging, and lowering blood fat [1–4]. As a precious traditional Chinese medicine, *Ranae Oviductus* has been in short supply because of its limited production [5]. Its high price and lucrative profits have tempted many counterfeit products, such as bullfrog oviduct, toad oviduct, or frog oviduct, to inundate the market, resulting in the uneven quality of *Ranae Oviductus* in the market [6,7]. Those counterfeits have a similar appearance but have less efficacy. To guarantee the quality of *Ranae Oviductus*, its authenticity identification has attracted more and more attention from the pharmacists, doctors, and medicinal scientists. The identification method of *Ranae Oviductus* is still under development. In the 2005 China Pharmacopoeia, the appearance and expansion degree were employed as discriminating items of *Ranae Oviductus* [8]. Our group has reported using UV spectra to identify *Ranae Oviductus* [9]. According to a previous study, it is difficult to identify the *Ranae Oviductus* and counterfeit products using traditional methods [10]. Therefore, it is essential to establish a highly reliable method for the identification of *Ranae Oviductus*.

More than 40% of the components in *Ranae Oviductus* are proteins and the proteins are the major bioactive components of *Ranae Oviductus* [11,12]. However, the identification of *Ranae Oviductus* and counterfeit products using HPLC based on protein has not been studied yet. In addition, reversed-phase high-performance liquid chromatography (RP-HPLC) is a simple, fast, and effective technique for protein separation and characterization, as used for protein in milk, wheat gliadin, and transgenic zein [13–15]. On the other hand, the fingerprint chromatogram is considered as a comprehensive qualitative and quantitative method for the identification of different species, especially in the quality assessment of traditional Chinese medicines [16]. The World Health Organization (WHO) has admitted the use of chromatographic fingerprints as an identification strategy for traditional Chinese medicinal preparations [17]. Many reports have employed HPLC fingerprint chromatograms to study the quality control of traditional Chinese medicines. For example, Lu et al. used the HPLC fingerprint to identify Chinese *Angelica* from related umbellifer herbs. Sun et al. analyzed polysaccharides from different *Ganoderma*. Li et al. established the fingerprint analysis of polyphenols, which were extracted from pomegranate peel, with reliable results [18–20].

In this work, the main proteins components of *Ranae Oviductus* were used as the study objects. We used RP-HPLC to establish a fingerprint method for the identification of *Ranae Oviductus*. Ten batches of *Ranae Oviductus* were collected from different main producing areas of the Changbai Mountains. A protein reference chromatogram was established using those *Ranae Oviductus*, based on protein composition similarity analysis. Furthermore, the difference between the authentic *Ranae Oviductus* and counterfeit products were investigated. The results were verified via a chemometric approach, utilizing principal component analysis and hierarchical clustering analysis. Both showed that the newly established *Ranae Oviductus* identification method was reliable.

2. Materials and Methods

2.1. Chemicals and Samples

The petroleum ether, guanidine hydrochloride, and ammonium sulfate analytical grade were purchased from Beijing Chemical Factory (Beijing, China). The dithiothreitol (DTT) and trifluoroacetic acid (TFA) were purchased from Sigma-Aldrich (St. Louis, MO, USA). The HPLC-grade acetonitrile (MeCN) and HPLC-grade methanol were purchased from Fisher (Fisher Scientific, USA). The ultrapure water was obtained from a gradient water purification system (Water Purifier, Sichuan, China).

Ranae Oviductus, bullfrog oviduct, toad oviduct and frog oviduct were provided by Jilin Province Rana Industry Association which were collected from the Changbai Mountain area in the Jilin province of China. The specific location is shown on the map in Figure 1. Ten batches of *Ranae Oviductus* samples were collected from different regions from the main producing area of the Changbai Mountain range. The specific collection information is shown in Table 1.

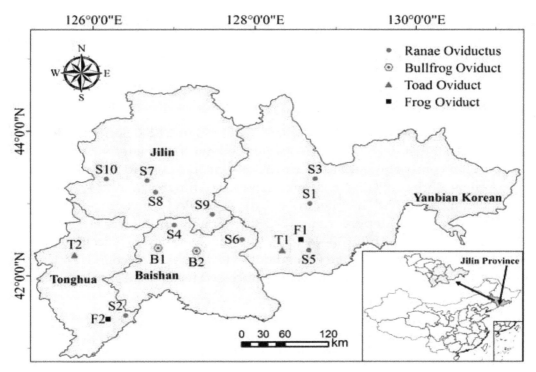

Figure 1. Distribution map of origins for *Ranae Oviductus* and its counterfeits in the Changbai mountain area.

Table 1. Origin and collecting date of the *Ranae Oviductus* samples and their counterfeits.

No.	Name of Medicine	Origin	Collection Date
S1	*Ranae Oviductus*	Yanbian, Jilin	2016.3
S2	*Ranae Oviductus*	Tonghua, Jilin	2016.1
S3	*Ranae Oviductus*	Yanbian, Jilin	2016.3
S4	*Ranae Oviductus*	Baishan, Jilin	2015.11
S5	*Ranae Oviductus*	Yanbian, Jilin	2016.3
S6	*Ranae Oviductus*	Baishan, Jilin	2015.11
S7	*Ranae Oviductus*	Jilin, Jilin	2016.12
S8	*Ranae Oviductus*	Jilin, Jilin	2016.12
S9	*Ranae Oviductus*	Jilin, Jilin	2015.11
S10	*Ranae Oviductus*	Jilin, Jilin	2015.11
B1	Bullfrog Oviduct	Baishan, Jilin	2016.12
B2	Bullfrog Oviduct	Baishan, Jilin	2016.12
T1	Toad Oviduct	Yanbian, Jilin	2016.10
T2	Toad Oviduct	Tonghua, Jilin	2016.11
F1	Frog Oviduct	Yanbian, Jilin	2016.10
F2	Frog Oviduct	Tonghua, Jilin	2016.11

2.2. Protein Extraction

The dried *Ranae Oviductus* was pulverized into a powder (passing through a 20-mesh sieve) and degreased with petroleum ether at room temperature. After filtration, the powder was placed in an oven at 55 °C for 1 h. Afterward, 0.50 g of the sample was added to PBS buffer (50 mL, 0.1 M pH 7.4). After continuously stirring for 8 h, the mixture was centrifuged at 5000 r/min for 15 min. The supernatant was collected and the precipitate was extracted again. The two centrifugal supernatants were combined. To the supernatant, an ammonium sulfate solid was slowly added until to 60% saturation [21,22]. The mixture was centrifuged at 8000 r/min for 20 min after standing at 4 °C for 1 h. The precipitate was dissolved in 6 M guanidine hydrochloride (containing 10 mM DTT) [23,24], and dialyzed in distilled water in a dialysis bag (molecular weight cutoff: 8000 Da) for 12 h [25]. The sample solution was finally scaled to 5 mL with 6 M guanidine hydrochloride (containing 10 mM DTT) in a volumetric flask and filtrated with a 0.45 µm filter membrane prior HPLC injection [26]. The preparations of bullfrog oviduct, toad oviduct and frog oviduct were the same as that for *Ranae Oviductus*.

2.3. RP-HPLC Chromatography Analysis

The samples were separated using an Agilent Technologies 1200 Series liquid chromatograph (Agilent Technologies, Pittsburgh, PA, USA) equipped with a quaternary pump, autosampler, thermostatted column compartment, diode array detector (DAD), and UV detector. The columns used were the Agilent Zorbax 300SB-C8 column (250 × 4.6 mm, 5 µm) and Agilent Zorbax SB-C18 column (250 × 4.6 mm, 5 µm) with mobile phase A (0.085% TFA in *v/v* with acetonitrile) and mobile phase B (0.1% TFA in *v/v* with ultrapure water) [27,28]. Gradient elution was adopted as follows, from 12–30% A in the first 52 min, and from 30–44% A in the next 28 min. The injection volume was 20 µL. The optimized separation conditions were tested under the different detection wavelengths, flow rates and temperatures [29]. The data were recorded and processed using the Agilent Chemstation software.

2.4. Validation of the RP-HPLC Method

Ranae Oviductus sample (S1) was used to verify the RP-HPLC method. A precision analysis was carried out by repeatedly injecting the same solution 5 times on the same day. The repeatability was assessed by injecting 5 separate solutions obtained from the same *Ranae Oviductus* sample. The stability was evaluated by analyzing the same sample solution at different time periods of 0, 2, 4, 8, 16 and 24 h at room temperature.

2.5. Establishment of the HPLC Fingerprint

The common characteristic peaks and similarities of fingerprint data of 10 batches of *Ranae Oviductus* were investigated using the professional software Similarity Evaluation System for the Chromatographic Fingerprint, according to the recommendations of the State Food and Drug Administration (SFDA). The HPLC fingerprint data of the samples were imported to the evaluation system (the solvent peaks in the first 4 min were removed and the time window was set at 0.2 s). The calibration method was multi-point calibration. The significant common peaks were labeled as mark peaks and the reference chromatogram fingerprint was generated with a mean value method. The similarity of the fingerprint data was represented by a correlation coefficient (similarity) and the higher similarity between the two samples resulted in a correlation coefficient value close to 1. The correlation coefficients of all chromatograms of 10 batches of *Ranae Oviductus* samples were calculated throughout the study and a correlation analysis was performed.

2.6. Data Analysis

Hierarchical clustering analysis (HCA) is a cluster analysis technique that reflects the similarities and differences between samples in the form of a hierarchical tree diagram [30,31]. This method is easier to observe than the complex raw data. Based on the clustering method between different groups and the Pearson correlation intervals, SPSS (version 25.0; SPSS Inc., Chicago, IL, USA) was used to group the different samples in this study.

Principal component analysis (PCA) is a classification method that uses dimensionality reduction techniques to simplify numerous original variables into several representative composite indicators [32,33]. According to the contribution rate of each comprehensive indicator, the information of the original data could be reflected when using appropriate numbers of principal components (PCs) [34]. In this study, PCA was performed using SPSS (version 25.0; SPSS Inc., Chicago, IL, USA) and the fractional scatter plot was interpreted by the relationship between PC1, PC2, and PC3 for visual analysis of the data matrix.

3. Results and Discussion

3.1. Optimization of the RP-HPLC Conditions

In order to improve the separation rate of the proteins in *Ranae Oviductus*, the *Ranae Oviductus* (S1) collected from the China Changbai mountain area were systematically investigated. The RP-HPLC chromatography method was optimized through the detection wavelength, separation column, flow rate and temperature. Three classical UV detection conditions were previously reported: 215 nm corresponding to the maximum absorption of peptide bonds; 254 nm corresponding to the maximum absorption of phenylalanine residues; and 280 nm corresponding to tyrosine and maximum absorption of tyrosine residues and tryptophan residues [35]. Figure 2a shows the UV absorption diagram of *Ranae Oviductus* using a diode array detector (DAD) with a wavelength range of 195–300 nm. The red region in the diagram indicated a larger absorption value. Although obvious solvent peaks around 215 nm were observed, the analysis of the core substance was not affected. The UV absorption diagram suggested that the separation effect at 215 nm was better than 254 nm and 280 nm.

Two types of columns (Agilent Zorbax SB-C18 column 250 × 4.6 mm, 5 μm, 80 Å and Agilent Zorbax 300SB-C8 column 250 × 4.6 mm, 5 μm, 300 Å) were used to examine the column effect on the protein separation of *Ranae Oviductus*. The results showed that the C8 column had a higher separation rate than the C18 column, which could be attributed to the large molecular weight of the proteins (Figure 2b). Therefore, the C8 column with a 300 Å pore diameter was selected for this study.

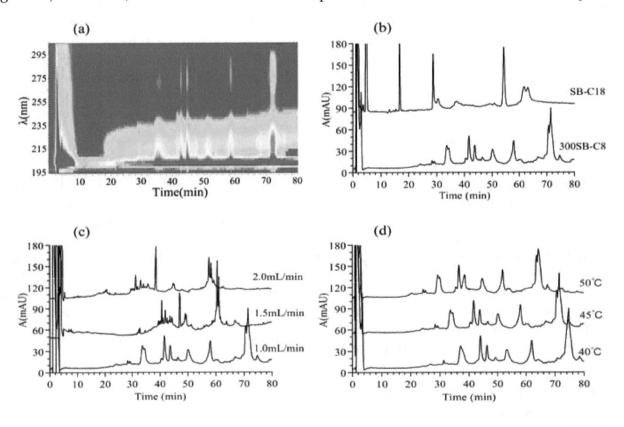

Figure 2. Optimization of reversed-phase high-performance liquid chromatography (RP-HPLC) separation method of the proteins from *Ranae Oviductus*. (**a**) The detection wavelength effect on the RP-HPLC chromatography of the *Ranae Oviductus* proteins. Diode array detector (DAD), 195–300 nm. (**b**) Column type effect on RP-HPLC chromatography of the *Ranae Oviductus* proteins (Agilent Zorbax 300SB-C8 column 250 × 4.6 mm, 5 μm, 300 Å and Agilent Zorbax SB-C18 column 250 × 4.6 mm, 5 μm, 80 Å). (**c**) Flow rate effect RP-HPLC chromatography of *Ranae Oviductus* (1.0 mL/min, 1.5 mL/min, 2.0 mL/min). (**d**) Temperature effect of RP-HPLC chromatography on the *Ranae Oviductus* proteins (40 °C, 45 °C, and 50 °C).

Since the flow rate of the mobile phase can affect the isolation efficiency, three flow rates (1.0, 1.5, 2.0 mL/min) were tested in this study. High flow rates showed that peaks overlapped (Figure 2c). The flow rate of 1.0 mL/min showed the highest separation effect and this, therefore, was chosen for the study.

On the other hand, the temperature played an important role in the RP-HPLC separation. Theoretically, high temperatures can increase the motion rate of proteins. In this study, three different temperatures (40, 45, and 50 °C) were investigated (Figure 2d). From the results, we could see that only one peak (t = 74.8 min) at 40 °C was observed, but two shoulder by shoulder peaks appeared at 45 and 50 °C. More proteins separated at 45 and 50 °C. Excessive temperature may damage the column's sorbent, therefore, 45 °C was selected as the optimum temperature.

3.2. RP-HPLC Methodology Validation

The accuracy of the RP-HPLC method was investigated through consecutive tests five times, using the same sample solution (*Ranae Oviductus* sample S1) within one day. The relative standard deviations (RSD) of the retention times and peak areas of the 12 common peaks were smaller than 2.02% and 4.23%, respectively. The repeatability was determined by injecting five separate sample solutions of the *Ranae Oviductus* sample. The results showed that the RSD of the retention time and peak area of the 12 common peaks were smaller than 2.96% and 5.62%, which suggested that the RP-HPLC method had good repeatability. The stability test was carried out at room temperature for 0, 2, 4, 8, 16 and 24 h. The RSD of the retention times and peak area were smaller than 2.62% and 5.22%. All tests indicated that the RP-HPLC method established in this work satisfied the requirements of protein fingerprinting analysis of *Ranae Oviductus*.

3.3. HPLC Fingerprint of Ranae Oviductus Protein

The protein chromatographic spectra of *Ranae Oviductus* collected from 10 sampling sites in Changbai Mountain area showed a similar profile using the optimized RP-HPLC method (Figure 3a). Based on the retention time, the 12 significant common-peaks were labeled with number 1 to 12. The 12 significant common-peaks in the *Ranae Oviductus* protein spectra were labeled as mark peaks according to the Chromatographic Fingerprint Similarity Evaluation System (2012 Edition) (Beijing, China). A reference fingerprint chromatographic spectrum of 10 batches of *Ranae Oviductus* was created (Figure 3b). The similarity was in the range of 0.779–0.980 (Table 2). The RSD value of the retention time of each common-peak was smaller than 4.70% and the RSD value of the relative peak area was smaller than 5.47%. This result pointed out that the common-peaks appearing in the chromatographic spectra were reliable in the analysis of *Ranae Oviductus*.

Table 2. Similarity values of 10 batches of *Ranae Oviductus* protein and reference chromatographic fingerprint spectra.

No.	Similarity	No.	Similarity
S1	0.779	S6	0.976
S2	0.906	S7	0.884
S3	0.967	S8	0.877
S4	0.970	S9	0.980
S5	0.970	S10	0.861

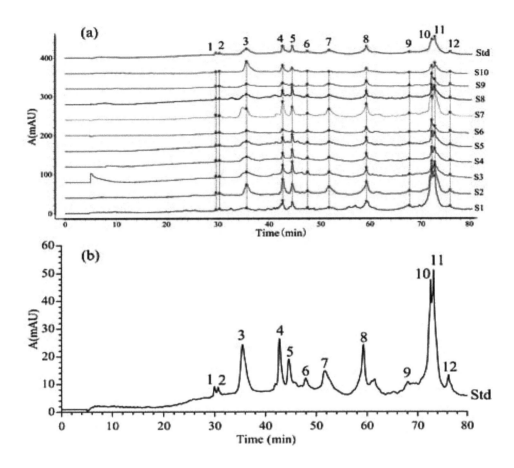

Figure 3. (a) HPLC fingerprint chromatographic spectra of 10 batches of *Ranae Oviductus* proteins.
(b) The reference protein chromatographic spectra of *Ranae Oviductus*.

3.4. Fingerprint Spectra Analysis

The fingerprint spectra analysis of *Ranae Oviductus* and counterfeit products (bullfrog oviduct, toad oviduct and frog oviduct) were performed depending on the aforementioned optimized RP-HPLC method. The results showed a significant difference. By comparing Figure 4a,b, we could see that the significant common-peaks appeared at around 30 min in the reference fingerprint of *Ranae Oviductus*. In contrary, the counterfeit products, including the bullfrog oviduct, showed four common-peaks (peak A, peak B, peak C and peak D) in 0–30 min and the toad oviduct, showed three common-peaks (peak J, peak K, peak L) in the same time period. *Ranae Oviductus* showed 12 common-peaks (peak1-peak12) in 30–80 min, whereas, the bullfrog oviduct and toad oviduct only showed five common-peaks. The frog oviduct only showed four tiny common-peaks (Figure 4c), which was a finding consistent with a previous report. Huang, et al. [36] reported that the protein types in frog oviduct were less than that of other species by using the SDS-PAGE method. Both the protein extraction method and the RP-HPLC conditions were optimized according to the *Ranae Oviductus* sample, which may have not been adequate for frog oviduct. Through the comparison, we noticed that even the three counterfeit products had a significant difference (Figure 4d). The bullfrog oviduct (nine peaks) and toad oviduct (eight peaks) had more peaks than the frog oviduct (four peaks), but the retention time was different. Therefore, although *Rana chensinensis*, bullfrog, toad, and frog are similar amphibians, they are not the same species. Their genetic differences cause the expression of different types of proteins in the fallopian tubes, so that in RP-HPLC chromatographic spectra, they showed significant differences. Those differences can be used to identify *Ranae Oviductus* and counterfeit products.

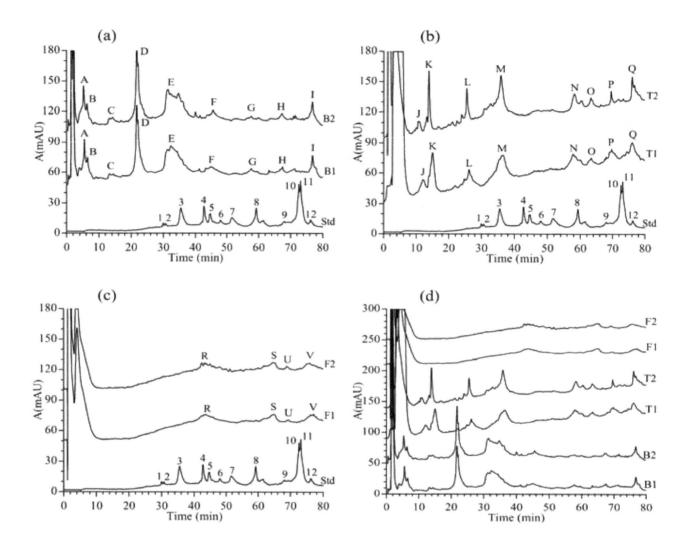

Figure 4. The comparison of *Ranae Oviductus* and counterfeit products. (**a**) Comparison of the protein HPLC fingerprint chromatogram of *Ranae Oviductus* (Std) and protein HPLC fingerprint chromatograms of the bullfrog oviduct (B1, B2). (**b**) Comparison of the protein HPLC fingerprint chromatogram of *Ranae Oviductus* (Std) and the protein HPLC fingerprint chromatograms of the toad oviduct (T1, T2). (**c**) Comparison of the protein HPLC fingerprint chromatogram of *Ranae Oviductus* (Std) and the protein HPLC fingerprint chromatograms of the frog oviduct (F1, F2). (**d**) Comparison of the protein HPLC fingerprint chromatograms of three counterfeits (bullfrog oviduct, toad oviduct, frog oviduct) of *Ranae Oviductus*.

3.5. Hierarchical Cluster Analysis (HCA)

Hierarchical cluster analysis was carried out using the relative peak areas of the characteristic peaks of *Ranae Oviductus* and counterfeit products. The 16 samples were analyzed using SPSS 25.0 software and the results are shown in Figure 5a. Obviously, there were four clusters when the interval of abscissa was 10. Cluster I, Cluster II and Cluster III were composed of the bullfrog oviduct sample, frog oviduct sample and toad oviduct sample, respectively. Cluster IV referred to the 10 samples of *Ranae Oviductus* used in the establishment of the fingerprint. The sample S1 with low similarity to *Ranae Oviductus* also showed a low correlation in Cluster IV. When the interval of abscissa was 25, the sample was divided into two clusters, one authentic and another one counterfeit.

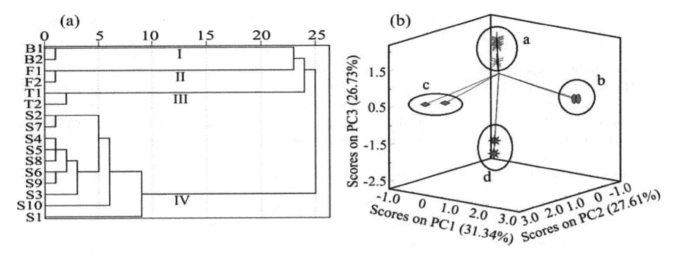

Figure 5. (a) The results of hierarchical cluster analysis of 10 batches of *Ranae Oveductus* and six counterfeit samples, (b) principal component analysis (PCA) score chart of 10 batches of *Ranae Oveductus* and six counterfeit samples in the first three principal components (PCs).

3.6. Principal Component Analysis(PCA)

As an effective data analysis technique, PCA has been used to study the classification of samples [37]. To directly reflect the difference between authentic and counterfeit products, 16 samples were used to perform the PCA analysis, based on the relative peak areas of the characteristic peaks of the samples. The variance contribution rates of the three main components (PC1, PC2, and PC3) were 31.34%, 27.61%, and 26.73%, respectively. The cumulative variance contribution rate of the three PCs was 85.68% and those variables reflected the majority of total information. To visualize the analysis results, the score charts were drawn using the three main components of PC1, PC2 and PC3 (Figure 5b). Four aggregation states are showed in Figure 5b. *Ranae Oviductus*, bullfrog oviduct, toad oviduct, and frog oviduct samples were classified in the a, b, c, and d regions, respectively. The *Ranae Oviductus* samples S1–10 could be classified in the same area (the a region), the bullfrog oviduct was classified in the b region, the toad oviduct was classified in the c region, and the frog oviduct was classified in the d region. The results were consistent with the HCA analysis, that both *Ranae Oviductus* and the counterfeit products were correctly classified. Comparing the similarity analysis with the HCA, PCA can provide a more visual comparison of the chromatograms.

4. Conclusions

This study used the RP-HPLC method and fingerprint technique to establish a chromatographic fingerprint of the proteins from *Ranae Oviductus*. Ten batches of *Ranae Oviductus* collected from the Changbai mountain area were used to analyze the protein components. The results showed 12 common-peaks in the reference fingerprint chromatographic spectrum. In combination with stoichiometry HCA and PCA, the results suggested that the method established in this work can satisfy the identification of *Ranae Oviductus* and counterfeit products. The method established in this work provides a promising approach for the identification of *Ranae Oviductus* and counterfeit products.

Author Contributions: Conceptualization, Y.W.; methodology, Y.W. and Y.G.; formal analysis, Y.G., Y.X., S.W. and H.G.; investigation, Y.G., Y.X., S.W., H.G. and M.L.; data curation, Y.G., S.W. and H.G.; writing—original draft preparation, Y.G., Y.X., S.W. and H.G.; writing—review and editing, Z.W., S.W. and Y.W.; visualization, Y.G., S.W., Z.W. and H.G.; supervision, Y.W.; project administration, S.W. and Y.W.; funding acquisition, Y.W.

References

1. Xie, C.; Zhang, L.J.; Zhang, W.Y.; Yang, X.; Fan, L.; Li, X. Immunomodulatory effect of *Oviductus Ranae* on the mice. *Chin. J. Gerontol.* **2010**, *30*, 3132–3133.

2. Li, Z.G.; Wang, C. The anti-fatigue effect of protein hydrolysate of *Oviductus Ranae* on mice and its physiological mechanism. *Chin. Sch. Phys. Educ.* **2017**, *4*, 81–87.

3.   Mo, Y.; Yu, M.J.; Mo, Y.L. Protective effect of *Oviductus Ranae* on D-galactose-induced aging mice. *Chin. J. Gerontol.* **2011**, *31*, 1603–1604.

4.   Peng, F.; Xu, F.; Liu, B.; Zhou, B.; Chen, X.; Zhao, Q. Effects of *Rana Temporaria Chensinensis David* egg oil on blood lipid in hyperlipemia rats. *Acad. J. Guangzhou Med. Coll.* **2003**, *31*, 57–59.

5.   He, Z.; Tang, X.; Liu, J.; Yuan, Y.; Jiang, C.; Zhao, Y.; Wang, Y. Application of rapid PCR to authenticate *Ranae Oviductus*. *China J. Chin. Mater. Med.* **2017**, *42*, 2467–2472.

6.   Hou, G.L.; Lu, B.Z.; Wang, C.L. Authentic identification of *Ranae Oviductus*. *Strait. Pharm. J.* **2007**, *19*, 63–64.

7.   Jin, P.; Zhang, Y.; Wang, H.; Lan, M.; Zhang, H.; Sun, J.M. Advances in identification of forest frog oil. *Jilin J. Chin. Med.* **2018**, *38*, 1179–1180.

8.   Liu, J.; Liu, S.; Liu, C. True or false identification of oviductus ranae. *Heilongjiang Med. Pharm.* **2009**, *32*, 20–21.

9.   Wang, Y.S.; Jiang, D.C.; Bai, X.X.; Wang, E.S. Identification Research *Rana temportva Chensinensis David*'s Quality with UASLG. *Lishizhen Med. Mater. Med. Res.* **2006**, *17*, 2125–2127.

10.   Zhang, W.; Wang, W.N.; Chen, F.F.; Zhang, L.; Yuan, D. Quality evaluation of *Oviductus Ranae* and similar products and fakes. *J. Shenyang Pharm. Univ.* **2012**, *29*, 951–958.

11.   Hu, X.; Liu, C.B.; Chen, X.P.; Wang, L.M. Main nourishment components of *Oviductus Ranae*. *J. Jilin Agric. Univ.* **2003**, *25*, 218–220.

12.   Hou, Z.H.; Zhao, H.; Yu, B.; Cui, B. Comprehensively analysis of components in *Oviductus Ranae*. *Sci. Technol. Food Ind.* **2017**, *38*, 348–352.

13.   Ma, L.; Yang, Y.; Chen, J.; Wang, J.; Bu, D. A rapid analytical method of major milk proteins by reversed-phase high-performance liquid chromatography. *Anim. Sci. J.* **2017**, *88*, 1623–1628. [CrossRef]

14.   Han, C.; Lu, X.; Yu, Z.; Li, X.; Ma, W.; Yan, Y. Rapid separation of seed gliadins by reversed-phase ultra performance liquid chromatography (RP-UPLC) and its application in wheat cultivar and germplasm identification. *Biosci. Biotechnol. Biochem.* **2015**, *79*, 808–815. [CrossRef]

15.   Rodríguez-Nogales, J.M.; Cifuentes, A.; García, M.C.; Marina, M.L. Characterization of Protein Fractions from Bt-Transgenic and Non-transgenic Maize Varieties Using Perfusion and Monolithic RP-HPLC. Maize Differentiation by Multivariate Analysis. *J. Agric. Food Chem.* **2007**, *55*, 3835–3842. [CrossRef] [PubMed]

16.   Xie, P.; Chen, S.; Liang, Y.-Z.; Wang, X.; Tian, R.; Upton, R. Chromatographic fingerprint analysis—a rational approach for quality assessment of traditional Chinese herbal medicine. *J. Chromatogr. A* **2006**, *1112*, 171–180. [CrossRef]

17.   Sun, J.; Chen, P. Chromatographic fingerprint analysis of yohimbe bark and related dietary supplements using UHPLC/UV/MS. *J. Pharm. Biomed. Anal.* **2012**, *61*, 142–149. [CrossRef]

18.   Lu, G.-H.; Chan, K.; Liang, Y.-Z.; Leung, K.; Chan, C.-L.; Jiang, Z.-H.; Zhao, Z.-Z. Development of high-performance liquid chromatographic fingerprints for distinguishing Chinese Angelica from related umbelliferae herbs. *J. Chromatogr. A* **2005**, *1073*, 383–392. [CrossRef]

19.   Sun, X.; Wang, H.; Han, X.; Chen, S.; Zhu, S.; Dai, J. Fingerprint analysis of polysaccharides from different Ganoderma by HPLC combined with chemometrics methods. *Carbohydr. Polym.* **2014**, *114*, 432–439. [CrossRef] [PubMed]

20.   Zhu, L.; Fang, L.; Li, Z.; Xie, X.; Zhang, L. A HPLC fingerprint study on Chaenomelis Fructus. *BMC Chem.* **2019**, *13*, 7. [CrossRef]

21.   Harrysson, H.; Hayes, M.; Eimer, F.; Carlsson, N.-G.; Toth, G.B.; Undeland, I. Production of protein extracts from Swedish red, green, and brown seaweeds, Porphyra umbilicalis Kützing, Ulva lactuca Linnaeus, and Saccharina latissima (Linnaeus) J. V. Lamouroux using three different methods. *J. Appl. Phycol.* **2018**, *30*, 3565–3580. [CrossRef]

22.   Huang, F.; Cockrell, D.C.; Stephenson, T.R.; Noyes, J.H.; Sasser, R.G. Isolation, purification, and characterization of pregnancy-specific protein B from elk and moose placenta. *Biol. Reprod.* **1999**, *61*, 1056–1061. [CrossRef] [PubMed]

23.   Takakura, D.; Hashii, N.; Kawasaki, N. An improved in-gel digestion method for efficient identification of protein and glycosylation analysis of glycoproteins using guanidine hydrochloride. *Proteomics* **2014**, *14*, 196–201. [CrossRef]

24.   Poulsen, J.W.; Madsen, C.T.; Young, C.; Poulsen, F.M.; Nielsen, M.L. Using Guanidine-Hydrochloride for Fast and Efficient Protein Digestion and Single-step Affinity-purification Mass Spectrometry. *J. Proteome Res.* **2013** *12*, 1020–1030. [CrossRef]

25. Mouecoucou, J.; Villaume, C.; Sanchez, C.; Mejean, L. Effects of gum arabic, low methoxy pectin and xylan on in vitro digestibility of peanut protein. *Food Res. Int.* **2004**, *37*, 777–783. [CrossRef]

26. Vincent, D.; Rochfort, S.; Spangenberg, G. Optimisation of Protein Extraction from Medicinal Cannabis Mature Buds for Bottom-Up Proteomics. *Molecules* **2019**, *24*, 659. [CrossRef]

27. Ali, I.; Aboul-Enein, H.Y.; Singh, P.; Singh, R.; Sharma, B. Separation of biological proteins by liquid chromatography. *Saudi Pharm. J.* **2010**, *18*, 59–73. [CrossRef]

28. Esteve, C.; Del Rio, C.; Marina, M.L.; Garcia, M.C. Development of an ultra-high performance liquid chromatography analytical methodology for the profiling of olive (*Olea europaea* L.) pulp proteins. *Anal. Chim. Acta* **2011**, *690*, 129–134. [CrossRef]

29. Gao, P.; Shi, B.; Li, Z.; Wang, P.; Yin, C.; Yin, Y.; Zan, L. Establishment and Application of Infant Formula Fingerprints by RP-HPLC. *Food Anal. Method.* **2018**, *11*, 23–33. [CrossRef]

30. Cui, L.L.; Zhang, Y.Y.; Shao, W.; Gao, D.M. Analysis of the HPLC fingerprint and QAMS from Pyrrosia species. *Ind. Crop Prod.* **2016**, *85*, 29–37. [CrossRef]

31. Kannel, P.R.; Lee, S.; Kanel, S.R.; Khan, S.P. Chemometric application in classification and assessment of monitoring locations of an urban river system. *Anal. Chim. Acta* **2007**, *582*, 390–399. [CrossRef] [PubMed]

32. Wang, C.; Zhang, C.-X.; Shao, C.-F.; Li, C.-W.; Liu, S.-H.; Peng, X.-P.; Xu, Y.-Q. Chemical Fingerprint Analysis for the Quality Evaluation of Deepure Instant Pu-erh Tea by HPLC Combined with Chemometrics. *Food Anal. Method.* **2016**, *9*, 3298–3309. [CrossRef]

33. Goodarzi, M.; Russell, P.J.; Vander Heyden, Y. Similarity analyses of chromatographic herbal fingerprints: A review. *Anal. Chim. Acta* **2013**, *804*, 16–28. [CrossRef]

34. Nelson, P.R.C.; MacGregor, J.F.; Taylor, P.A. The impact of missing measurements on PCA and PLS prediction and monitoring applications. *Chemom. Intell. Lab* **2006**, *80*, 1–12. [CrossRef]

35. Esteve, C.; Del Río, C.; Marina, M.L.; García, M.C. First Ultraperformance Liquid Chromatography Based Strategy for Profiling Intact Proteins in Complex Matrices: Application to the Evaluation of the Performance of Olive (*Olea europaea* L.) Stone Proteins for Cultivar Fingerprinting. *J. Agric. Food Chem.* **2010**, *58*, 8176–8182. [CrossRef] [PubMed]

36. Huang, Y.; Chang, L.; Zhang, S.W.; Yuan, D. Electrophoresis methods for the characterizaiton of *Ranae Oviductus* and its adulterants. *J. Shenyang Pharm. Univ.* **2017**, *34*, 1049–1054.

37. Fraige, K.; Pereira-Filho, E.R.; Carrilho, E. Fingerprinting of anthocyanins from grapes produced in Brazil using HPLC–DAD–MS and exploratory analysis by principal component analysis. *Food Chem.* **2014**, *145*, 395–403. [CrossRef] [PubMed]

9

An Update on Isolation Methods for Proteomic Studies of Extracellular Vesicles in Biofluids

Jing Li [1], Xianqing He [1], Yuanyuan Deng [2] and Chenxi Yang [2,*]

[1] School of Chemistry, Chemical Engineering and Life Science, Wuhan University of Technology, 122 luoshilu, Wuhan 430070, China; lij@whut.edu.cn (J.L.); 714769257@whut.edu.cn (X.H.)
[2] School of Biological Science & Medical Engineering, Southeast University, No.2 Sipailou, Nanjing 210096, China; 213150742@seu.edu.cn
* Correspondence: yangchenxi@seu.edu.cn

Academic Editors: Marcello Locatelli, Angela Tartaglia, Dora Melucci, Abuzar Kabir, Halil Ibrahim Ulusoy and Victoria Samanidou

Abstract: Extracellular vesicles (EVs) are lipid bilayer enclosed particles which present in almost all types of biofluids and contain specific proteins, lipids, and RNA. Increasing evidence has demonstrated the tremendous clinical potential of EVs as diagnostic and therapeutic tools, especially in biofluids, since they can be detected without invasive surgery. With the advanced mass spectrometry (MS), it is possible to decipher the protein content of EVs under different physiological and pathological conditions. Therefore, MS-based EV proteomic studies have grown rapidly in the past decade for biomarker discovery. This review focuses on the studies that isolate EVs from different biofluids and contain MS-based proteomic analysis. Literature published in the past decade (2009.1–2019.7) were selected and summarized with emphasis on isolation methods of EVs and MS analysis strategies, with the aim to give an overview of MS-based EV proteomic studies and provide a reference for future research.

Keywords: extracellular vesicles; isolation methods; biofluid; proteomics; mass spectrometry

1. Introduction

Although extracellular vesicles (EVs) were first described as 'platelet dust' in the late 1960s, it is now widely accepted that EVs are novel and important mediators for cellular communication by delivering bioactive molecules from donor to recipient cells [1,2]. Growing evidence has indicated that the cargo of EVs can reflect the content of their cells of origin and regulate physiological and pathological processes [3]. To date, EVs are considered as a novel source for biomarker discovery. With the benefits of liquid biopsy, analysis of EVs in biofluids has emerged as a promising diagnostic and monitoring tool for many diseases including cancer, neurodegenerative, kidney, and cardiovascular diseases [1,4,5].

EVs are membrane-enclosed particles that carry many bioactive molecules, including nucleic acids, proteins, and lipids, from their cells of origin. Based on their intracellular origin, EVs can be classified into three categories: exosomes, microvesicles (MVs), and apoptotic bodies. Exosomes are classically defined as the nanoparticles with sizes from 30–100 nm and formed by the fusion of multivesicular bodies with the plasma membranes; microvesicles, also called ectosomes, are usually described as the particles with sizes from 100–1000 nm and directly budded from the plasma membrane; apoptotic bodies (>1000 nm) are often considered as the particles that are released by apoptotic cells [6,7]. Despite apparent differences from their definition, it is difficult to differentiate the types of EVs after their release. It has been shown that the size of exosomes and microvesicles has a considerable overlap [7]. Currently, most of the isolation methods described in this review result in the mixed population of EVs. In addition to the physical heterogeneity, EVs are also highly heterogeneous in their cargo composition. Significant efforts have been made with the aim to comprehensively categorize EV subtypes, such as building an extensive and up–to–date database for EVs including ExoCarta, Vesiclepedia, and EVpedia [8–11]. However, consensus regarding the molecular markers to unambiguously distinguish the types of EVs remains to be a problem. Therefore, 'extracellular vesicle', which is suggested by the International Society of Extracellular Vesicles (ISEV), is used here for all the secreted vesicles [12].

Due to their tremendously diagnostic and therapeutic potential, EVs have gained increasing attention in the past decade, as shown by the number of publications (Figure 1). However, most of the studies focus on the nucleic acid content of EVs, such as microRNA or messenger RNA. With its improvements on sensitivity and high-throughput, mass spectrometry (MS) has become the fundamental technique of proteomics in recent years. Nowadays, MS has the capability to identify and characterize the protein content of EVs [6]. In the past decades, MS has been utilized to study EV proteome in various diseases, such as cancer and cardiovascular diseases [13,14]. This review will focus on publications within ten years that contain MS-based studies for EV proteins in human biofluids, such as urine, plasma, and saliva, rather than studies of EVs from laboratory animals or cell cultures and without any MS characterization. The references may be not comprehensive, but we try to highlight the recent improvements on isolation and MS strategies used in studies of EV proteome.

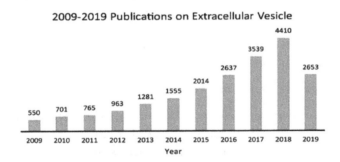

Figure 1. Publication trends on extracellular vesicle studies in the past decade (2009.1 to 2019.7). Publications were selected by searching the keyword "extracellular vesicle" in the Web of Science from the year of 2009.1 to 2019.7. x axis: year; y axis: number of publications.

2. Isolation Strategies for Extracellular Vesicles in MS-Based Proteomic Studies

EVs in biofluids are several orders of magnitude lower than other abundant components, such as lipoprotein particles, protein aggregates, and soluble proteins, including albumin in blood and Tamm-Horsfall protein (THP) in urine, which could interfere with the characterization of EVs [15,16]. Thus, the isolation step is required for all EV studies. In a typical MS-based bottom-up proteomic workflow, an additional isolation step for EVs is applied before the protein extraction and digestion (Figure 2). The commonly used isolation methods are either through the physical property of EVs, such as density and size, or based on the chemical property of EVs, such as through interacting with surface proteins of EVs, to achieve isolation [15]. Even though microfluidics-based devices hold promising potential for rapid and efficient isolation of EVs from biofluids, their low processing capacity greatly limits the downstream analysis due to the insufficient amounts of proteins [17]. Hence, this review will discuss the isolation methods, which could provide successful downstream MS-based proteomic EV studies and give an update for the ten-year improvements on isolation methods which are used in MS-based workflow studies.

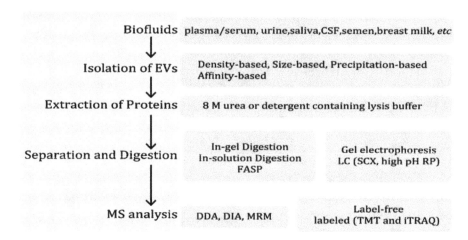

Figure 2. A general workflow of mass spectrometry (MS)-based proteomic extracellular vesicle (EV) study. EVs are firstly isolated from various biofluids, and EV proteins are extracted by adding detergent or non-detergent containing lysis buffer. The extracted EV proteins can be separated by gel electrophoresis and digested in-gel before MS analysis. Alternatively, digestion can be performed after protein extraction, and the generated peptides are either fractionated by liquid chromatography (LC) before MS analysis or directly subjected to MS analysis. The MS analysis can be conducted in data-dependent acquisition (DDA) or data-independent acquisition (DIA) for discovery EV studies or multiple reaction monitoring (MRM) for target EV studies. Differential expressed EV proteins also can be revealed by quantitative MS analysis via label-free or labeled quantitative proteomics. CSF: cerebrospinal fluid; FASP: filter aided sample preparation; SCX: strong cation exchange chromatography; RP: reverse phase chromatography; TMT: tandem mass tag; iTRAQ: isobaric tag for relative and absolute quantitation.

2.1. Sample Storage and Processing Conditions

Inappropriate storing and processing conditions can significantly affect the EV characteristics and recovery from biofluids, thus increasing pre-analytical variances or bringing artificial results. However, this aspect is not the focus of this review, and several comprehensive review or research papers have covered this topic [11,15,18–20]. Herein, some suggestions which are important and have been universally understood by the community are listed. In general, samples should be processed immediately after collection and in minimal waiting periods between each processing stages. Aliquots of samples are recommended in order to avoid multiple freezing–thawing cycles during whole processes. To obtain better EV recovery and preserve their characteristics in the biofluids, storing samples at −80 °C before EV isolation is important for long time storage [18,21–23]. However, one should be aware that there are no strict standards regarding sample storage and processing conditions for now. Most studies focus on the effects on concentration, size, RNA content, or some of the marker proteins of EVs under different conditions [18,21,24]. The comprehensive proteomic studies are still needed for evaluating the effects on protein content. In addition, each type of biofluid has special considerations which should be noticed before starting experiments.

2.2. Density-Based Isolation

Differential ultracentrifugation (dUC) as the current gold standard is the most commonly used isolation method of EVs. A recent worldwide survey of ISEV members has reported that 80% of EV isolation was conducted by dUC [25]. Biofluids typically contain a multicomponent mixture of particles that differ in sizes and densities, thus resulting in different sedimentation rates. During dUC, smaller particles can be isolated from larger ones according to their sedimentation rates by a successive increase of centrifugation forces and durations [26]. Although the details of protocols used by different groups are different to some extent, the general steps should be similar which usually include consecutively pelleting the apoptotic bodies and cell debris, the MVs, and the exosomes, as shown in Figure 3.

In most cases, samples are usually diluted by phosphate-buffered saline (PBS) before centrifugation to decrease their viscosity [27]. This dilution not only can increase the purity of EVs by decreasing the co-isolated contaminants, such as protein aggregates, but also can improve the efficiency of EV isolation since higher viscosity resulted in lower sedimentation efficiency [16,18,28]. After dilution, one or more centrifugation steps at 1000–3000× g are applied to remove dead cells and cell debris [15]. For example, a 30 min centrifugation at 2000× g can be used for viscous fluids according to one of the most cited protocols from Théry et al. [27].

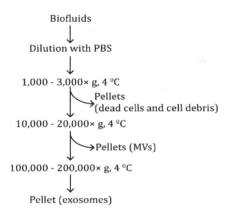

Figure 3. A basic differential ultracentrifugation (dUC) workflow for isolation of MVs and exosomes. Biofluids are diluted by phosphate-buffered saline (PBS) before centrifugation. Dead cells and cell debris are removed as pellets during the centrifugation at 1000–3000× g. Further centrifugation of supernatant at 10,000–20,000× g facilitates the isolation of MVs from exosomes. Finally, the recovery of exosomes is achieved by ultracentrifuging the 10,000–20,000× g-derived supernatant at 100,000–200,000× g.

Afterward, higher speed centrifugation, such as 10,000–20,000× g, typically follows to isolate MVs in the biofluids (Figure 3) [29,30]. The so-called ultracentrifugation at 100,000–200,000× g for hours is normally used to isolate exosomes from samples (Figure 3) [15,31]. Chutipongtanate et al. collected urinary MVs at a 20 min-centrifugation of 10,000× g before proceeding to prepare urinary exosomes at 100,000× g for 1 h [32]. Sun et al. also isolated MVs and exosomes from saliva samples by sequentially centrifuging at 10,000× or 20,000× g for 1 h and 100,000× or 125,000× g for 2.5 h, with 785 proteins identified from MVs and 910 proteins from exosomes [33]. Table 1 lists the details of centrifugation force and time from the selected EV studies for future reference. Their corresponding MS strategies and results are also included in Table 1. Rather than using common gel-based bottom-up proteomics, different methodologies on MS-based workflow were also developed and applied to EV studies as summarized in Table 1, such as different liquid chromatography (LC) fractionation methods, digestion strategies, and MS acquisition approaches, which will be discussed in Section 4. Many exosomes studies discarded the pellets resulted from 10,000–20,000× g before ultracentrifugation at 100,000–200,000× g (Table 1). However, Whitham et al. recently isolated EVs at 20,000× g for 1 h to study the exercise-induced EV proteome and found that a host of small-vesicle and exosomal markers, such as SDCBP, TSG101, PDCD6IP (ALIX), CD63, and CD9, identified in 20,000× g-derived EV lysates. Further quantitative studies revealed that no significant differences were observed in any EV markers between samples subjected to 20,000× or 100,000× g centrifugation. They claimed that a quantitative proteomic analysis of small-vesicle and exosomal protein cargo was possible with the 20,000× g centrifugation for 1 h rather than prolonged centrifugation at 100,000× g [34]. Besides, Kim et al. claimed that centrifugation at 40,000× g could provide comparable or improved results relative to ultracentrifugation at 110,000× g [35]. Those studies may imply that the purity of exosome samples yielded by dUC are obtained with the cost of exosome loss during centrifugation at 10,000–20,000× g.

The pellets of interest are usually washed once at the final steps by resuspension and centrifugationagain. It has been demonstrated that less washing can result in a higher EV yield, but also have more contaminants [36]. Therefore, the balance between yield and purity should be judged when adopting protocols. It is also worth noting that the efficiency of isolation is not only dependent on the viscosity of the samples, centrifugation force, and time, but also on rotor type since sedimentation path lengths are dependent on the type of rotors used and different distances from the rotational axis could result in differences in the g-force. Cvjetkovic et al. applied a 70 min centrifugation at $100,000\times g$ for exosome isolation on three different rotors and found that the yield and purity of exosomes obtained were significantly different [37]. To address this issue, a web-calculator was developed by Livshits et al. to adjust the common dUC protocol to the "individual" dUC protocol [26]. Therefore, one should be aware that proper modifications are necessary when adopting dUC for different types of biofluids and laboratory settings in order to achieve optimal isolation.

Table 1. Selected MS analysis for EVs obtained from centrifugation-based isolation.

Isolation	Proteomic Sample Preparation	Mass Spectrometry	Sample Origin	Number of Proteins	Year	Study
$19,000\times g$ for 120 min	2D-LC/MS: SCX as 1st dimensional	LTQ ion trap	plasma	1806 proteins	2017	[30]
Sucrose cushion at $100,000\times g$ for 90 min	2D-LC/MS: C18-SCX stage-tip as 1st dimensional	Q-Exactive	serum	702 proteins	2017	[38]
$100,000\times g$ for 90 min incubation with DTT	iTRAQ 2D-LC/MS	LTQ-Orbitrap Velos Elite	urine	4710 proteins in total and 3528 proteins for quantification	2017	[39]
Sucrose cushion at $100,000\times g$ for 90 min	iTRAQ 2D-LC/MS: high pH as 1st dimensional	Orbitrap Fusion Lumos	semen	3699 proteins in total	2018	[40]
$110,000\times g$ for 70 min	FASP	Q Exactive	serum	655 proteins	2018	[41]
$10,000\times g$, 20 min for MVs and at $100,000\times g$, 1 h for exosomes	in-solution digestion	SWATH-MS TripleTof 5600+	urine	Targeted data analysis for 888 proteins	2018	[32]
Density ultracentrifugation at $270,000\times g$, 1 h and incubation with DTT	in-solution digestion	MS^E	urine	1877 proteins	2011	[42]
$100,000\times g$ for 180 min	in-solution digestion	L Q-Exactive Orbitrap	umbilical cord blood	211 proteins	2015	[43]
$200,000\times g$, 1 h and incubation with DTT	in-gel digestion	LTQ Orbitrap XL and LTQ Orbitrap Velos	urine	1989 proteins in total	2012	[44]
$100,000\times g$ for 90 min	in-solution digestion	LTQ Orbitrap Velos	saliva	381 proteins	2015	[45]
$200,000\times g$ for 90 min and incubation with KBr	iTRAQ LC off-line separation	MALDI * tandem mass spectrometry	plasma	not report	2010	[46]
Sucrose cushion at $192,000\times g$ for 15–18 h	in-gel digestion	Q-Exactive	breast milk	1963 proteins	2016	[47]
$20,000\times g$ for 1 h for MVs	in-solution digestion	Q-Exactive/Plus	plasma	3294 proteins in 4 h LC/MS	2015	[29]
10,000 or $20,000\times g$, 1 h for MVs; 100,000 or $125,000\times g$, 2.5 h for exosomes	SDS-PAGE FASP	Q-Exactive	saliva	785 proteins for MVs; 910 proteins for exosomes	2018	[33]
$20,000\times g$, 1 h for MVs; $100,000\times g$, 1 h for exosomes	in-solution digestion	LTQ-Orbitrap Velos Pro	plasma	9225 phosphopeptides in MVs; 1014 phosphopeptides in exosomes	2017	[48]
$100,000\times g$ for 70 min	in-gel digestion	LTQ-XL	CSF	91 proteins identified from control466 proteins identified from disease	2018	[49]

* MALDI: Matrix-assisted laser desorption/ionization.

dUC has been utilized to isolate MVs and exosomes from different types of biofluids, such as plasma, urine, saliva, breast milk, and semen, as listed in Table 1. But the EV pellets obtained from dUC are usually contaminated with some co-sediment high abundant components in the biofluids including lipoprotein particles, protein aggregates, and high abundant soluble proteins, which significantly affect the downstream MS analysis. To improve the purity of isolated EVs, density gradient (DG) flotation, such as the sucrose gradient or OptiPrep velocity gradient (iodixanol gradient), is developed and incorporated into the dUC protocol [15,50]. Although the density of MVs remains unclear, the density of exosomes is 1.13–1.19 g/mL [14]. Upon centrifugation, EVs migrate to the surrounding medium if their densities are same, resulting in further purification of the EVs from other contaminants. For example, the purified exosome pellets from dUC are resuspended into PBS and overlaid on a 30% sucrose cushion with centrifugation at 100,000× g [27]. The EV samples can be further fractionated by a step DG using a series of solutions with different density. Iwai et al. used a series of sucrose solutions with concentrations at 2.0, 1.6, 1.18, and 0.8 M and iodixanol solutions with concentration at 50%, 40%, 30%, and 20% to separately isolate exosomes from saliva and collect fractions from different densities [51]. A recent proteomic comparative study was performed to evaluate the dUC and DG and found that DG reduced the presence of co-isolated proteins aggregates and other membranous particles [52]. In comparison to the sucrose gradient, the OptiPrep velocity gradient is reported to perform better at removing some lipoproteins and preserving the size of the vesicles in the gradient [15]. One of the reasons is that the osmotic pressure of sucrose is higher than iodixanol, which could damage EVs in the samples [51].

Some additional strategies are also included in the dUC workflow to increase the purity of EVs for different types of biofluids. THP (also called uromodulin) is a highly abundant protein in urine and can form a polymeric network to trap exosomes during centrifugation at 10,000–20,000× g. To alleviate this effect and increase the yield of exosomes, incubation of the crude exosome pellets with dithiothreitol (DTT) or 3-[(3-cholamidopropyl)dimethylammonio]-1-propanesulfonic (CHAPS) were developed. DTT could denature THP, thus inhibiting aggregation and allowing THP to be removed from the supernatant. Moon et al. resuspended the 200,000× g-derived urinary pellets in the sucrose solution and incubated with 60 mg/mL DTT at 60 °C for 10 min before DG. A total of 1877 urinary exosome proteins were identified in MS^E analyses [42]. But one of the side effects caused by DTT is that exosomal protein remodeling as DTT is a strong reducing agent and may reduce the exosomal proteins, thus resulting in detrimental effects on their biological activity. Musante et al. used CHAPS which is a mild detergent and known to solubilize THP to replace DTT. They found that CHAPS did not affect vesicle morphology or exosomal marker distribution and preserved better biological activity. Further MS analysis revealed that 76.2% of proteins recovered by CHAPS were identified in those treated by DTT [53]. In addition, Barrachina et al. used KBr in a similar mechanism for plasma samples to reduce lipoproteins in EV samples by solubilizing them [54]. Alternative strategies to improve dUC can be achieved by combinational usage with other types of isolation methods, such as filter device or size exclusion chromatography (SEC). Those combinational methods not only can improve the purity of EVs, but also can dramatically reduce the overall processing time. Details will be presented in the following subsections.

2.3. Size-Based Isolation

Size-based isolation, such as filtration and size exclusion chromatography (SEC), is another type of isolation method, which can be used alone or with other methods to isolate EVs from biofluids. For filtration, samples are passed through a membrane with a specific pore size by centrifugation or pressure. Centrifugation-based filter devices have been reported to yield approximately three-fold greater EVs than that prepared by pressure-driven filter devices [55]. Filters made by different materials have been demonstrated as a fast and simple alternative to dUC. Merchant et al. applied a pore size 0.1 μm of commercially available VVLP (hydrophilized polyvinylidene difluoride) disc membranes to isolate urinary exosomes before MALDI (Matrix-assisted laser desorption/ionization) TOF analysis, and filtration of 50 mL urine samples was achieved within 15 min [56]. Musante et al. developed a "hydrostatic filtration dialysis" process to isolate urinary EVs. Urine samples were centrifuged at 2000× g before loaded onto a dialysis membrane with a molecular weight cut-off (MWCO) of 1000 kDa. They found that centrifugation at 2000× g allowed to remove the bulk of THP without losing exosomes. By using the dialysis membrane with MWCO of 1000 kDa, solvent, together with all the analytes below 1000 kDa were pushed through the mesh of the membrane due to the hydrostatic pressure of the urine. This method avoided the laborious and time-consuming steps of dUC, while the yield of EVs from this dialysis membrane was reported to outperform the dUC [57,58]. Sequential usage of different types of filters was also explored to isolate EVs. A three-step protocol was established based on sequential steps of dead-end pre-filtration, tangential flow filtration, and low-pressure track-etched membrane filtration. But this sequential filtration step was tested for cell culture, not for biofluids [59]. Instead of used alone, filtration is more commonly used with other types of methods for EV isolation, such as with dUC as a concentration/enrichment step with the aim to concentrate the samples and reduce the processing duration. For example, a 0.22 μm filter device is the most used filter device in EV studies to remove components with a diameter exceeding ca. 200 nm and as one of the processing steps in the dUC [16,60]. In the protocol of Théry et al., the pellets yielded by 2 h of centrifugation at 110,000× g were resuspended in PBS and passed through a 0.22 μm filter before another round of centrifugation at 110,000× g [27]. Shiromizu et al. further simplified the steps by initially using a 300× g centrifugation followed by a filtration step with a 0.22 μm filter to obtain the exosomes crude before a 30% sucrose DG in colorectal cancer biomarker studies [38]. The hydrostatic filtration dialysis can also be used as a pre-enrichment step for dUC to isolate urinary EVs [61].

Despite that the filtration is fast and has the capability of high throughput for EV isolation, the filters can be easily blocked resulted from trapping vesicles or other contaminant aggregates. SEC as another type of size-based isolation strategy has not been normally reported with this limitation posed by filtration [16]. For SEC, samples are loaded onto a column packed with heterogeneous polymeric beads, such as Sepharose, with diverse pore size. In general, the larger molecules are eluted earlier than the smaller ones since the smaller molecules can enter more pores than the larger ones, thus eluted later. Menezes-Neto et al. used SEC as a stand-alone methodology for isolation of EVs. They packed Sepharose CL-2B into a syringe and isolated exosomes from a 1 ml plasma after centrifugation at 500× g for 10 min. A total of 269 proteins were identified from the plasma of one healthy donor on an LTQ Orbitrap Velos mass spectrometer [62]. However, Karimi et al. also packed Sepharose CL-2B beads into a Telos solid phase extraction column and found that this SEC column failed to separate EVs from lipoproteins. Instead of using SEC alone, they overlaid a 6 mL plasma on top of an OptiPrep cushion and centrifuged at 178,000× g before SEC separation. The combinational usage of the density cushion and SEC reduced about 100-fold lipoprotein particles in the EV samples with 1187 proteins identified. [63]. SEC was also reported as an alternative step to replace the final step of dUC. Smolarz et al. used the SEC to isolate exosomes instead of ultracentrifugation at 100,000–200,000× g. Briefly, serum was centrifuged at 1000× g and 10,000× g for 10 and 30 min, respectively. The generated supernatant was filtrated using a 0.22 μm syringe filter unit before loading onto the micro-SEC column to isolate exosomes. A total of 267 proteins were identified by the downstream LC/MS analysis [64]. A commercial size-exclusion chromatography column, qEV, was also used to extract EVs from saliva and tears to study primary Sjögren's syndrome [65]. One of the problems faced by SEC is the increased sample volume obtained after elution, resulting in an extra concentration step for the downstream EV analysis. Foers et al. compared ultracentrifugation and ultrafiltration for the concentration of the SEC eluent. They loaded 10,000× g supernatant of human synovial fluid into a HiPrep 26/60 Sephacryl S-500 HR prepacked gel filtration column. This column contains a hydrophilic, rigid allyl dextran/bisacrylamide matrix and allows for large sample volume input and small EV infiltration. SEC fractions were concentrated by either ultracentrifugation at 100,000× g for 90 min or passing an Amicon Ultra-15 100 kDa cellulose ultrafiltration device. They found ultrafiltration could avoid artifactual aggregation of EVs with contaminants, such as extracellular debris, which were typically observed in samples prepared by ultracentrifugation [66].

2.4. Precipitation-Based Isolation

Polymer precipitation-based isolation has the benefits of commercial availability and easy processing and is now widely applied to isolate EVs from the biofluids under many disease statuses, such as colorectal cancer, epithelial ovarian cancer, and rheumatoid arthritis [67–69]. This type of isolation method is initially used in viral studies by forming a polymer network to decrease the solubility of all components present in the sample [70]. The whole procedure includes mixing an appropriate volume of a polymer solution with samples and incubation. Then, the precipitated EVs are recovered by low-speed centrifugation. The polymer solution could be from a commercial kit, such as ExoQuick, Total Exosome Isolation, and ExoSpin, or home-made polyethylene glycol (PEG) solution [14]. Comparative studies have been conducted to evaluate the EVs isolated by different commercial kits in order to facilitate the choice of isolation methods. Ding et al. compared three commonly used commercial kits for EV isolation, including Total Exosome Isolation, ExoQuick, and RIBO Exosome Isolation Reagent. They found that the size of the majority of particles isolated by those kits was from 30–150 nm, while RIBO generated the highest particle yields. Further western blot (WB) results revealed that ExoQuick was the most efficient method by evaluating the marker proteins of CD63 and TSG101 [71]. Lobb et al. found that ExoSpin performed significantly better in avoiding co-isolation of contaminating proteins and yielded higher levels of EV markers compared to ExoQuick [55].

Although easy–to–use EV commercial kits are now widely used, home-made PEG has relative low-cost of EV preparation. Weng et al. added PEG into samples with a final concentration of 10% and incubated the samples at 4 °C for 2 h before recovery at centrifugation of 3000× g. Then a second-round of PEG precipitation was followed in order to improve the purity of EVs. The downstream MS analysis identified a total of 6299 protein groups from HeLa cell culture supernatant. Unfortunately, they did not test any biofluid sample in the study [72]. PEG has also been demonstrated to be used together with ultracentrifugation. Rider et al. purified the EVs resulted from one-round of PEG precipitation by further centrifugation at 100,000× g for 70 min [73]. Instead of isolating EVs by precipitation, aqueous two-phase systems (ATPSs) were proposed by Shin et al. They used a PEG/dextran ATPS to isolate EVs from the tumor interstitial fluid based on the mechanism that different kinds of particles are effectively partitioned to different phases in a short time. Their comparative studies showed that ATPSs could recovery about 70% of EVs from the EV protein mixtures, whereas the recovery for dUC and ExoQuick were about 16% and 40% [74]. But one should notice that EVs isolated by precipitation may be contaminated by polymer molecules, such as PEG, which is well-known for interfering in MS-based proteomic analysis. Therefore, it is necessary to remove those polymer molecules before MS analysis.

2.5. Affinity-Based Isolation

Apart from size and density, EVs share some common characteristics, like general protein composition and lipid bilayer structure. By utilizing those common characteristics, affinity-based isolation could achieve the isolation of EVs from complex biological samples. The main principle of affinity-based isolation is via the interaction between the surface markers of EVs with the antibody, molecules, or function group immobilized onto various carriers to separate EVs from the analyzed biofluids. Among those methods, immuno-based isolation is the most widely available and used method [15,75]. Some proteins have often been used as exosome-associated markers including the tetraspanin family (such as CD8, CD9, CD61, CD63, CD81, and CD82), cytoplasmic proteins (such as tubulin, actin, actin-binding proteins, annexins, and Rab proteins), and heat shock proteins (such as Hsp70, and Hsp90). Therefore, the antibodies against those common proteins coupled to different carriers have been utilized to isolate EVs [76–78]. Hildonen at el. isolated urinary exosomes from healthy subjects by immunocapture on magnetic beads. They coupled the antibody cocktail against CD8, CD61, and CD81 to magnetic beads. By digestions on beads in non-detergent containing buffer, they studied the outer membrane-associated proteins of exosomes and found 49 proteins associated or bound to membranes [76]. Antibody against tetraspanins was also shown to immobilize on highly porous monolithic silica microtips and applied to investigate lung cancer biomarker proteins on exosomes in serum samples. The subsequent MS analysis had identified 1369 proteins [77]. In addition to those common markers of EVs, immuno-based isolation was also explored to isolate the desired groups of EVs because the function of EVs appears to be determined by its specific protein content. For example, anti-EpCAM-coupled microbeads were employed to extract epithelial tumor-derived

EVs from plasma since it has been demonstrated that exosomes from epithelial tumors express EpCAM (epithelial cell adhesion molecule) on their surface [78,79]. Tauro et al. isolated two distinct populations of exosomes released from organoids derived from the human colon carcinoma cell line LIM1863EVs, via sequential immunocapture using anti-A33- and anti-EpCAM-coupled magnetic beads [80].

In addition to antibodies, some EV-binding molecules, such as specific peptides including venceremin or Vn, and heparin, were also investigated to isolate EVs [14]. Vn, a novel class of peptides, which exhibit the specific affinity for heat shock proteins were selected for isolation of EVs from breast cancer [81]. Bijnsdorp et al. compared the urinary EVs isolated by Vn-96 and dUC and found that more than 85% of the proteins were identified both in EVs isolated by Vn and dUC. But the Vn96-peptide offered easier and time convenient methods in comparison with dUC [82]. Heparin is a highly sulfated glycosaminoglycan and has recently been used to isolate the EVs in which the surface contains the cell surface receptor, heparan sulfate proteoglycans. Balaj et al. incubated plasma with heparin-coated beads overnight and further processed the enriched samples by ultracentrifuging at 100,000× g for 90 min or a 100 kDa MWCO filter. The EVs isolated by heparin-affinity beads were detected to contain the EV marker of Alix and lower level of protein contamination [83].

Affinity for targeted proteins on the surface of EVs can be problematic for general EV studies since an unreliable analysis could be obtained due to the exclusion of EVs without targeted proteins.

Therefore, an affinity for the lipid membrane structures of EVs is utilized. Gao et al. recently adopted the TiO_2 material, which is commonly used for the enrichment of phosphopeptides to isolate EVs. Through the interaction with the phosphate groups on the lipid bilayer of EVs, TiO_2 can enrich EVs from serum within 5 min [84]. Tan et al. also focused on the membrane lipid as the target and used phospholipid-binding ligands to extract plasma EVs. Based on previous studies, EVs could be differentiated by their membrane phospholipid composition, specifically GM1 gangliosides and phosphatidylserines. They found two distinct groups of EVs by using cholera toxin B chain (CTB) and annexin V (AV), which, respectively, binds GM1 ganglioside and phosphatidylserine [85]. Nakai et al. developed a novel method for EV purification by using Tim4 proteins. Tim4 proteins can capture EVs via the specific interaction with the phosphatidylserine displayed on the surface of EVs and release the EVs by adding Ca^{2+} chelators. They claimed that the lower contaminations were found in the EV samples isolated by Tim4 proteins [86].

3. Comparative Studies for Isolation Methods of EVs

Among the isolation methods discussed above, it is generally thought that dUC is time-consuming. Filtration has the risk of stuck EVs in the membrane pores, while SEC is not ideal for large scale isolation. Although precipitation-based and immuno-based methods usually involve easy processing, the purity of EVs from precipitation is often problematic and affinity-based isolation is often considered as a good technique for isolation of sub-populations of EVs [16]. However, it is more reasonable to evaluate each isolation method based on the detailed protocol used and criteria of evaluation in each study. Otherwise, purity, efficiency, and reproducibility of different isolations could easily confound literature. For example, Kalra et al. performed a comparative evaluation of three exosome isolation techniques: dUC, anti-EpCAM conjugated microbeads, and OptiPrep DG. Their results suggested that the OptiPrep DG was superior in isolating pure exosomal populations by comparing the level of highly abundant plasma proteins which were detected by MS in the isolated plasma EV samples [79]. Those three methods were also compared by Greening et al. in a cell model. Based on the quantitative MS results for the identified exosome markers and proteins associated with EV biogenesis, trafficking, and release, anti-EpCAM was shown to be the most effective method to isolate exosomes [50]. Results from those two comparative studies can be explained by the differences in the sample types, details of protocols, and criteria of evaluation used in each study. Therefore, the selected studies for evaluation of different EV isolation methods are listed in Table 2 for better interpretation of each isolation. One thing to be mentioned is that the comparative studies listed in Table 2 also include the studies based on cell cultures, animals, and characterization of EVs by other methods, and are not just based on biofluid samples and analyses of MS.

Table 2. Selected comparative studies for EV isolation.

Isolation Methods	Characterization Techniques	Samples	Study
dUC, SEC	NTA, Dissociation-enhanced lanthanide fluorescence immunoassay, WB, TEM	rat plasma, cell culture	[87]
dUC, SEC	TEM, AFM, WB, MS	cell culture	[88]
Affinity-based (exoEasy kit) and SEC (qEV)	WB, TEM, NTA, lipid quantification kit, RNA quality	plasma	[89]
dUC and Commercial kit from Invitrogen, 101Bio, Wako and iZON	Dynamic Light Scattering, immunoblot analysis, qRT-PCR, MS, Cell Proliferation Assay	cell culture	[90]
dUC, precipitation (ExoQuick, Total Exosome Isolation Reafent, Exo-PREP) and SEC (qEV)	TEM, NTA, WB	cell culture	[91]
Lectin-based, Exoquick, Total exosome Isolation and in-house modified procedure	WB, Reverse transcriptase and qPCR, EM	urine	[92]
dUC, precipitation (ExoQuick, Total exosome isolation, PEG, Exo-spin), filtration (ExoMir)	NTA, Flow cytometry, WB, PCR,	serum	[93]
dUC, filtration (Stirred cell and Centricon), OptiPrep DG, ExoQuick, Exo-spin, SEC	Tunable resistive pulse sensing, EM, WB	cell culture and plasma	[55]
SEC and Exo-Spin	NTA, Flow cytometry, MS	plasma	[62]
dUC, anti-EpCAM, OptiPrep DG	MS, WB, TEM	plasma	[79]
Nanomembrane ultrafiltration, dUC and dUC-SEC	MS, TEM, WB	urine	[94]
dUC, anti-EpCAM, OptiPrep DG	TEM, CryoEM, MS	cell culture	[50]
Sucrose DG and ExoQuick	TEM, NTA, WB	serum	[95]

* EM: electron microscopy; TEM: transmission electron microscopy; NTA: nanoparticle tracking analysis AFM: atomic force microscopy; WB: western blot.

As shown in Table 2, many studies have compared the EV isolation by different techniques; thus, according to different criteria. Different criteria were also applied, even if the same technique was used for assessment [55,88,92,94]. WB for EV marker proteins is one of the commonly used methods to compare the efficiency of EV isolation. But how many and which marker proteins should be chosen for the good comparison has not been well established. Lobb et al. provided a comparative analysis of four EV isolation techniques. dUC, ultrafiltration, SEC, OptiPrep DG, and precipitation (ExoQuick and ExoSpin) were used to isolate EVs from cell culture and plasma. By comparing the levels of exosomal markers of HSP70, Flotillin-1, and TSG 101 in WB, precipitation protocols provided the least pure preparations of EVs, whereas SEC isolation was comparable to DG purification of EVs [55]. In a similar way, Royo et al. tested the EV isolation of lectin-based purification, Exoquick, Total Exosome Isolation, and an in-house modified EV isolation procedure via WB of eight EV protein markers including CD9, CD10, CD63, TSG101, CD10, AIP1/Alix, AQP2, and FLT1. They observed that the levels of different EV marker proteins varied by different isolations and, thus, suggested that different methods isolated a different mixture of urinary EV marker proteins [92]. Evaluation of EV isolation by MS also lacks criteria to make a universal, comprehensive comparison. Rood et al. centrifugated the urine samples at 17,000× g for 15 min and then isolated the EVs by further centrifuging at 200,000× g for 110 min or filtering with 100 kDa Vivaspin 20 polyethersulfone nanomembrane concentrators. They found that either ultracentrifugation or ultrafiltration was difficult to isolate EVs from urine since highly abundant proteins, especially albumin and α-1-antitrypsin, were present in large amounts, which significantly limited the detection of MALDI-TOF. Additional SEC following ultracentrifugation was suggested to use in order to improve the purity of EVs [94]. Based on the gene ontology analysis for the identified proteins by MS, Davis et al. believed that dUC and SEC did not isolate equivalent EV population profiles [88]. Altogether, cautions should be taken when interpreting each EV isolation.

Rather than focus on the performance in yield or purity of each isolation, the functional activity of EVs was also reported to depend on the isolation method used [87,91]. Antounians et al. noticed that amniotic fluid stem cell-derived EVs isolated by dUC, precipitation (ExoQuick, Total Exosome Isolation Reagent, and Exo-PREP), and SEC (qEV column) had different effects on a model of damaged lung epithelium [91]. It suggests the necessity of evaluating the isolation methods within the content of biology.

4. MS Strategies Used in Proteomic Studies of Extracellular Vesicles

4.1. Sample Preparation and Separation

To date, proteomic studies of EVs are mainly conducted based on the bottom-up MS strategy. As shown in Figure 2, protein should be extracted from the isolated EVs and digested before MS analysis. For proteomic analysis, EV proteins are commonly extracted using the lysis buffer with detergent (such as sodium dodecyl sulfate (SDS)) or without detergent (such as 8 M urea). TRIzol reagent, which is often used in isolation of nucleic acid from EVs, has been recently reported to extract proteins from EVs. Joy et al. compared the EV protein extraction between Laemmli and TRIzol. Laemmli buffer typically contains 2% SDS, 10% glycerol in Tris-HCl with pH 6.8, which is an effective protein-extraction for EVs. They found that these two methods gave similar results in their ability to extract proteins and ~60% of proteins were identified in the samples prepared by both methods. However, they did not apply TRIzol reagent on any EV samples from biofluids [96]. Special extraction methods are also investigated to facilitate studies of sub-populations of proteins in the EVs, such as membrane proteins. Hu et al. optimized the Triton X-114 detergent partitioning protocol to target the analysis of membrane proteins of urinary EVs. Dried EV pellets were dissolved in 1% SDS containing lysis buffer for 1 h before adding 2.2% pre-condensed Triton X-114 buffer. A lower detergent phase, with an oily appearance, and an upper aqueous phase were formed when the temperature was above the clouding point of Triton X-114. Proteins in each phase were recovered by acetone precipitation before MS analysis. Most of the membrane proteins of urinary EVs were found in the detergent fraction [58].

As shown in Table 1, filter aided sample preparation (FASP) was utilized in some EV studies to achieve an easy process for buffer exchange and protein digestion [97]. In FASP, the extracted EV proteins are transferred into a molecular weight cut-off filter. This filter can retain most of the proteins on the membrane after simple centrifugation. Meanwhile, peptides can freely pass through the membrane during centrifugation. By using this kind of filter, the denaturing detergent-based buffer used for protein extraction can be easily changed to a digestion buffer, and the sample can be digested on the filter without extra transferring steps. FASP, with easy processing and minimal sample loss, has become the method of choice in many EV studies, especially in the limited amount of samples available [16]. Fel et al. improved the FASP by using multi-enzyme digestion to prepare EV samples obtained by precipitation. In their studies, serum samples from polycythemia vera patients were centrifuged at 2000× *g* for 30 min to remove cells and debris before incubation with the reagent from the Total Exosome Isolation kit. Afterward, the proteins were extracted from EVs and digested sequentially by Lys C, trypsin, and chymotrypsin in a Micron 30 kDa filter (Figure 4). A total of 706 proteins were identified with thirty-eight proteins showing significant differences in the patients' group [97].

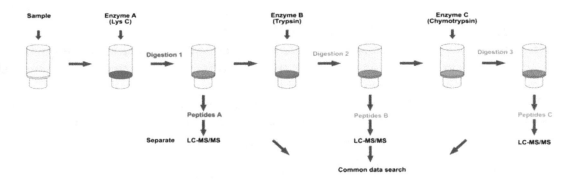

Figure 4. The schematic workflow for multi-enzyme digestion filter-aided sample preparation. This figure was adopted from Ref. [97].

To perform in-depth proteomic analysis, additional separation before LC/MS analysis can be performed by either gel electrophoresis or liquid chromatography. Gel electrophoresis can effectively remove the most common contaminants in the samples according to the molecular weight of proteins, which could benefit the downstream MS analysis. Both Tsuno et al. and Xie et al. isolated EVs from serum using ExoQuick and separated the protein content through two-dimensional gel electrophoresis before MALDI-TOF analysis to study rheumatoid arthritis and coronary artery aneurysms, respectively [69,98]. Gel electrophoresis has also been applied to study EVs from urine, breast milk, and saliva [45,47,99]. Apart from separation based on gel electrophoresis, two-dimensional liquid chromatography (2D-LC) is utilized to analyze EV samples [30,38–40,100]. Antwi-Baffour et al. isolated MVs from the plasma of malaria patients and used a microcapillary strong cation exchange (SCX) column to fractionate the digested MVs samples. A total of 1729 proteins were identified in malaria samples, while only 234 proteins were identified in healthy control samples [30]. Their finding may imply that MVs in disease status could result in more protein identification than in healthy. Shiromizu et al. further simplified the fractionation of EV samples by using a C18-SCX Stage-tip. Using this strategy, they identified 702 proteins from the serum of colorectal cancer patients [38]. Instead of SCX as the first-dimensional separation, Lin et al. performed a high pH reverse phase chromatography to fractionate EVs from semen and study asthenozoospermia with 3699 protein identified by MS [40].

In addition to the typical proteomic studies, separation methods vary according to different studies, such as the studying of post-translational modifications of EV proteins. The electrostatic repulsion-hydrophilic interaction chromatography (ERLIC) was employed to facilitate the study of glycoproteins from EVs. Cheow et al. centrifuged plasma at 100,000× g for 2 h and 200,000× g for 18 h. They recovered a visible yellow suspension that was highly enriched in soluble glycoproteins and EVs. After protein extraction and digestion, an ERLIC column was used to simultaneously enrich secretory and EV-enriched glycoproteins and further fractionate the sample. A total of 127 plasma glycoproteins were identified with high confidence [101]. In order to study N-linked glycoproteomics of urinary exosomes, Saraswat et al. isolated urinary EVs by centrifugation at 200,000× g for 2 h and applied SNA affinity chromatography or SEC to enrich glycopeptides in the urinary EVs after tryptic digestion. In total, 126 N-glycopeptides from 51 N-glycosylation sites belonging to 37 glycoproteins were found [102].

4.2. MS Acquisition

During MS analysis, data-dependent acquisition (DDA) are normally used. Recently, data-independent acquisitions (DIA), such as SWATH (sequential window acquisition of all theoretical fragment ion), MS^E, and multiplexed MS/MS, are used in EV studies to satisfy different purposes. Unlike DDA, DIA simultaneously fragments all precursor ions present in a wide isolation window. Braga-Lagache et al. analyzed MV proteins from plasma samples by both

DDA and multiplexed DIA on a quadrupole orbitrap instrument. In each cycle of multiplexed DIA, data is usually acquired with one full MS scan followed by a series of MS2, such as ten MS2 scans. Each MS2 scan records all the fragment ions generated by precursor ions that are isolated from multiple different isolation windows with a fixed m/z range, such as isolated from three randomly combined 10 m/z isolation windows. A targeted approach is used to analyze the DIA data by using spectral libraries from formerly acquired fragment spectra with exact mass and retention time of precursors. They found that a multiplexed DIA approach only consumed one third of the DDA acquisition time when data was extracted by a targeted approach. Their results suggested that multiplexed DIA was a valuable alternative to DDA [103]. Moon et al. and Chutipongtanate et al. also applied DIA to analyze the protein content of EVs [32,42]. In the study of Moon et al., crude exosomes prepared by sucrose density ultracentrifugation were digested in-gel and analyzed by MS^E on a Waters Q-TOF mass spectrometer. In MS^E, alternating low- and high-energy collision-induced dissociation are used. The low-energy scan is used to obtain precursor information, while the high-energy scan is to collect fragment ions. A total of 1877 urinary exosome proteins were identified from IgA nephropathy and thin basement membrane nephropathy patients [42]. Chutipongtanate et al. utilized SWATH to analyze urinary EV proteins. In SWATH, the mass range of interest is divided into several segments with a fixed m/z range, such as 25 m/z. Then, precursor ions within each segment are fragmented together until all the segments are analyzed. They achieve a label-free DIA quantitative analysis for EV and MV proteins with a curated spectral library of 1145 targets, suggesting their potential clinical use [32].

Quantitative MS based on label and label-free have been demonstrated to study various diseases, such as prostate cancer, asthenozoospermia and venous thrombosis [39,40,46,104]. Fujita et al. labeled the urinary EV proteins with isobaric tag for relative and absolute quantitation (iTRAQ). A total of 4710 proteins were identified by MS, including 3528 proteins quantified [39]. Lin et al. quantified seminal EV proteins with iTRAQ labeling and revealed 91 proteins with significant changes [40]. 2D-LC and tandem mass tag (TMT) were also used to quantitative analysis of EVs in HIV-infected alcohol drinkers and cigarette smokers through precipitation-based isolation [104]. Although stable isotope labeling by amino acids in cell culture (SILAC) cannot label EV proteins from human biofluids, a PROMIS-Quan method which based on SILAC quantification was developed in order to gain a comprehensive quantification for potential clinical EV protein analysis. In PROMIS-Quan, EV lysates were spiked with super-SILAC which was prepared from cell cultures and served as an internal standard. Then, the same set of super-SILAC mix was quantified relative to purified proteins of interest, with known absolute amounts. By this way, EV proteins can be quantified not only in large-scale but also retrospectively only relative to the same set of super-SILAC standard [29]. Quantitative MS is not only applied to the EV studies with the aim of biomarker discovery but also developed as an evaluation method to assess the EV isolation. Wang et al. established a multiple reaction monitoring (MRM) based method to assess the purity of EVs. MRM is often used for target quantitative analysis as a validation method for biomarkers reported in discovery MS analysis. They first generated ^{15}N-labeled quantification concatamers (QconCATs) for a pattern of targeted EV proteins and abundant serum proteins (non-EV proteins or contaminants) as the internal standards for quantification of those proteins in MRM. QconCATs were artificial proteins composed of concatenated tryptic peptides from targeted proteins. The purity of EVs was then assessed by the quantitative results of the targeted EV proteins and abundant serum proteins in MRM [105]. They further expanded this method to separate EVs and lipoprotein particles by adding QconCAT for apolipoproteins into the previous MRM assay [106].

5. Conclusions

With a greater understanding of the roles of EVs in the regulation of physiological and pathological processes, an increased need to use that knowledge for diagnosis and therapy of diseases has emerged. To satisfy that increased need, establishing an EV isolation method that provides rapid, efficient, and high throughput isolation and enables assessment of the full spectrum of EVs is required. Unfortunately, the currently available isolation methods only partially meet the requirement. MS is a powerful tool for the characterization of the protein content of EVs, which is crucial to decipher the biological role of EVs and explore their potential use as diagnostic, monitoring, and therapeutic tools. Currently, the application of MS in EV studies is largely limited by the imperfections of EV isolation methods.

The increasing number of studies have pointed out the EV samples prepared by current isolation methods containing different sub-populations of EVs and contaminants from surroundings. Contaminants in the isolated EV samples may not only cover the signal of lower abundant EV proteins during MS analysis but also increase the difficulty of MS data analysis, since there is no current standard to clearly distinguish EV proteins from contaminants, especially the uncommon contaminants, in the MS-generated list. To address those problems, future improvements on EV isolation and MS analysis are urgently required.

Author Contributions: Conceptualization: C.Y.; writing and original draft preparation: C.Y., J.L., X.H., and Y.D.; review and editing: C.Y. and J.L.

References

1. Merchant, M.L.; Rood, I.M.; Deegens, J.K.J.; Klein, J.B. Isolation and Characterization of Urinary Extracellular Vesicles: Implications for Biomarker Discovery. *Nat. Rev. Nephrol.* **2017**, *13*, 731–749. [CrossRef] [PubMed]
2. Wolf, P. The Nature and Significance of Platelet Products in Human Plasma. *Br. J. Haematol.* **1967**, *13*, 269–288. [CrossRef] [PubMed]
3. Yuana, Y.; Sturk, A.; Nieuwland, R. Extracellular Vesicles in Physiological and Pathological Conditions. *Blood Rev.* **2013**, *27*, 31–39. [CrossRef] [PubMed]
4. Quinn, J.F.; Patel, T.; Wong, D.; Das, S.; Freedman, J.E.; Laurent, L.C.; Carter, B.S.; Hochberg, F.; Van Keuren-Jensen, K.; Huentelman, M.; et al. Extracellular Rnas: Development as Biomarkers of Human Disease. *J. Extracell Vesicles* **2015**, *4*, 27495. [CrossRef] [PubMed]
5. Loyer, X.; Vion, A.C.; Tedgui, A.; Boulanger, C.M. Microvesicles as Cell-Cell Messengers in Cardiovascular Diseases. *Circ. Res.* **2014**, *114*, 345–353. [CrossRef] [PubMed]
6. Pocsfalvi, G.; Stanly, C.; Vilasi, A.; Fiume, I.; Capasso, G.; Turiák, L.; Buzas, E.I.; Vékey, K. Mass Spectrometry of Extracellular Vesicles. *Mass Spectrom. Rev.* **2016**, *35*, 3–21. [CrossRef] [PubMed]
7. Simonsen, J.B. What Are We Looking At? Extracellular Vesicles, Lipoproteins, or Both. *Circ. Res.* **2017**, *121*, 920–922. [CrossRef] [PubMed]
8. Simpson, R.J.; Kalra, H.; Mathivanan, S. Exocarta as a Resource for Exosomal Research. *J. Extracell Vesicles* **2012**, *1*, 18374. [CrossRef] [PubMed]
9. Kalra, H.; Simpson, R.J.; Ji, H.; Aikawa, E.; Altevogt, P.; Askenase, P.; Bond, V.C.; Borras, F.E.; Breakefield, X.; Budnik, V.; et al. Vesiclepedia: A Compendium for Extracellular Vesicles with Continuous Community Annotation. *PLoS Biol.* **2012**, *10*, e1001450. [CrossRef] [PubMed]
10. Kim, D.K.; Lee, J.; Kim, S.R.; Choi, D.S.; Yoon, Y.J.; Kim, J.H.; Go, G.; Nhung, D.; Hong, K.; Jang, S.C.; et al. Evpedia: A Community Web Portal for Extracellular Vesicles Research. *Bioinformatics* **2015**, *31*, 933–939. [CrossRef] [PubMed]
11. Witwer, K.W.; Buzas, E.; Bemis, L.T.; Bora, A.; Lasser, C.; Lotvall, J.; Nolte-t Hoen, E.N.; Piper, M.G.; Sivaraman, S.; Skog, J.; et al. Standardization of Sample Collection, Isolation and Analysis Methods in Extracellular Vesicle Research. *J. Extracell Vesicles* **2013**, *2*, 20360. [CrossRef] [PubMed]
12. Gould, S.J.; Raposo, G. As We Wait: Coping with an Imperfect Nomenclature for Extracellular Vesicles. *J. Extracell Vesicles* **2013**, *2*, 20389. [CrossRef] [PubMed]
13. Barrachina, M.N.; Calderon-Cruz, B.; Fernandez-Rocca, L.; Garcia, A. Application of Extracellular Vesicles Proteomics to Cardiovascular Disease: Guidelines, Data Analysis, and Future Perspectives. *Proteomics* **2019**, *19*, 1800247. [CrossRef] [PubMed]
14. Wang, W.; Luo, J.; Wang, S. Recent Progress in Isolation and Detection of Extracellular Vesicles for Cancer Diagnostics. *Adv. Healthc. Mater.* **2018**, *7*, e1800484. [CrossRef] [PubMed]
15. Szatanek, R.; Baran, J.; Siedlar, M.; Baj-Krzyworzeka, M. Isolation of Extracellular Vesicles: Determining the Correct Approach. *Int. J. Mol. Med.* **2015**, *36*, 11–17. [CrossRef] [PubMed]

16. Abramowicz, A.; Widlak, P.; Pietrowska, M. Proteomic Analysis of Exosomal Cargo: The Challenge of High Purity Vesicle Isolation. *Mol. Biosyst.* **2016**, *12*, 1407–1419. [CrossRef] [PubMed]

17. Tzouanas, C.; Lim, J.S.Y.; Wen, Y.; Thiery, J.P.; Khoo, B.L. Microdevices for Non-Invasive Detection of Bladder Cancer. *Chemosensors* **2017**, *5*, 30. [CrossRef]

18. Yuana, Y.; Boing, A.N.; Grootemaat, A.E.; van der Pol, E.; Hau, C.M.; Cizmar, P.; Buhr, E.; Sturk, A.; Nieuwland, R. Handling and Storage of Human Body Fluids for Analysis of Extracellular Vesicles. *J. Extracell. Vesicles* **2015**, *4*, 29260. [CrossRef] [PubMed]

19. Lacroix, R.; Judicone, C.; Mooberry, M.; Boucekine, M.; Key, N.S.; Dignat-George, F.; The ISTH SSC Workshop. Standardization of Pre-Analytical Variables in Plasma Microparticle Determination: Results of the International Society on Thrombosis and Haemostasis Ssc Collaborative Workshop. *J. Thromb. Haemost.* **2013**, *11*, 1190–1193. [CrossRef] [PubMed]

20. Yuana, Y.; Bertina, R.M.; Osanto, S. Pre-Analytical and Analytical Issues in the Analysis of Blood Microparticles. *Thromb. Haemost.* **2011**, *105*, 396–408. [CrossRef] [PubMed]

21. Akers, J.C.; Ramakrishnan, V.; Yang, I.; Hua, W.; Mao, Y.; Carter, B.S.; Chen, C.C. Optimizing Preservation of Extracellular Vesicular Mirnas Derived from Clinical Cerebrospinal Fluid. *Cancer Biomark.* **2016**, *17*, 125–132. [CrossRef] [PubMed]

22. Jamaly, S.; Ramberg, C.; Olsen, R.; Latysheva, N.; Webster, P.; Sovershaev, T.; Braekkan, S.K.; Hansen, J.B. Impact of Preanalytical Conditions on Plasma Concentration and Size Distribution of Extracellular Vesicles Using Nanoparticle Tracking Analysis. *Sci. Rep.* **2018**, *8*, 17216. [CrossRef] [PubMed]

23. Jeyaram, A.; Jay, S.M. Preservation and Storage Stability of Extracellular Vesicles for Therapeutic Applications. *AAPS J.* **2017**, *20*, 1. [CrossRef] [PubMed]

24. Ge, Q.Y.; Zhou, Y.X.; Lu, J.F.; Bai, Y.F.; Xie, X.Y.; Lu, Z.H. Mirna in Plasma Exosome Is Stable under Different Storage Conditions. *Molecules* **2014**, *19*, 1568–1575. [CrossRef] [PubMed]

25. Gardiner, C.; Di Vizio, D.; Sahoo, S.; Thery, C.; Witwer, K.W.; Wauben, M.; Hill, A.F. Techniques Used for the Isolation and Characterization of Extracellular Vesicles: Results of a Worldwide Survey. *J. Extracell. Vesicles* **2016**, *5*, 32945. [CrossRef] [PubMed]

26. Livshts, M.A.; Khomyakova, E.; Evtushenko, E.G.; Lazarev, V.N.; Kulemin, N.A.; Semina, S.E.; Generozov, E.V.; Govorun, V.M. Isolation of Exosomes by Differential Centrifugation: Theoretical Analysis of a Commonly Used Protocol. *Sci Rep.* **2015**, *5*, 17319. [CrossRef] [PubMed]

27. Thery, C.; Amigorena, S.; Raposo, G.; Clayton, A. Isolation and Characterization of Exosomes from Cell Culture Supernatants and Biological Fluids. *Curr. Protoc. Cell Biol.* **2006**, *30*, 3–22. [CrossRef] [PubMed]

28. Momen-Heravi, F.; Balaj, L.; Alian, S.; Trachtenberg, A.J.; Hochberg, F.H.; Skog, J.; Kuo, W.P. Impact of Biofluid Viscosity on Size and Sedimentation Efficiency of the Isolated Microvesicles. *Front. Physiol.* **2012**, *3*, 162. [CrossRef]

29. Harel, M.; Oren-Giladi, P.; Kaidar-Person, O.; Shaked, Y.; Geiger, T. Proteomics of Microparticles with Silac Quantification (Promis-Quan): A Novel Proteomic Method for Plasma Biomarker Quantification. *Mol. Cell Proteom.* **2015**, *14*, 1127–1136. [CrossRef]

30. Antwi-Baffour, S.; Adjei, J.K.; Agyemang-Yeboah, F.; Annani-Akollor, M.; Kyeremeh, R.; Asare, G.A.; Gyan, B. Proteomic Analysis of Microparticles Isolated from Malaria Positive Blood Samples. *Proteome Sci.* **2017**, *15*, 5. [CrossRef] [PubMed]

31. Raposo, G.; Nijman, H.W.; Stoorvogel, W.; Liejendekker, R.; Harding, C.V.; Melief, C.J.; Geuze, H.J. B Lymphocytes Secrete Antigen-Presenting Vesicles. *J. Exp. Med.* **1996**, *183*, 1161–1172. [CrossRef] [PubMed]

32. Chutipongtanate, S.; Greis, K.D. Multiplex Biomarker Screening Assay for Urinary Extracellular Vesicles Study: A Targeted Labelfree Proteomic Approach. *Sci. Rep.* **2018**, *8*, 15039. [CrossRef] [PubMed]

33. Sun, Y.; Huo, C.; Qiao, Z.; Shang, Z.; Uzzaman, A.; Liu, S.; Jiang, X.; Fan, L.; Ji, L.; Guan, X.; et al. Comparative Proteomic Analysis of Exosomes and Microvesicles in Human Saliva for Lung Cancer. *J. Proteome Res.* **2018**, *17*, 1101–1107. [CrossRef] [PubMed]

34. Whitham, M.; Parker, B.L.; Friedrichsen, M.; Hingst, J.R.; Hjorth, M.; Hughes, W.E.; Egan, C.L.; Cron, L.; Watt, K.I.; Kuchel, R.P.; et al. Extracellular Vesicles Provide a Means for Tissue Crosstalk During Exercise. *Cell Metab.* **2018**, *27*, 237–251. [CrossRef] [PubMed]

35. Kim, J.; Tan, Z.; Lubman, D.M. Exosome Enrichment of Human Serum Using Multiple Cycles of Centrifugation. *Electrophoresis* **2015**, *36*, 2017–2026. [CrossRef] [PubMed]

36. Langevin, S.M.; Kuhnell, D.; Orr-Asman, M.A.; Biesiada, J.; Zhang, X.; Medvedovic, M.; Thomas, H.E.

Balancing Yield, Purity and Practicality: A Modified Differential Ultracentrifugation Protocol for Efficient Isolation of Small Extracellular Vesicles from Human Serum. *RNA Biol.* **2019**, *16*, 5–12. [CrossRef] [PubMed]

37. Cvjetkovic, A.; Lotvall, J.; Lasser, C. The Influence of Rotor Type and Centrifugation Time on the Yield and Purity of Extracellular Vesicles. *J. Extracell. Vesicles* **2014**, *3*, 23111. [CrossRef] [PubMed]

38. Shiromizu, T.; Kume, H.; Ishida, M.; Adachi, J.; Kano, M.; Matsubara, H.; Tomonaga, T. Quantitation of Putative Colorectal Cancer Biomarker Candidates in Serum Extracellular Vesicles by Targeted Proteomics. *Sci. Rep.* **2017**, *7*, 12782. [CrossRef] [PubMed]

39. Fujita, K.; Kume, H.; Matsuzaki, K.; Kawashima, A.; Ujike, T.; Nagahara, A.; Uemura, M.; Miyagawa, Y.; Tomonaga, T.; Nonomura, N. Proteomic Analysis of Urinary Extracellular Vesicles from High Gleason Score Prostate Cancer. *Sci. Rep.* **2017**, *7*, 42961. [CrossRef]

40. Lin, Y.; Liang, A.; He, Y.; Li, Z.; Li, Z.; Wang, G.; Sun, F. Proteomic Analysis of Seminal Extracellular Vesicle Proteins Involved in Asthenozoospermia by iTRAQ. *Mol. Reprod Dev.* **2019**. [CrossRef]

41. Jiao, Y.J.; Jin, D.D.; Jiang, F.; Liu, J.X.; Qu, L.S.; Ni, W.K.; Liu, Z.X.; Lu, C.H.; Ni, R.Z.; Zhu, J.; et al. Characterization and Proteomic Profiling of Pancreatic Cancer-Derived Serum Exosomes. *J. Cell Biochem.* **2019**, *120*, 988–999. [CrossRef] [PubMed]

42. Moon, P.G.; Lee, J.E.; You, S.; Kim, T.K.; Cho, J.H.; Kim, I.S.; Kwon, T.H.; Kim, C.D.; Park, S.H.; Hwang, D.; et al. Proteomic Analysis of Urinary Exosomes from Patients of Early Iga Nephropathy and Thin Basement Membrane Nephropathy. *Proteomics* **2011**, *11*, 2459–2475. [CrossRef] [PubMed]

43. Jia, R.; Li, J.; Rui, C.; Ji, H.; Ding, H.; Lu, Y.; De, W.; Sun, L. Comparative Proteomic Profile of the Human Umbilical Cord Blood Exosomes between Normal and Preeclampsia Pregnancies with High-Resolution Mass Spectrometry. *Cell Physiol. Biochem.* **2015**, *36*, 2299–2306. [CrossRef] [PubMed]

44. Pisitkun, T.; Gandolfo, M.T.; Das, S.; Knepper, M.A.; Bagnasco, S.M. Application of Systems Biology Principles to Protein Biomarker Discovery: Urinary Exosomal Proteome in Renal Transplantation. *Proteomics Clin. Appl.* **2012**, *6*, 268–278. [CrossRef] [PubMed]

45. Winck, F.V.; Ribeiro, A.C.P.; Domingues, R.R.; Ling, L.Y.; Riano-Pachon, D.M.; Rivera, C.; Brandao, T.B.; Gouvea, A.F.; Santos-Silva, A.R.; Coletta, R.D.; et al. Insights into Immune Responses in Oral Cancer through Proteomic Analysis of Saliva and Salivary Extracellular Vesicles. *Sci. Rep.* **2015**, *5*, 16305. [CrossRef] [PubMed]

46. Ramacciotti, E.; Hawley, A.E.; Wrobleski, S.K.; Myers, D.D., Jr.; Strahler, J.R.; Andrews, P.C.; Guire, K.E.; Henke, P.K.; Wakefield, T.W. Proteomics of Microparticles after Deep Venous Thrombosis. *Thromb. Res.* **2010**, *125*, e269–e274. [CrossRef] [PubMed]

47. Van Herwijnen, M.J.; Zonneveld, M.I.; Goerdayal, S.; Nolte-'t Hoen, E.N.; Garssen, J.; Stahl, B.; Maarten Altelaar, A.F.; Redegeld, F.A.; Wauben, M.H. Comprehensive Proteomic Analysis of Human Milk-Derived Extracellular Vesicles Unveils a Novel Functional Proteome Distinct from Other Milk Components. *Mol. Cell. Proteom.* **2016**, *15*, 3412–3423. [CrossRef]

48. Chen, I.H.; Xue, L.; Hsu, C.C.; Paez, J.S.; Pan, L.; Andaluz, H.; Wendt, M.K.; Iliuk, A.B.; Zhu, J.K.; Tao, W.A. Phosphoproteins in Extracellular Vesicles as Candidate Markers for Breast Cancer. *Proc. Natl. Acad. Sci. USA* **2017**, *114*, 3175–3180. [CrossRef]

49. Manek, R.; Moghieb, A.; Yang, Z.; Kumar, D.; Kobessiy, F.; Sarkis, G.A.; Raghavan, V.; Wang, K.K.W. Protein Biomarkers and Neuroproteomics Characterization of Microvesicles/Exosomes from Human Cerebrospinal Fluid Following Traumatic Brain Injury. *Mol. Neurobiol.* **2018**, *55*, 6112–6128. [CrossRef]

50. Greening, D.W.; Xu, R.; Ji, H.; Tauro, B.J.; Simpson, R.J. A Protocol for Exosome Isolation and Characterization: Evaluation of Ultracentrifugation, Density-Gradient Separation, and Immunoaffinity Capture Methods. *Methods Mol. Biol.* **2015**, *1295*, 179–209.

51. Iwai, K.; Minamisawa, T.; Suga, K.; Yajima, Y.; Shiba, K. Isolation of Human Salivary Extracellular Vesicles by Iodixanol Density Gradient Ultracentrifugation and Their Characterizations. *J. Extracell. Vesicles* **2016**, *5*, 30829. [CrossRef] [PubMed]

52. Arab, T.; Raffo-Romero, A.; Van Camp, C.; Lemaire, Q.; Le Marrec-Croq, F.; Drago, F.; Aboulouard, S.; Slomianny, C.; Lacoste, A.S.; Guigon, I.; et al. Proteomic Characterisation of Leech Microglia Extracellular Vesicles (Evs): Comparison between Differential Ultracentrifugation and Optiprep (Tm) Density Gradient Isolation. *J. Extracell. Vesicles* **2019**, *8*, 1603048. [CrossRef] [PubMed]

53. Musante, L.; Saraswat, M.; Duriez, E.; Byrne, B.; Ravida, A.; Domon, B.; Holthofer, H. Biochemical and Physical Characterisation of Urinary Nanovesicles Following Chaps Treatment. *PLoS ONE* **2012**, *7*, e37279. [CrossRef] [PubMed]

54. Barrachina, M.N.; Sueiro, A.M.; Casas, V.; Izquierdo, I.; Hermida-Nogueira, L.; Guitian, E.; Casanueva, F.F.; Abian, J.; Carrascal, M.; Pardo, M.; et al. A Combination of Proteomic Approaches Identifies a Panel of Circulating Extracellular Vesicle Proteins Related to the Risk of Suffering Cardiovascular Disease in Obese Patients. *Proteomics* **2019**, *19*, e1800248. [CrossRef] [PubMed]

55. Lobb, R.J.; Becker, M.; Wen, S.W.; Wong, C.S.F.; Wiegmans, A.P.; Leimgruber, A.; Moller, A. Optimized Exosome Isolation Protocol for Cell Culture Supernatant and Human Plasma. *J. Extracell. Vesicles* **2015**, *4*, 27031. [CrossRef] [PubMed]

56. Merchant, M.L.; Powell, D.W.; Wilkey, D.W.; Cummins, T.D.; Deegens, J.K.; Rood, I.M.; McAfee, K.J.; Fleischer, C.; Klein, E.; Klein, J.B. Microfiltration Isolation of Human Urinary Exosomes for Characterization by MS. *Proteom. Clin. Appl.* **2010**, *4*, 84–96. [CrossRef]

57. Musante, L.; Tataruch, D.; Gu, D.F.; Benito-Martin, A.; Calzaferri, G.; Aherne, S.; Holthofer, H. A Simplified Method to Recover Urinary Vesicles for Clinical Applications, and Sample Banking. *Sci. Rep.* **2014**, *4*, 7532. [CrossRef]

58. Hu, S.; Musante, L.; Tataruch, D.; Xu, X.; Kretz, O.; Henry, M.; Meleady, P.; Luo, H.; Zou, H.; Jiang, Y.; et al. Purification and Identification of Membrane Proteins from Urinary Extracellular Vesicles Using Triton X-114 Phase Partitioning. *J. Proteome Res.* **2018**, *17*, 86–96. [CrossRef]

59. Heinemann, M.L.; Ilmer, M.; Silva, L.P.; Hawke, D.H.; Recio, A.; Vorontsova, M.A.; Alt, E.; Vykoukal, J. Benchtop Isolation and Characterization of Functional Exosomes by Sequential Filtration. *J. Chromatogr. A* **2014**, *1371*, 125–135. [CrossRef]

60. Osti, D.; Del Bene, M.; Rappa, G.; Santos, M.; Matafora, V.; Richichi, C.; Faletti, S.; Beznoussenko, G.V.; Mironov, A.; Bachi, A.; et al. Clinical Significance of Extracellular Vesicles in Plasma from Glioblastoma Patients. *Clin. Cancer Res.* **2019**, *25*, 266–276. [CrossRef]

61. Musante, L.; Tataruch-Weinert, D.; Kerjaschki, D.; Henry, M.; Meleady, P.; Holthofer, H. Residual Urinary Extracellular Vesicles in Ultracentrifugation Supernatants after Hydrostatic Filtration Dialysis Enrichment. *J. Extracell. Vesicles* **2017**, *6*, 1267896. [CrossRef] [PubMed]

62. De Menezes-Neto, A.; Saez, M.J.F.; Lozano-Ramos, I.; Segui-Barber, J.; Martin-Jaular, L.; Ullate, J.M.E.; Fernandez-Becerra, C.; Borras, F.E.; del Portillo, H.A. Size-Exclusion Chromatography as a Stand-Alone Methodology Identifies Novel Markers in Mass Spectrometry Analyses of Plasma-Derived Vesicles from Healthy Individuals. *J. Extracell. Vesicles* **2015**, *4*, 27378. [CrossRef] [PubMed]

63. Karimi, N.; Cvjetkovic, A.; Jang, S.C.; Crescitelli, R.; Hosseinpour Feizi, M.A.; Nieuwland, R.; Lotvall, J.; Lasser, C. Detailed Analysis of the Plasma Extracellular Vesicle Proteome after Separation from Lipoproteins. *Cell Mol. Life Sci.* **2018**, *75*, 2873–2886. [CrossRef] [PubMed]

64. Smolarz, M.; Pietrowska, M.; Matysiak, N.; Mielanczyk, L.; Widlak, P. Proteome Profiling of Exosomes Purified from a Small Amount of Human Serum: The Problem of Co-Purified Serum Components. *Proteomes* **2019**, *7*, 18. [CrossRef] [PubMed]

65. Aqrawi, L.A.; Galtung, H.K.; Vestad, B.; Ovstebo, R.; Thiede, B.; Rusthen, S.; Young, A.; Guerreiro, E.M.; Utheim, T.P.; Chen, X.; et al. Identification of Potential Saliva and Tear Biomarkers in Primary Sjogren's Syndrome, Utilising the Extraction of Extracellular Vesicles and Proteomics Analysis. *Arthritis Res. Ther.* **2017**, *19*, 14. [CrossRef]

66. Foers, A.D.; Chatfield, S.; Dagley, L.F.; Scicluna, B.J.; Webb, A.I.; Cheng, L.; Hill, A.F.; Wicks, I.P.; Pang, K.C. Enrichment of Extracellular Vesicles from Human Synovial Fluid Using Size Exclusion Chromatography. *J. Extracell. Vesicles* **2018**, *7*, 1490145. [CrossRef] [PubMed]

67. Chen, Y.Y.; Xie, Y.; Xu, L.; Zhan, S.H.; Xiao, Y.; Gao, Y.P.; Wu, B.; Ge, W. Protein Content and Functional Characteristics of Serum-Purified Exosomes from Patients with Colorectal Cancer Revealed by Quantitative Proteomics. *J. Extracell. Vesicles* **2017**, *140*, 900–913. [CrossRef]

68. Zhang, W.; Ou, X.; Wu, X. Proteomics Profiling of Plasma Exosomes in Epithelial Ovarian Cancer: A Potential Role in the Coagulation Cascade, Diagnosis and Prognosis. *Int. J. Oncol.* **2019**, *54*, 1719–1733. [CrossRef]

69. Tsuno, H.; Arito, M.; Suematsu, N.; Sato, T.; Hashimoto, A.; Matsui, T.; Omoteyama, K.; Sato, M.; Okamoto, K.; Tohma, S.; et al. A Proteomic Analysis of Serum-Derived Exosomes in Rheumatoid Arthritis. *BMC Rheumatol.* **2018**, *2*, 35. [CrossRef]

70. Leberman, R. The Isolation of Plant Viruses by Means of "Simple" Coacervates. *Virology* **1966**, *30*, 341–347. [CrossRef]

71. Ding, M.; Wang, C.; Lu, X.; Zhang, C.; Zhou, Z.; Chen, X.; Zhang, C.Y.; Zen, K.; Zhang, C. Comparison of Commercial Exosome Isolation Kits for Circulating Exosomal Microrna Profiling. *Anal. Bioanal. Chem.* **2018**, *410*, 3805–3814. [CrossRef] [PubMed]

72. Weng, Y.J.; Sui, Z.G.; Shan, Y.C.; Hu, Y.C.; Chen, Y.B.; Zhang, L.H.; Zhang, Y.K. Effective Isolation of Exosomes with Polyethylene Glycol from Cell Culture Supernatant for in-Depth Proteome Profiling. *Analyst* **2016**, *141*, 4640–4646. [CrossRef] [PubMed]

73. Rider, M.A.; Hurwitz, S.N.; Meckes, D.G. Extrapeg: A Polyethylene Glycol-Based Method for Enrichment of Extracellular Vesicles. *Sci. Rep.* **2016**, *6*, 23978. [CrossRef] [PubMed]

74. Shin, H.; Han, C.; Labuz, J.M.; Kim, J.; Kim, J.; Cho, S.; Gho, Y.S.; Takayama, S.; Park, J. High-Yield Isolation of Extracellular Vesicles Using Aqueous Two-Phase System. *Sci. Rep.* **2015**, *5*, 13103. [CrossRef] [PubMed]

75. Wang, D.; Sun, W. Urinary Extracellular Microvesicles: Isolation Methods and Prospects for Urinary Proteome. *Proteomics* **2014**, *14*, 1922–1932. [CrossRef] [PubMed]

76. Hildonen, S.; Skarpen, E.; Halvorsen, T.G.; Reubsaet, L. Isolation and Mass Spectrometry Analysis of Urinary Extraexosomal Proteins. *Sci. Rep.* **2016**, *6*, 36331. [CrossRef] [PubMed]

77. Ueda, K.; Ishikawa, N.; Tatsuguchi, A.; Saichi, N.; Fujii, R.; Nakagawa, H. Antibody-Coupled Monolithic Silica Microtips for Highthroughput Molecular Profiling of Circulating Exosomes. *Sci. Rep.* **2014**, *4*, 6232. [CrossRef] [PubMed]

78. Taylor, D.D.; Gercel-Taylor, C. Microrna Signatures of Tumor-Derived Exosomes as Diagnostic Biomarkers of Ovarian Cancer. *Gynecol. Oncol.* **2008**, *110*, 13–21. [CrossRef]

79. Kalra, H.; Adda, C.G.; Liem, M.; Ang, C.S.; Mechler, A.; Simpson, R.J.; Hulett, M.D.; Mathivanan, S. Comparative Proteomics Evaluation of Plasma Exosome Isolation Techniques and Assessment of the Stability of Exosomes in Normal Human Blood Plasma. *Proteomics* **2013**, *13*, 3354–3364. [CrossRef]

80. Tauro, B.J.; Greening, D.W.; Mathias, R.A.; Mathivanan, S.; Ji, H.; Simpson, R.J. Two Distinct Populations of Exosomes Are Released from Lim1863 Colon Carcinoma Cell-Derived Organoids. *Mol. Cell. Proteom.* **2013**, *12*, 587–598. [CrossRef]

81. Ghosh, A.; Davey, M.; Chute, I.C.; Griffiths, S.G.; Lewis, S.; Chacko, S.; Barnett, D.; Crapoulet, N.; Fournier, S.; Joy, A.; et al. Rapid Isolation of Extracellular Vesicles from Cell Culture and Biological Fluids Using a Synthetic Peptide with Specific Affinity for Heat Shock Proteins. *PLoS ONE* **2014**, *9*, e110443. [CrossRef] [PubMed]

82. Bijnsdorp, I.V.; Maxouri, O.; Kardar, A.; Schelfhorst, T.; Piersma, S.R.; Pham, T.V.; Vis, A.; van Moorselaar, R.J.; Jimenez, C.R. Feasibility of Urinary Extracellular Vesicle Proteome Profiling Using a Robust and Simple, Clinically Applicable Isolation Method. *J. Extracell. Vesicles* **2017**, *6*, 1313091. [CrossRef] [PubMed]

83. Balaj, L.; Atai, N.A.; Chen, W.L.; Mu, D.; Tannous, B.A.; Breakefield, X.O.; Skog, J.; Maguire, C.A. Heparin Affinity Purification of Extracellular Vesicles. *Sci. Rep.* **2015**, *5*, 10266. [CrossRef] [PubMed]

84. Gao, F.Y.; Jiao, F.L.; Xia, C.S.; Zhao, Y.; Ying, W.T.; Xie, Y.P.; Guan, X.Y.; Tao, M.; Zhang, Y.J.; Qin, W.J.; et al. A Novel Strategy for Facile Serum Exosome Isolation Based on Specific Interactions between Phospholipid Bilayers and TiO_2. *Chem. Sci.* **2019**, *10*, 1579–1588. [CrossRef] [PubMed]

85. Tan, K.H.; Tan, S.S.; Sze, S.K.; Lee, W.K.R.; Ng, M.J.; Lim, S.K. Plasma Biomarker Discovery in Preeclampsia Using a Novel Differential Isolation Technology for Circulating Extracellular Vesicles. *Am. J. Obstet. Gynecol.* **2014**, *211*, 380-e1. [PubMed]

86. Nakai, W.; Yoshida, T.; Diez, D.; Miyatake, Y.; Nishibu, T.; Imawaka, N.; Naruse, K.; Sadamura, Y.; Hanayama, R. A Novel Affinity-Based Method for the Isolation of Highly Purified Extracellular Vesicles. *Sci. Rep.* **2016**, *6*, 33935. [CrossRef] [PubMed]

87. Takov, K.; Yellon, D.M.; Davidson, S.M. Comparison of Small Extracellular Vesicles Isolated from Plasma by Ultracentrifugation or Size-Exclusion Chromatography: Yield, Purity and Functional Potential. *J. Extracell. Vesicles* **2019**, *8*, 1560809. [CrossRef] [PubMed]

88. Davis, C.N.; Phillips, H.; Tomes, J.J.; Swain, M.T.; Wilkinson, T.J.; Brophy, P.M.; Morphew, R.M. The Importance of Extracellular Vesicle Purification for Downstream Analysis: A Comparison of Differential Centrifugation and Size Exclusion Chromatography for Helminth Pathogens. *PLoS Negl. Trop. Dis.* **2019**, *13*, e0007191. [CrossRef]

89. Stranska, R.; Gysbrechts, L.; Wouters, J.; Vermeersch, P.; Bloch, K.; Dierickx, D.; Andrei, G.; Snoeck, R. Comparison of Membrane Affinity-Based Method with Size-Exclusion Chromatography for Isolation of Exosome-Like Vesicles from Human Plasma. *J. Transl. Med.* **2018**, *16*, 1. [CrossRef]

90. Patel, G.K.; Khan, M.A.; Zubair, H.; Srivastava, S.K.; Khushman, M.; Singh, S.; Singh, A.P. Comparative Analysis of Exosome Isolation Methods Using Culture Supernatant for Optimum Yield, Purity and Downstream Applications. *Sci. Rep.* **2019**, *9*, 5335. [CrossRef]

91. Antounians, L.; Tzanetakis, A.; Pellerito, O.; Catania, V.D.; Sulistyo, A.; Montalva, L.; McVey, M.J.; Zani, A. The Regenerative Potential of Amniotic Fluid Stem Cell Extracellular Vesicles: Lessons Learned by Comparing Different Isolation Techniques. *Sci. Rep.* **2019**, *9*, 1837. [CrossRef] [PubMed]

92. Royo, F.; Zuniga-Garcia, P.; Sanchez-Mosquera, P.; Egia, A.; Perez, A.; Loizaga, A.; Arceo, R.; Lacasa, I.; Rabade, A.; Arrieta, E.; et al. Different Ev Enrichment Methods Suitable for Clinical Settings Yield Different Subpopulations of Urinary Extracellular Vesicles from Human Samples. *J. Extracell. Vesicles* **2016**, *5*, 29497. [CrossRef] [PubMed]

93. Andreu, Z.; Rivas, E.; Sanguino-Pascual, A.; Lamana, A.; Marazuela, M.; Gonzalez-Alvaro, I.; Sanchez-Madrid, F.; de la Fuente, H.; Yanez-Mo, M. Comparative Analysis of Ev Isolation Procedures for Mirnas Detection in Serum Samples. *J. Extracell. Vesicles* **2016**, *5*, 31655. [CrossRef] [PubMed]

94. Rood, I.M.; Deegens, J.K.; Merchant, M.L.; Tamboer, W.P.; Wilkey, D.W.; Wetzels, J.F.; Klein, J.B. Comparison of Three Methods for Isolation of Urinary Microvesicles to Identify Biomarkers of Nephrotic Syndrome. *Kidney Int.* **2010**, *78*, 810–816. [CrossRef] [PubMed]

95. Caradec, J.; Kharmate, G.; Hosseini-Beheshti, E.; Adomat, H.; Gleave, M.; Guns, E. Reproducibility and Efficiency of Serum-Derived Exosome Extraction Methods. *Clin. Biochem.* **2014**, *47*, 1286–1292. [CrossRef] [PubMed]

96. Joy, A.P.; Ayre, D.C.; Chute, I.C.; Beauregard, A.P.; Wajnberg, G.; Ghosh, A.; Lewis, S.M.; Ouellette, R.J.; Barnett, D.A. Proteome Profiling of Extracellular Vesicles Captured with the Affinity Peptide Vn96: Comparison of Laemmli and Trizol (C) Protein-Extraction Methods. *J. Extracell. Vesicles* **2018**, *7*, 1438727. [CrossRef] [PubMed]

97. Fel, A.; Lewandowska, A.E.; Petrides, P.E.; Wisniewski, J.R. Comparison of Proteome Composition of Serum Enriched in Extracellular Vesicles Isolated from Polycythemia Vera Patients and Healthy Controls. *Proteomes* **2019**, *7*, 20. [CrossRef]

98. Xie, X.F.; Chu, H.J.; Xu, Y.F.; Hua, L.; Wang, Z.P.; Huang, P.; Jia, H.L.; Zhang, L. Proteomics Study of Serum Exosomes in Kawasaki Disease Patients with Coronary Artery Aneurysms. *Cardiol. J.* **2018**. [CrossRef]

99. Gonzales, P.A.; Pisitkun, T.; Hoffert, J.D.; Tchapyjnikov, D.; Star, R.A.; Kleta, R.; Wang, N.S.; Knepper, M.A. Large-Scale Proteomics and Phosphoproteomics of Urinary Exosomes. *J. Am. Soc. Nephrol.* **2009**, *20*, 363–379. [CrossRef]

100. Kittivorapart, J.; Crew, V.K.; Wilson, M.C.; Heesom, K.J.; Siritanaratkul, N.; Toye, A.M. Quantitative Proteomics of Plasma Vesicles Identify Novel Biomarkers for Hemoglobin E/Beta-Thalassemic Patients. *Blood Adv.* **2018**, *2*, 95–104. [CrossRef]

101. Sok Hwee Cheow, E.; Hwan Sim, K.; de Kleijn, D.; Neng Lee, C.; Sorokin, V.; Sze, S.K. Simultaneous Enrichment of Plasma Soluble and Extracellular Vesicular Glycoproteins Using Prolonged Ultracentrifugation-Electrostatic Repulsion-Hydrophilic Interaction Chromatography (Puc-Erlic) Approach. *Mol. Cell. Proteom.* **2015**, *14*, 1657–1671. [CrossRef] [PubMed]

102. Saraswat, M.; Joenvaara, S.; Musante, L.; Peltoniemi, H.; Holthofer, H.; Renkonen, R. N-Linked (N-) Glycoproteomics of Urinary Exosomes. *Mol. Cell. Proteom.* **2015**, *14*, 263–276. [CrossRef] [PubMed]

103. Braga-Lagache, S.; Buchs, N.; Iacovache, M.I.; Zuber, B.; Jackson, C.B.; Heller, M. Robust Label-Free, Quantitative Profiling of Circulating Plasma Microparticle (Mp) Associated Proteins. *Mol. Cell. Proteom.* **2016**, *15*, 3640–3652. [CrossRef] [PubMed]

104. Kodidela, S.; Wang, Y.; Patters, B.J.; Gong, Y.; Sinha, N.; Ranjit, S.; Gerth, K.; Haque, S.; Cory, T.; McArthur, C.; et al. Proteomic Profiling of Exosomes Derived from Plasma of Hiv-Infected Alcohol Drinkers and Cigarette Smokers. *J. Neuroimmune Pharmacol.* **2019**. [CrossRef] [PubMed]

105. Wang, T.T.; Anderson, K.W.; Turko, I.V. Assessment of Extracellular Vesicles Purity Using Proteomic Standards. *Analyt. Chem.* **2017**, *89*, 11070–11075. [CrossRef]

106. Wang, T.; Turko, I.V. Proteomic Toolbox to Standardize the Separation of Extracellular Vesicles and Lipoprotein Particles. *J. Proteome Res.* **2018**, *17*, 3104–3113. [CrossRef]

10

Artificial Neural Network Prediction of Retention of Amino Acids in Reversed-Phase HPLC under Application of Linear Organic Modifier Gradients and/or pH Gradients

Angelo Antonio D'Archivio🆔

Dipartimento di Scienze Fisiche e Chimiche, Università degli Studi dell'Aquila, Via Vetoio, 67100 Coppito, L'Aquila, Italy; angeloantonio.darchivio@univaq.it

Abstract: A multi-layer artificial neural network (ANN) was used to model the retention behavior of 16 o-phthalaldehyde derivatives of amino acids in reversed-phase liquid chromatography under application of various gradient elution modes. The retention data, taken from literature, were collected in acetonitrile–water eluents under application of linear organic modifier gradients (φ gradients), pH gradients, or double pH/φ gradients. At first, retention data collected in φ gradients and pH gradients were modeled separately, while these were successively combined in one dataset and fitted simultaneously. Specific ANN-based models were generated by combining the descriptors of the gradient profiles with 16 inputs representing the amino acids and providing the retention time of these solutes as the response. Categorical "bit-string" descriptors were adopted to identify the solutes, which allowed simultaneously modeling the retention times of all 16 target amino acids. The ANN-based models tested on external gradients provided mean errors for the predicted retention times of 1.1% (φ gradients), 1.4% (pH gradients), 2.5% (combined φ and pH gradients), and 2.5% (double pH/φ gradients). The accuracy of ANN prediction was better than that previously obtained by fitting of the same data with retention models based on the solution of the fundamental equation of gradient elution.

Keywords: amino acids; reversed-phase liquid chromatography; gradient elution; retention prediction; artificial neural network

1. Introduction

Reversed-phase high-performance liquid chromatography (RP-HPLC) is an extensively applied technique in the separation and determination of a wide range of multi-class compounds, including biomolecules, pharmaceuticals, and industrial chemicals, in human, environmental, and food samples [1–4]. Separation of complex mixtures by RP-HPLC generally requires the application of mobile-phase gradients to overcome the typical disadvantages of isocratic elution, such as poor resolution of early peaks, broadening of late peaks, band tailing, and long separation times [5,6]. In organic modifier mobile-phase gradients (φ gradients), the concentration of organic solvent in the mobile phase is increased, determining a progressive increase of the elution power of the eluent during the gradient run and a consequent decrease in solute retention. A similar effect occurs in the pH gradient of the mobile phase [7], where an increase or decrease in pH in the case of weak bases or acids, respectively, produces a progressive increase of the ionized form of the analyte and a consequent decrease in its retention time.

In the last few decades, various predictive models [8–11] were proposed with the aim of supporting the empirical strategies commonly utilized in the development of the chromatographic methods, which can be particularly slow and inefficient when a large number of parameters have to be fixed, such as in the case of programmed elution analysis.

Many attempts to describe the retention of solutes in RP-HPLC under the application of mobile-phase gradients are based on the solution of the fundamental equation of gradient elution [12–16],

$$\int_0^{t_R - t_0} \frac{dt}{t_0 k} = 1 \tag{1}$$

where t_R is the retention time, t_0 is the column hold-up time, and k is the retention factor. Analytical or numerical solutions of Equation (1) require the dependence of k upon the mobile-phase composition. To this end, popular relationships relating k and φ, or empirical models arising from the experimental properties of the system are often used, where the adjustable eluent- and sometimes solute-dependent parameters associated with these relationships are determined by appropriate fitting algorithms applied to the retention data.

Artificial neural networks (ANNs), since their introduction in 1990s, are used as regression tools to address various complex issues in chromatography. The main advantage of ANN regression is that both multilinear and non-linear phenomena can be handled without the need of prior definition of a fitting function. The ANN-based applications in retention prediction include the development of quantitative structure–retention relationships (QSSRs) [17,18], modeling of the combined effects of solute structure and separation conditions (column, eluent, or both) [19,20], and transfer of retention data between different columns or eluent types [21–23]. ANN models based simultaneously on molecular descriptors and instrumental conditions associated with the elution mode were used to predict the retention times of diverse sets of organic compounds in gradient RP-HPLC [24–27]. We previously used ANN regression to model the retention times of 16 selected purines, pyrimidines, and nucleosides under the application of multilinear φ gradients [28]. With this aim, a network was trained to associate the retention times with both gradient profiles and solutes, the latter being represented by "bit-string" categorical descriptors, which, unlike the aforementioned QSSR-inspired approaches, did not require any assumption of the chemical structure of the analytes. The generalization ability of the so-obtained model was tested on external multilinear gradients, providing an accurate prediction of the solute retention times (within 2–3% on average). This approach was here extended to the RP-HPLC retention of ionizable solutes of biological relevance, such as amino acids, analyzed under the application of linear φ gradients, pH gradients, or combined φ/pH gradients, whereby the target compounds were previously derivatized with o-phthalaldehyde (OPA) to allow their fluorescence detection. The data investigated in the present study were taken from three works of Pappa-Louisi and co-workers [14–16], who collected the experimental data and developed retention models based on the solution of the fundamental equation of gradient elution to verify the accuracy of the predicted retention by different equations or fitting algorithms.

The present study is aimed at exploring the capability of ANN regression calibrated with the retention data collected in representative gradients to predict the chromatographic behavior of ionizable solutes in external separation conditions. Retention in gradient RP-HPLC is governed by several factors, such as the chemical structure of solutes, their acid–base properties, the polarity/acidity of the mobile phase, and how these properties change during the chromatographic run. While, on the one hand, ANN is potentially able to treat such complexity, on the other, the network does not provide a fitting equation that could be useful for getting information about the relative role of the different factors in the retention process. Nevertheless, finding the optimal condition for the chromatographic separation of a complex mixture, which is anyway a multivariate problem, can be handled by statistical retention models, but their predictive performance is more important than the knowledge of their physical meaning. In this view, a network was trained to associate the experimental parameters describing the gradient elution profile with the retention times of a mixture of target analytes to be separated. The ANN-based model, once calibrated on a sufficiently large set of representative separation conditions, was later applied to simultaneously predict the retention times of all the solutes in external elution conditions. In the end, the ANN response can be useful for optimization purposes, because it allows deducing the retention of the target solutes at any point of the experimental domain explored in calibration, and it may replace or support inefficient trial-and-error empirical approaches usually adopted to search the optimal separation conditions. At first, two separate ANN-based retention models were generated to predict the data collected under application of linear φ gradients or pH gradients. In addition, the retention behavior of the amino acids under the independent or simultaneous application of linear φ and pH gradients was modeled by ANN. The predictive performance of the various ANN-based models developed in this work was compared with the prediction ability of the retention models based on the solution of the fundamental equation of gradient elution.

2. Results

2.1. Identification of Model Variables and Data Subsets

In this paper, ANN regression was used to model the RP-HPLC retention of o-phthaladehyde (OPA) derivatives of 16 amino acids collected under the application of φ gradients, pH gradients, or combined pH/φ gradients. The retention datasets (A, B, and C, respectively), taken from the literature [14–16], are described in Section 3.1. The following variables were considered to describe the linear φ gradients of dataset A: the starting pH (pH$_i$), the starting organic solvent content (φ_i), and the φ-gradient slope ($\Delta\varphi/t_g = (\varphi_f - \varphi_i)/t_g$), where φ_f is the φ value at the end of gradient run and t_g is the gradient time. The pH gradients of dataset B were described by φ_i, pH$_i$ and the gradient slope ($\Delta pH/t_g = (pH_f - pH_i)/t_g$, where pH$_f$ is the final pH value). The respective values of the above quantities, determined from the experimental conditions reported in the original papers [14,16], are collected in Table 1. Among these parameters, the constant ones ($\Delta pH/t_g$ and φ_i in dataset A, and $\Delta\varphi/t_g$ in dataset B) were not considered as network inputs in ANN modeling of φ gradients or pH gradients. Datasets A and B were successively fused in one comprehensive dataset, hereafter indicated as A+B, to attempt ANN modeling of retention data collected under independent applications of φ gradients and pH gradients. In this case, all four gradient descriptors reported in Table 1 are informative and were considered as ANN inputs. To describe the 27 double pH/φ gradients of dataset C (referring to double pH/φ gradients), the three non-constant experimental quantities (pH$_f$, φ_f, and t_g) varying according to a three-level experimental design in Reference [15] were assumed as ANN inputs. The level values selected for these variables are given in Section 3.1.

As described in Section 3.2, ANN regression requires a training set, which is processed to update the network weights and biases; however, the network performance must also be monitored during learning using unknown data (validation set) to avoid overfitting. Moreover, the real generalization ability of the learned network must be finally evaluated on external data (test set) neither used in training nor in validation. To design these three datasets, the various φ gradients of dataset A, the pH gradients of dataset B, and the φ/pH gradients of dataset C were graphically represented in the space of the variables previously selected to describe the changes in the eluent composition (Figure 1). These plots helped us generate three well-balanced subsets in terms of representativeness; the data samples assigned to each subset were selected to cover the investigated experimental domain as much as homogeneously possible. Regardless of the dataset, six gradients were selected for the final test; three gradients (dataset A) or four gradients (datasets B and C) were selected for the internal validation and the remaining elution conditions were used to train the networks (Table 1 and Figure 1). The training, validation, and test sets for dataset A+B were designed by fusing the respective subsets of the A and B matrices. Considering that the retention data of 16 amino acids are associated with each experimental elution mode, the training, validation, and test data points were 160, 48, and 96, respectively, for dataset A; 192, 64, and 96, respectively, for dataset B; 352, 112, and 196, respectively, for dataset A+B; and 272, 64, and 96 for dataset C. Rather than representing the solutes by molecular descriptors, according to conventional QSRR approach, each of the 16 amino acids was identified by a

16-bit string, consisting of all "0" values except the n-th bit, which was set to "1", where n corresponds to the position of that solute in an arbitrary and predefined sequence of the investigated analytes. In this condition, the network was trained to properly associate the retention times to both solutes and gradient modes, without any explicit reference to the solute molecular structure.

2.2. ANN Modeling of Retention

The distinct networks handling the retention datasets A, B, A + B, and C were optimized following a usual procedure aimed at founding the combination of the ANN adjustable parameters providing the lowest validation error. A range-scaling between 0 and 1 was always applied to both input and output variables. Retention time (t_R(min)) values and their logarithmic values were alternatively considered as the ANN responses. Both options provided good ANN models and a random distribution of absolute residuals; however, logarithmic transformation of retention times was preferred to the unscaled values because it gave lower relative errors for the less retained amino acids.

Table 1. Descriptors of the linear φ gradients (dataset A) and pH gradients (dataset B).

Dataset	Gradient Code	Subset [a]	pH_i	$\Delta pH/t_g$	φ_i	$\Delta\varphi/t_g$
A	1A	train	2.8	0	0.20	0.06
	2A	val	2.8	0	0.20	0.03
	3A	train	2.8	0	0.20	0.015
	4A	test	3.3	0	0.20	0.03
	5A	test	3.3	0	0.20	0.02
	6A	train	3.3	0	0.20	0.015
	7A	test	3.3	0	0.20	0.01
	8A	train	3.82	0	0.20	0.06
	9A	test	3.82	0	0.20	0.018
	10A	train	3.82	0	0.20	0.012
	11A	train	4.2	0	0.20	0.03
	12A	train	4.2	0	0.20	0.015
	13A	val	4.2	0	0.20	0.01
	14A	val	5.85	0	0.20	0.015
	15A	train	5.85	0	0.20	0.01
	16A	test	5.85	0	0.20	0.0075
	17A	train	7.8	0	0.20	0.015
	18A	test	7.8	0	0.20	0.01
	19A	train	7.8	0	0.20	0.0075
B	1B	train	2.8	0.79	0.35	0
	2B	val	2.8	0.527	0.35	0
	3B	test	2.8	0.395	0.35	0
	4B	train	2.8	0.263	0.35	0
	5B	val	2.8	0.527	0.25	0
	6B	train	2.8	0.527	0.27	0
	7B	test	2.8	0.527	0.30	0
	8B	train	2.8	0.263	0.25	0
	9B	test	2.8	0.263	0.27	0
	10B	train	2.8	0.263	0.30	0
	11B	train	3.2	0.580	0.25	0
	12B	train	3.2	0.387	0.25	0
	13B	val	3.2	0.290	0.25	0
	14B	train	3.2	0.193	0.25	0
	15B	train	3.2	0.387	0.27	0
	16B	test	3.2	0.387	0.30	0
	17B	test	3.2	0.387	0.35	0
	18B	train	3.2	0.290	0.30	0

Table 1. *Cont.*

Dataset	Gradient Code	Subset [a]	pH_i	$\Delta pH/t_g$	φ_i	$\Delta\varphi/t_g$
	19B	val	3.2	0.290	0.35	0
B	20B	test	3.2	0.193	0.27	0
	21B	train	3.2	0.193	0.30	0
	22B	train	3.2	0.193	0.35	0

[a] Training set (train), validation set (val), test set (test).

Based on the results of preliminary ANN runs, in which a sigmoid or a tangent hyperbolic activation function was applied to the hidden neurons, the latter was preferred, while application of a non-linear transformation in the output neuron was not required because it did not produce any improvement in the model performance. The number of hidden neurons was varied in the range between $N - 6$ and $N + 6$, where N is the number of inputs, and each tested network was trained until the validation error reached a minimum value.

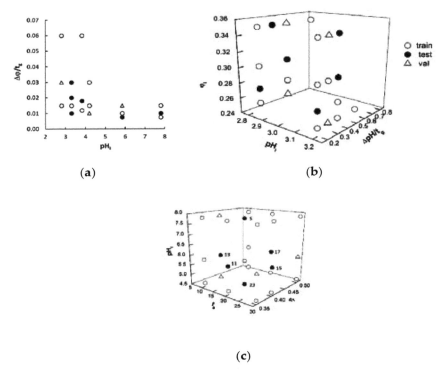

(a)   (b)

(c)

Figure 1. Gradients used in artificial neural network (ANN) training, validation, and test data projected in the space of the variables adopted as network inputs for datasets A (**a**), B (**b**), and C (**c**). Test samples of dataset C are labeled according to Reference [15].

The best ANN architectures and learning durations are presented in Table 2. Because of a randomization of the starting weights, here generated between -0.1 and 0.1, the optimal network produced slightly different responses upon being re-trained several times. To minimize the influence of the initial weights on the ANN-based model performance, the network was re-trained 100 times and the outputs were averaged. The agreement between computed or predicted ANN responses and the experimental t_R values for each retention dataset are graphically shown in Figure 2. Table 2 displays the determination coefficients in calibration and prediction (R^2 and Q^2) and the related standard errors (SEC and SEP, respectively) associated with the ANN-based models, where Q^2 was determined according to Todeschini et al. [29]. The average and maximum absolute percentage errors (mean(%) and max(%), respectively) in each subset are also reported. All the above statistical parameters refer to the unscaled t_R values.

Table 2. Description of the ANN models developed on the retention datasets A, B, A+B, and C: architecture and learning duration of the optimal network, coefficients of determination in training (R^2) and prediction (Q^2) and respective standard errors (SEC and SEP), and the mean and maximum percentage errors (mean(%) and max(%)).

Data Set	Network Topology	Learning Epochs	Training				Validation				Test			
			R^2	SEC	mean(%)	max(%)	Q^2	SEP	mean(%)	max(%)	Q^2	SEP	mean(%)	max(%)
A	18-14-1[a]	251	0.9999	0.06	0.3	1.6	0.9906	0.37	1.6	7.4	0.9984	0.22	1.4	6.4
B	19-21-1	63	0.9980	0.46	0.7	4.1	0.9778	0.78	1.4	4.1	0.9949	0.48	1.1	5.3
A+B	20-23-1	286	0.9993	0.23	1.2	6.3	0.9939	0.65	3.3	12.6	0.9799	0.48	2.5	10.4
C	19-18-1	125	0.9994	0.22	1.0	4.2	0.9938	0.72	2.6	6.9	0.9958	0.59	2.5	6.8

[a] Number of neurons in the input, hidden and output layer, respectively.

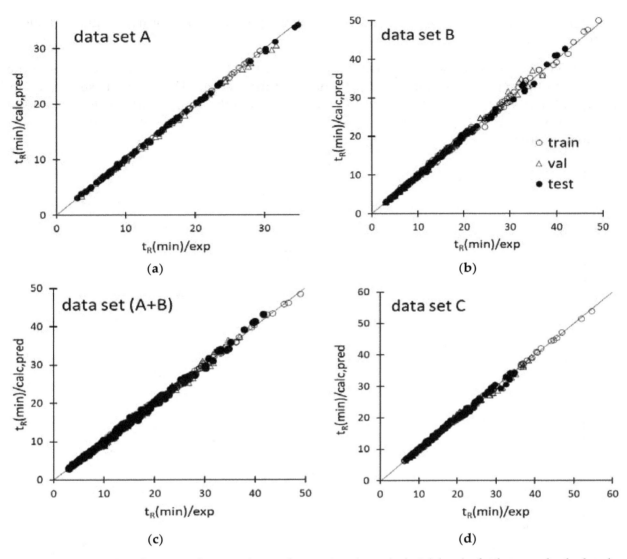

Figure 2. Agreement between the experimental retention times (t_R(min)/exp) of solutes and calculated or predicted ANN responses (t_R(min)/calc,pred) of datasets A (**a**), B (**b**), A+B (**c**), and D (**d**).

2.3. Predictive Performance of the ANN-Based Models

Inspection of the agreement plots for the various retention datasets modeled by ANN (Figure 2) reveals that both computed and predicted responses were very close to the ideal line, ensuring an accurate prediction of the retention times of the amino acids within the respective experimental domains. As expected, the training data samples were better modeled than the validation and test data; nonetheless, worsening of the predictive performance as compared to the fitting ability was slight, as confirmed by the small differences among the statistical parameters of training, validation, and test sets, reported in Table 2. The data samples were also randomly distributed around the ideal line of the agreement plots, suggesting the absence of systematic errors, except for dataset C, for which a small group of validation cases in the t_R range between 30 and 40 min were all underestimated (Figure 2d). Most of these data samples were associated with the most retained amino acids analyzed under the application of a same gradient ($\varphi_f = 0.5$, $pH_f = 5.86$, $t_g = 30$ min), but the errors were anyway acceptable (within 4–7%). The retention data collected under the application of φ gradients and pH gradients were very well modeled according to the mean errors, which were smaller than 1% and 1.5% for the training and test data, respectively (Table 2). Only a slight worsening of the descriptive/predictive ANN performance was observed when the network was called to model the retention times of the amino acids under the independent application of φ gradients and pH gradients (dataset A+B) or double pH/φ gradients; the mean error in both cases was just above 1% for the training set and 2.5% for the test set (Table 2).

Figure 3 displays the trend of the relative (%) errors (err (%)) for the retention times of the 16 amino acids in the φ gradients and/or pH gradients of the test set. Therefore, these data quantify the ability of the ANN-based models to predict the retention of the amino acids in elution conditions external to those used in calibration, and give a measure of the applicability of this approach in optimization problems.

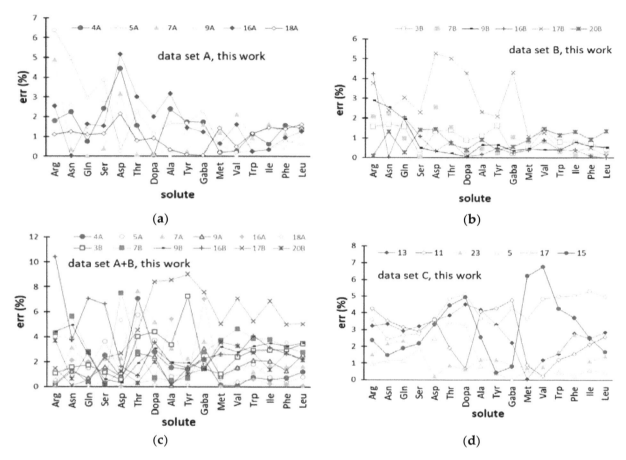

Figure 3. Percentage errors (err (%)) for the retention times of the amino acids provided by the ANN-based models in the external gradients (test set) of datasets A (**a**), B (**b**), A+B (**c**), and C (**d**). Abbreviations used for the amino acids are reported in Section 3.1. Gradient codes are specified in Table 1 (datasets A, B, and A+B) and Figure 1 (dataset C).

For most gradients of dataset A (φ gradients) and B (pH gradients), err (%) almost regularly decreased, passing from the less retained (Arg) to the most retained amino acid (Leu), seen from the left to the right of the plots displayed in Figure 3a,b. This arose from the fact that the absolute errors were homogeneously distributed over the target amino acids and, therefore, the relative errors were inversely related to the t_R value. Most of the predicted errors associated with the 16 amino acids in the external gradients of datasets A and B were smaller than 3%, while ANN modeling of datasets A+B (Figure 3c) and C (Figure 3d) provided slightly greater residuals, although generally below 5%. It can be noted that the retention times of most amino acids were less accurately predicted in the pH gradient 17B, when the data referring to pH gradients were modeled separately (dataset B) and when pH gradients and φ gradients were combined (dataset A+B). The moderately worse performance of the ANN model in this experimental condition can be due to the fact that the values of the two eluent descriptors (φ_i and pH_i) of pH gradient 17B were the greatest within the respective variability ranges (Table 1) and, therefore, the network was called to extrapolate the response.

2.4. Comparison of the ANN-Based Models with Retention Models Based on the Solution of the Fundamental Equation of Gradient Elution

The error trends provided by the retention models based on the solution of the fundamental equation of gradient elution that Pappa-Louisi and co-workers applied to datasets A [16] and B [14] are displayed in Figure 4 for comparison purposes. With regards to dataset A, the ANN-based model gave a lower number of errors above 3% as compared with the retention model developed in Reference [16]. Concerning dataset B, it must be noted that the pH gradient retention data collected in the pH ranges of 2.8–10.7 and 3.2–9 (in Table 1, gradients 1B–10B and 11B–22B, respectively) were fitted by two separate models in Reference [14], while, in this work, all 22 elution conditions were modeled by the same network. Nevertheless, the comprehensive ANN-based model built here seems to give a better prediction of the retention times, whereby the number of errors above 2% was lower as compared with the results provided by the two separate retention models generated from the solution of the fundamental equation of gradient elution. Although t_R values of the most retained amino acids (Val, Trp, Ile, Phe, and Leu) were predicted by the two alternative approaches with a comparable accuracy (errors were close to 1% or lower), the behavior of the less retained solutes was better described by the ANN model.

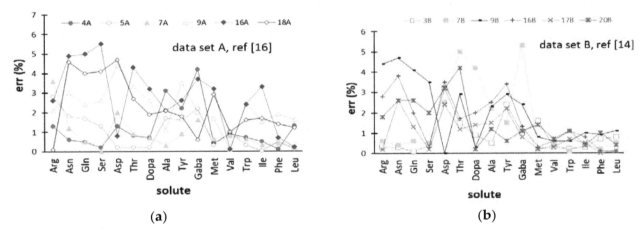

(a)   **(b)**

Figure 4. Percentage errors (err (%)) for the retention times of the amino acids provided by the retention models developed in References [14,16] for the external gradients (test set) of datasets A (**a**) and B (**b**). Abbreviations used for the amino acids are reported in Section 3.1. Gradient codes are specified in Table 1.

The comparison of Figures 3c and 4a,b reveals that the accuracy of prediction in the external φ gradients and pH gradients of dataset A+B was substantially equivalent to that provided by the solution of the fundamental equation of gradient elution. However, it should be remarked that a single ANN-based model was required to fit these data, while the data collected in pH gradients and φ gradients covering two different pH ranges were interpolated with three different retention models in References [14,16].

The ANN model describing retention under the application of double pH/φ gradients (dataset C) exhibited individual t_R errors in the external gradients that surpassed 5% only in a limited number of cases (Figure 3d). For this dataset, instead of the detailed trend of errors, not given in Reference [15], the mean errors provided by the retention model obtained from the solution of the fundamental equation of gradient elution could be considered for comparison. The mean and maximum errors reported for the model calibrated with all 27 gradients of dataset C were 2.9% and 18.9%, respectively. Moreover, the mean error associated with individual amino acids over the 27 gradients monotonically grew with the increase in retention time, from 1.5% (Arg) up to 6.5% (Leu) (Figure 5 of Reference [15]), revealing a poor modeling of the retention behavior of the most retained solutes. In the present work, the mean and maximum errors for the 17 gradients used to train the network were noticeably lower (1.0 and 4.2%, Table 2), and we observed a substantial independence of the training errors from the kind of amino acid. In Reference [15], the model initially developed using all the 27 gradients was recalibrated with 18 gradients and applied to the remaining nine gradients providing mean and maximum errors of 3.5 and 11.8%, respectively. In the present work, the network trained with 17 gradients gave lower mean and maximum errors both in internal validation (2.6 and 6.9%) and external prediction (2.5 and 6.8%). In summary, the ANN-based model, as compared with the retention models generated from the solution of the fundamental equation of gradient elution, provided a more accurate prediction of the retention times of the amino acids in double pH/φ gradients, as well as a more homogenous error distribution.

3. Methods

3.1. Retention Data

The data here analyzed were taken from three papers of Pappa-Louisi and co-workers [14–16] regarding the RP-HPLC retention of OPA derivatives of amino acids collected under the application of φ gradients, pH gradients, or combined pH/φ gradients. The mobile phases consisted of mixtures of aqueous phosphate buffer with a total ionic strength of 0.02 M and acetonitrile. In the first paper [16], 19 chromatographic runs were performed in different fixed eluent pHs (between 2.80 and 7.80), while the organic solvent volume fraction φ was linearly varied between 0.2 and 0.5 in different gradient durations (t_g, ranging between 5 and 40 min). In the second paper [14], φ was kept fixed (at 0.25, 0.27, 0.3, or 0.35), and 22 different linear pH gradients were applied in the pH ranges of 2.8–10.7 or 3.2–9, where t_g was varied between 10 and 30 min. A third retention dataset (dataset C) was collected by Zisi et al. [15] under the application of a double organic solvent and pH gradient, in which both φ and pH were linearly changed from initial values (φ_i and pH_i) to final values (φ_f and pH_f). This consisted of 27 different runs performed at fixed values of φ_i (0.25) and pH_i (3.21), while pH_f, φ_f, and t_g were varied according to a three-level experimental design. The selected levels were 4.68, 5.86, and 7.86 for pH_f; 0.35, 0.40, and 0.50 for φ_f; and 10, 20, and 30 min for t_g.

The amino acids analyzed in the above conditions were as follows: L-arginine (Arg), L-asparagine (Asn), L-glutamine (Gln), L-serine (Ser), L-aspartic acid (Asp), L-threonine (Thr), beta-(3,4-dihydroxyphenyl)-L-alanine (Dopa), L-alanine (Ala), L-tyrosine (Tyr), 4-aminobutyric acid (GABA), L-methionine (Met), L-valine (Val), L-tryptophan (Trp), L-isoleucine (Ile), L-phenylanine (Phe), and L-leucine (Leu). The amino acid L-glutamic acid (Glu), which was analyzed only in some experimental conditions, was not considered here. Apart from the different gradient profiles, all the retention data were collected with the same column, detector, and eluent flow rate. A 250 mm × 4.6 mm MZ-PerfectSil Target ODS-3HD analytical column with a 5-μm particle size kept at 25 °C was used, and the spectrofluorometric detector worked at 455 nm after excitation at 340 nm. Further experimental details can be found in the original papers [14–16].

3.2. Artificial Neural Network Modelling

A three-layer feed-forward ANN [30,31] was used in this work. The network consisted of one layer of input neurons, one output neuron, and an adjustable number of neurons in the hidden layer, fully connected to both the input and output neurons. Weights were associated to the connections, which modulated the information flowing from the input layer collecting the independent variables to the output neuron providing the network response. The weighted input variables entering each neuron of the hidden layer were summed and added to a bias value, and the result was transformed by a non-linear activation function, providing an output signal. The output neuron operated in a similar way on the weighted outputs of the hidden neurons producing the final response. A starting set of weights and biases, randomly generated, was sequentially updated in a learning (or training) procedure in which the network evaluated several input/output pairs (training set) to produce the best agreement between the target and computed responses. The optimized set of weights and biases, which represented a sort of memory of the learned network, could later be recalled, making predictions of the unknown response when the predictors were known. In this work, the network was trained by a quasi-Newton method [31], which incorporates second-order information about the error surface shape, ensuring faster convergence and a greater probability of avoiding local minima as compared to the classical error backpropagation learning algorithm. To avoid overfitting, the ANN performance during the learning step was monitored on unknown data samples (validation set), and the weight update was interrupted when the validation error started increasing. Minimization of the validation error was the criterion also adopted to select among alternative ANN models, differing in their network architecture, kind of activation function, kind of data scaling, and so on, the one with the best expected generalization ability. The real predictive performance of the final ANN-based model was finally evaluated on data samples (test set) external to both the training and validation sets. Software OpenNN [32] was used to perform ANN modeling.

4. Conclusions

In this paper, a three-layer artificial neural network was used to model the retention times of 16 amino acids under the separate or simultaneous application of linear organic modifier and pH gradients. We focused on the ANN's capability to predict the retention data of the target solutes in external gradients, which is a useful response for optimization purposes. Using a "bit-string" representation of solutes allowed simultaneously modeling the retention behavior of all 16 amino acids with no explicit reference to their chemical structure or properties. It follows that the approach presented in this work can be transferred to chemical classes or heterogeneous groups of solutes different from those investigated. Moreover, the model generation did not require any assumption concerning the dependence of the retention factors on the eluent pH and composition, which is, by contrast, a prerequisite to attempt the solution of the fundamental equation of gradient elution. The predictive ability of the ANN-based models tested on external gradients was very good, whereby the mean errors for the retention times were 1.1% for φ gradients, 1.4% for pH gradients, and 2.5% for pH/φ gradients, and better than that provided by retention models based on the solution of the fundamental equation of gradient elution. In summary, ANN modeling seems a powerful and flexible regression tool to describe the effect of the experimental conditions in linear gradient elution on the retention of ionizable solutes and, in combination with experimental design, can be applied to optimize HPLC methods.

References

1. Fekete, S.; Veuthey, J.L.; Guillarme, D. New trends in reversed-phase liquid chromatographic separations of therapeutic peptides and proteins: Theory and applications. *J. Pharm. Biomed. Anal.* **2012**, *69*, 9–27. [CrossRef] [PubMed]

2. Domínguez-Álvarez, J.; Mateos-Vivas, M.; Rodríguez-Gonzalo, E.; García-Gómez, D.; Bustamante-Rangel, M.; Delgado Zamarreño, M.M.; Carabias-Martínez, R. Determination of nucleosides and nucleotides in food samples by using liquid chromatography and capillary electrophoresis. *TrAC Trends Anal. Chem.* **2017**, *92*, 12–31. [CrossRef]

3. Mazzeo, P.; Di Pasquale, D.; Ruggieri, F.; Fanelli, M.; D'Archivio, A.A.; Carlucci, G. HPLC with diode-array detection for the simultaneous determination of di(2-ethylhexyl)phthalate and mono(2-ethylhexyl)phthalate in seminal plasma. *Biomed. Chromatogr.* **2007**, *21*, 1166–1171. [CrossRef] [PubMed]

4. D'Archivio, A.A.; Maggi, M.A.; Ruggieri, F.; Carlucci, M.; Ferrone, V.; Carlucci, G. Optimisation by response surface methodology of microextraction by packed sorbent of non steroidal anti-inflammatory drugs and ultra-high performance liquid chromatography analysis of dialyzed samples. *J. Pharm. Biomed. Anal.* **2016**, *125*, 114–121. [CrossRef] [PubMed]

5. Fanali, S.; Haddad, P.R.; Poole, C.F.; Schoenmakers, P.; Lloyd, D. *Liquid Chromatography: Fundamentals and Instrumentation*; Elsevier: Amsterdam, The Netherlands, 2013; ISBN 9780124158078.

6. Jandera, P.; Churáček, J. Gradient elution in liquid chromatography. II. Retention characteristics (retention volume, band width, resolution, plate number) in solvent-programmed chromatography—Theoretical considerations. *J. Chromatogr. A* **1974**, *91*, 223–235. [CrossRef]

7.   Kaliszan, R.; Wiczling, P.; Markuszewski, M.J. pH Gradient Reversed-Phase HPLC. *Anal. Chem.* **2004**, *76*, 749–760. [CrossRef]

8.   Poole, C.F.; Lenca, N. Applications of the solvation parameter model in reversed-phase liquid chromatography. *J. Chromatogr. A* **2017**, *1486*, 2–19.

9.   Vitha, M.; Carr, P.W. The chemical interpretation and practice of linear solvation energy relationships in chromatography. *J. Chromatogr. A* **2006**, *1126*, 143–194. [CrossRef]

10. Torres-Lapasió, J.R.; García-Alvarez-Coque, M.C.; Rosés, M.; Bosch, E.; Zissimos, A.M.; Abraham, M.H. Analysis of a solute polarity parameter in reversed-phase liquid chromatography on a linear solvation relationship basis. *Anal. Chim. Acta* **2004**, *515*, 209–227. [CrossRef]

11. Cela, R.; Ordoñez, E.Y.; Quintana, J.B.; Rodil, R. Chemometric-assisted method development in reversed-phase liquid chromatography. *J. Chromatogr. A* **2013**, *1287*, 2–22. [CrossRef]

12. Andrés, A.; Téllez, A.; Rosés, M.; Bosch, E. Chromatographic models to predict the elution of ionizable analytes by organic modifier gradient in reversed phase liquid chromatography. *J. Chromatogr. A* **2012**, *1247*, 71–80. [CrossRef] [PubMed]

13. Fasoula, S.; Zisi, C.; Gika, H.; Pappa-Louisi, A.; Nikitas, P. Retention prediction and separation optimization under multilinear gradient elution in liquid chromatography with Microsoft Excel macros. *J. Chromatogr. A* **2015**, *1395*, 109–115. [CrossRef] [PubMed]

14. Pappa-Louisi, A.; Zisi, C. A simple approach for retention prediction in the pH-gradient reversed-phase liquid chromatography. *Talanta* **2012**, *93*, 279–284. [CrossRef]

15. Zisi, C.; Fasoula, S.; Nikitas, P.; Pappa-Louisi, A. Retention modeling in combined pH/organic solvent gradient reversed-phase HPLC. *Analyst* **2013**, *138*, 3771–3777. [CrossRef] [PubMed]

16. Fasoula, S.; Zisi, C.; Nikitas, P.; Pappa-Louisi, A. Retention prediction and separation optimization of ionizable analytes in reversed-phase liquid chromatography under organic modifier gradients in different eluent pHs. *J. Chromatogr. A* **2013**, *1305*, 131–138. [CrossRef] [PubMed]

17. Héberger, K. Quantitative structure-(chromatographic) retention relationships. *J. Chromatogr. A* **2007**, *1158*, 273–305. [CrossRef] [PubMed]

18. D'Archivio, A.A.; Incani, A.; Ruggieri, F. Retention modelling of polychlorinated biphenyls in comprehensive two-dimensional gas chromatography. *Anal. Bioanal. Chem.* **2011**, *399*, 903–913. [CrossRef]

19. D'Archivio, A.A.; Maggi, M.A.; Mazzeo, P.; Ruggieri, F. Quantitative structure-retention relationships of pesticides in reversed-phase high-performance liquid chromatography based on WHIM and GETAWAY molecular descriptors. *Anal. Chim. Acta* **2008**, *628*, 162–172. [CrossRef]

20. D'Archivio, A.A.; Maggi, M.A.; Ruggieri, F. Multiple-column RP-HPLC retention modelling based on solvatochromic or theoretical solute descriptors. *J. Sep. Sci.* **2010**, *33*, 155–166. [CrossRef]

21. D'Archivio, A.A.; Giannitto, A.; Maggi, M.A. Cross-column prediction of gas-chromatographic retention of polybrominated diphenyl ethers. *J. Chromatogr. A* **2013**, *1298*, 118–131. [CrossRef]

22. D'Archivio, A.A.; Incani, A.; Ruggieri, F. Cross-column prediction of gas-chromatographic retention of polychlorinated biphenyls by artificial neural networks. *J. Chromatogr. A* **2011**, *1218*, 8679–8690. [CrossRef] [PubMed]

23. D'Archivio, A.A.; Giannitto, A.; Maggi, M.A.; Ruggieri, F. Cross-column retention prediction in reversed-phase high-performance liquid chromatography by artificial neural network modelling. *Anal. Chim. Acta* **2012**, *717*, 52–60. [CrossRef] [PubMed]

24. Fatemi, M.H.; Abraham, M.H.; Poole, C.F. Combination of artificial neural network technique and linear free energy relationship parameters in the prediction of gradient retention times in liquid chromatography. *J. Chromatogr. A* **2008**, *1190*, 241–252. [CrossRef] [PubMed]

25. Golubović, J.; Protić, A.; Otašević, B.; Zečević, M. Quantitative structure-retention relationships applied to development of liquid chromatography gradient-elution method for the separation of sartans. *Talanta* **2016**, *150*, 190–197. [CrossRef] [PubMed]

26. Barron, L.P.; McEneff, G.L. Gradient liquid chromatographic retention time prediction for suspect screening applications: A critical assessment of a generalised artificial neural network-based approach across 10 multi-residue reversed-phase analytical methods. *Talanta* **2016**, *147*, 261–270. [CrossRef] [PubMed]

27. D'Archivio, A.A.; Maggi, M.A.; Ruggieri, F. Prediction of the retention of s-triazines in reversed-phase high-performance liquid chromatography under linear gradient-elution conditions. *J. Sep. Sci.* **2014**, *37*, 1930–1936. [CrossRef] [PubMed]

28. D'Archivio, A.A.; Maggi, M.A.; Ruggieri, F. Artificial neural network prediction of multilinear gradient retention in reversed-phase HPLC: Comprehensive QSRR-based models combining categorical or structural solute descriptors and gradient profile parameters. *Anal. Bioanal. Chem.* **2015**, *407*, 1181–1190. [CrossRef]

29. Todeschini, R.; Ballabio, D.; Grisoni, F. Beware of Unreliable Q2! A Comparative Study of Regression Metrics for Predictivity Assessment of QSAR Models. *J. Chem. Inf. Model.* **2016**, *56*, 1905–1913. [CrossRef]

30. Marini, F.; Bucci, R.; Magrì, A.L.; Magrì, A.D. Artificial neural networks in chemometrics: History, examples and perspectives. *Microchem. J.* **2008**, *88*, 178–185. [CrossRef]

31. Svozil, D.; Kvasnička, V.; Pospíchal, J. Introduction to multi-layer feed-forward neural networks. *Chemometr. Intell. Lab. Syst.* **1997**, *39*, 43–62. [CrossRef]

32. Lopez, R. Open NN: An Open Source Neural Networks C++ Library. 2014. Available online: http://opennn. cimne.com/ (accessed on 20 January 2019).

11

Synthesis of Graphene Oxide Based Sponges and their Study as Sorbents for Sample Preparation of Cow Milk Prior to HPLC Determination of Sulfonamides

Martha Maggira [1], **Eleni A. Deliyanni** [2] and **Victoria F. Samanidou** [2,*]

[1] Laboratory of Analytical Chemistry, Department of Chemistry, Aristotle University of Thessaloniki, GR-541 24 Thessaloniki, Greece; marthamaggira@gmail.com

[2] Laboratory of General and Environmental Technology, Department of Chemistry, Aristotle University of Thessaloniki, GR-541 24 Thessaloniki, Greece; lenadj@chem.auth.gr

* Correspondence: samanidu@chem.auth.gr

Abstract: In the present study, a novel, simple, and fast sample preparation technique is described for the determination of four sulfonamides (SAs), namely Sulfathiazole (STZ), sulfamethizole (SMT), sulfadiazine (SDZ), and sulfanilamide (SN) in cow milk prior to HPLC. This method takes advantage of a novel material that combines the extractive properties of graphene oxide (GO) and the known properties of common polyurethane sponge (PU) and that makes sample preparation easy, fast, cheap and efficient. The PU-GO sponge was prepared by an easy and fast procedure and was characterized with FTIR spectroscopy. After the preparation of the sorbent material, a specific extraction protocol was optimized and combined with HPLC-UV determination could be applied for the sensitive analysis of trace SAs in milk. The proposed method showed good linearity while the coefficients of determination (R^2) were found to be high (0.991–0.998). Accuracy observed was within the range 90.2–112.1% and precision was less than 12.5%. Limit of quantification for all analytes in milk was $50~\mu g~kg^{-1}$. Furthermore, the PU-GO sponge as sorbent material offered a very clean extract, since no matrix effect was observed.

Keywords: sulfonamides; HPLC; graphene oxide; sponge; milk

1. Introduction

Sulfonamides are a group of synthetic antibacterial agents, which are widely used in veterinary practice for prophylactic and therapeutic purposes and as feed additives. Due to their ability to inhibit folic acid synthesis in microorganisms, they are commonly used against a wide range of bacteria, protozoa, parasites, and fungi [1–3].

However, the improper administration of sulfa drugs in dairy husbandry and the insufficient withdrawal periods can lead to noncompliant residues in animal originated foods, a fact which can contribute to several concerns in the dairy industry and public health [4].

In humans, such concerns comprise the rise of allergic or toxic reactions and the development of drug-resistance, whereas in the dairy industry they provoke the inhibition of bacterial fermentation in cheese and yoghurt production [5]. In order to safeguard public health and ensure food safety, monitoring of such residues in products designated for human consumption is considered mandatory.

For this reason, the European Union has established a maximum residue level (MRL) for sulfonamides in foodstuffs of animal origin, which in the case of milk is 100 μg kg^{-1} [6]

Additionally, several methods have been described for the detection and/or determination of sulfonamides in foods of animal origin such as microbial inhibition assays, immunochemical methods, capillary electrophoresis (CE), gas chromatography (GC), and HPLC [5,7].

Sample preparation is a key step prior to the detection of sulfonamides present in different kinds of samples. The clean-up procedure of various matrices can be accomplished by either traditional techniques, such as liquid-liquid extraction (LLE) [8], or modern methods, like solid phase extraction (SPE) [9], solid phase micro extraction (SPME) [1,10], fabric phase solid extraction [11], matrix solid phase dispersion (MSPD) [12] and Quick, Easy, Cheap, Effective, Rugged and Safe (QuEChERS) method [13,14]. Most of the aforementioned techniques depend on an absorbent material to achieve high analytical specificity and selectivity.

However, in the analysis of complex matrices, many innovative materials have emerged as valuable tools to enhance the efficiency of the extraction and isolation of the target analytes. As such, graphene-based materials are preferred to other carbon-based nanomaterials due to their great potential on the sample preparation procedure. Graphene (G) is a two dimensional nanomaterial with extraordinary physicochemical properties such as thermal and chemical stability, thermal conductivity, hydrophobicity, and large specific surface area [15]. Graphene oxide (GO) is a single-atomic layered material, an important derivative of graphene with similar structure, which is composed easily from the oxidation of graphite. However, GO is more polar than G because of the hydroxyl (–OH) and carboxyl (–COOH) groups, a characteristic that facilitates GO bonds into other compounds such as aminopropyl silica [16].

Graphene based materials are extensively applied in SPE procedure as they offer high sorption efficiency for organic compounds and metal ions mainly in environmental samples [17–19]. Although G and GO demonstrate excellent sorbent characteristics, many limitations have been reported concerning their isolation from well dispersed solutions and their sheets' restacking or escaping from the SPE column [20,21].

In order to surpass the problems having occurred during the elution and sample loading in SPE, new sample preparation techniques have been developed such as the use of graphene-based materials in dispersive solid phase extraction (DSPE) and MSPD. In DSPE the absorbent is mainly utilized in food [22] and environmental samples [23–26], whereas MSPD has been performed for the extraction of sulfonamides in milk samples [27].

Recently, melamine sponge was functionalized with graphene, via a microwave-assisted hydrothermal process, in order to be used as adsorbent for SAs extraction from milk, egg, and environmental water [28]. The proposed method was highly accurate and sensitive for the analysis of nine SA's. However, it is not referred to the determination of sulfathiazole (STZ), sulfamethizole (SMT), and sulfanilamide (SN). In the current study, commercial polyurethane (PU) sponges, a kind of cheap porous material, were examined for SAs extraction from milk. PU sponges, compared with other sponge materials, such as melamine [29,30], and chitosan sponge [31] present certain advantages like easy access, low cost, and high resilience, excellent flexibility, and reuse [32]. Moreover, the surface of the PU sponge was used as a skeleton for hydrophobic modifiers. Hence, in the current study, surface modification was achieved via a green route at ambient conditions.

Polyurethane (PU) sponges with a unique 3D structure have a potential application as absorbents due to their advantages of easy access, low cost, and high resilience compared to other porous materials, such as melamine foam and chitosan sponge. Although PU sponge is hydrophilic, modifications or physical coating like functionalization with graphene are required to increase the hydrophobicity and are usually used to achieve higher efficiency in separations [32].

Consequently, the objective of this study was to combine the unique properties of PU sponge being functionalized with GO in order to serve as an innovative absorbent material in the sample preparation procedures. Due to its properties of low cost, time saving, and simplicity, the GO-PU material was further used for the determination of sulfonamides in cow milk samples prior to HPLC-DAD method.

2. Results and Discussion

2.1. Characterization

Polyurethane sponge was used as a base material in order to be functionalized with graphene oxide. Polyurethane presents an open-hole structure, with a high porosity as well as a rich surface chemistry with surface-groups that can attract and react with different molecules. Graphene oxide was embodied in the PU skeleton after the dispersion of GO in water. Graphene oxide was connected to polyurethane after chemical interactions between the GO (epoxy-groups) and polyurethane surface groups (C=O and –N–H groups). After the polyurethane functionalization with graphene oxide, the sponge prepared appeared with a black color and presented hydrophobicity that was further increased after the coating with PVA.

The XRD diffraction patterns of the prepared graphite oxide (GO) as well as of the GO impregnated sponge before (PU-GO) and after the PVA coating (PU-GO-PVA) are presented in Figure 1. Graphite presents a sharp diffraction peak at 26.6° in the XRD pattern (not presented), attributed to interlayer (002) spacing (d = 0.33 nm). The characteristic XRD peak of graphite oxide appeared at 2θ = 10.9°; as estimated by the Bragg's law, the interlayer distance between the carbon layers, increased from 0.33 nm for graphite to 0.81 nm for GO [33]. In the XRD pattern of the GO impregnated sponge (PU-GO) the characteristic XRD peak of graphite oxide, at 2θ = 10.9°, was not present, indicating that the layered structure of GO was destroyed. A diffraction peak at 2θ = 21° could be due to PVA while the broad peaks at around 11.6° and 19.8° indicated some degree of crystallinity of the PU [34–36]. The XRD pattern for the sample after the sulfonamide adsorption (PU-GO-SA), which is also presented in Figure 2, reveals that a decrease of crystallinity was observed, evidenced by the disappearance of the peak at 2θ = 11.6°.

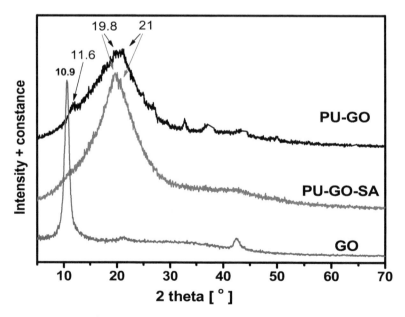

Figure 1. X-ray diffraction (XRD) patterns of the graphite oxide (GO), the graphene oxide impregnated sponge (PU-GO), and the sponge after the adsorption of sulfonamides (PU-GO-SA).

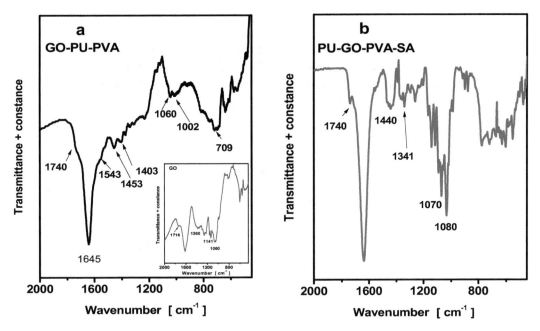

Figure 2. Fourier-transform infrared (FTIR) spectra for (**a**) polyurethane-graphene oxide- polyvinyl alcohol (PU-GO-PVA) sponge raw and after (**b**) the absorption of sulfonamide's (SA's) (PU-GO-PVA-SA)-(in the inset the spectrum of GO).

FTIR spectroscopy was used in this study to identify the possible interactions between GO and PU (PU-GO), between PU-GO and PVA (PU-GO-PVA sponge) as well as between the sponge and the sulfonamides (PU-GO-PVA-SA) in order for the adsorption mechanism to be revealed. The FTIR spectra of PU-GO-PVA as well as of PU-GO-PVA after the sorption of sulfonamide (PU-GO-PVA-SA), are presented in Figure 2. The FTIR spectra of GO is presented in the inset of Figure 2. GO contains polar groups on the edges of graphite layers such as carbonyl, carboxyl, and epoxide, as well as hydroxyl groups within the basal planes of the graphene sheets. In the spectrum of GO (Figure 2a), the bands at 1050–1100 cm^{-1} and ~1716 cm^{-1} can be attributed to carboxylic groups whereas the band at ~1600 cm^{-1} can be attributed to C=C stretching mode of the sp^2 carbon skeletal network and/or to epoxy groups. The band at 1356 cm^{-1} is due to C–OH stretching of O–H groups, while the band at 1045 and at 1141 cm^{-1} can be also attributed to epoxy and alkoxy C–O groups, respectively.

Polyurethane (PU) is a polymer obtained after the polymerization of diisocyanate and polyol that contains C=O and –NH groups (electron donating sites); these groups are able to form hydrogen bonds with graphene oxide during the complexation. The spectra of PU-GO-PVA sponge presented peaks at 1740 and 1060 cm^{-1} attributed to carboxyl and epoxy groups, respectively, at a lower intensity compared to the relative peaks of the spectra of GO, indicating the involvement of these groups in the composite synthesis. The peaks at 1543 cm^{-1} could be attributed to amide II formation after reaction of the carboxylic groups of GO with –NH groups of PU while the peaks at about 1453 cm^{-1} could be attributed to –CH$_3$ groups of PVA indicating the covering [37–39].

The most significant spectra alterations for the GO-PU-PVA after the SA adsorption (GO-PU-PVA-SA sample), are the new bands appearing at 1260 and 1070 cm^{-1} in addition to the diminishing of the peaks at 1191, 1130 and 1740 cm^{-1} (carbonyl) absorption bands (Figure 2). The new band at 1440 cm^{-1} can be attributed to amide I formation due to interactions between the SA amines and the sponge carboxylates, causing the diminishing of the band at 1740 cm^{-1}. The new band at 1260 cm^{-1}, can be attributed to hydrogen bond interaction between the GO-PU-PVA carboxyl groups and the sulfones/O=S=O groups of SA which are strong hydrogen-bond acceptors. It is obvious that the grafting of PU with extra carboxyl groups enhanced the SA adsorption owning to their reactions with the amines and the hydrogen bond with the sulfones/O=S=O groups of the SA. This was also reported for dorzolamine encapsulation to chitosan, as well as for pramipexole adsorption on activated carbon.

2.2. Synthesis Optimization

The mass of the material retained in the sponge was initially studied, keeping its second mass at 0.04 g. After selecting three different levels (0.12, 0.24, and 0.32 g), the procedure of the sponge preparation was followed. The sample preparation was performed in standard solutions with all three materials. From the results as presented in Figure 3, it seems that the mass of 0.12 g is more effective for the adsorption.

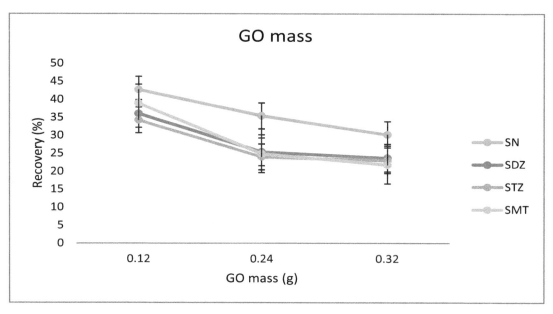

Figure 3. Effect of the graphene oxide (GO) mass on the adsorption efficiency of the sulfonamides.

The size of the sponge was optimized after the testing of two different sizes. Particularly 0.04 g and 0.07 g sponge were dipped in the dispersed solution. The results showed that the bigger sponge is sufficient to achieve the optimum adsorption.

For the PU-GO sponge formation, the GO molecules should be immobilized during its preparation. This is accomplished with the adding of a solvent like water or some polymer, of which polyvinyl alcohol (PVA) is more common due to its low cost. In the present research, two such solvents were tested, water and PVA. As shown in the results (Figure 4), PVA helps in the sample preparation procedure.

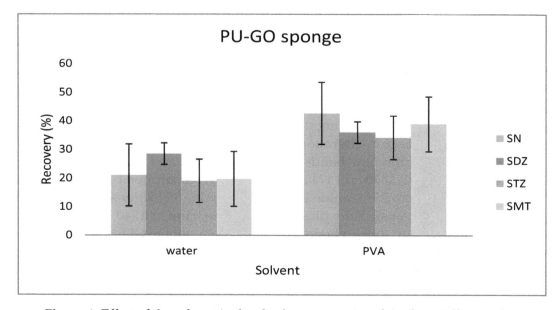

Figure 4. Effect of the solvent in the absolute recoveries of the four sulfonamides.

Different solutions of NH_3/EtOH containing 60 mL of the mixture were prepared in three different volume ratios (4:1, 1:1 and 1:4) and were further applied in the functionalization of the GO-PU material. The results revealed that the quantity of NH_3 was crucial to the absorption and that the volume ratio 4:1 achieved higher efficiency.

2.3. Chromatography

The target analytes were separated by gradient elution. Optimum gradient program was chosen as providing good analytes' resolution, at the shortest analysis. A typical chromatogram is shown in Figure 5. The retention times were observed at 6.345, 7.566, 8.748, and 12.899 min for SN, SDZ, STZ, and SMT respectively.

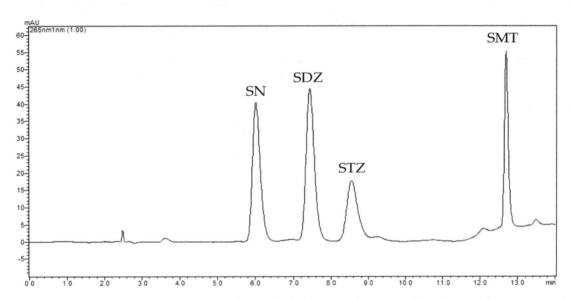

Figure 5. A typical HPLC chromatogram of standard solution of examined analytes at the concentration of 5 ng μL^{-1}. Peaks are as follows: SN: 6.345 min, SDZ: 7.566 min, STZ: 8.748 min, and SMT: 12.899 min.

2.4. Sample Preparation Optimization

All initial optimization experiments were performed using standard solutions of sulfonamides. The optimum conditions established were further checked for their appropriateness to the milk matrix.

In the loading and elution step different methods were tested. Although stirring showed the best results in the tests with the standard solutions, as shown in Table 1 the extraction declined sharply when the milk samples were tested and the recovery rates ranged from 7 to 14%. Thus, centrifugation in low rates was selected. Centrifugation at low rates had two purposes: (1) sufficient sample interaction with the material, and (2) preventing the adsorbent from escaping from the structure of the sponge. High centrifugation rates hindered the extraction process. With regards to sonication, GO particles were released from the sponge and sample handling was difficult.

Table 1. Effect of the loading/elution time and the extraction procedure on the efficiency of the method. (SN = sulfanilamide, SDZ = sulfadiazine, STZ = sulfathiazole, SMT = sulfamethizole)

Loading/Elution Time (min)		Absolute Recovery Rates (R%)			
		SN	SDZ	STZ	SMT
Rest	15/15	21.1	23.5	29.9	29.3
Sonication	7/7	15.6	21.2	29.7	33.7
Stirring	15/15	22.3	28.7	34.8	36.4
Centrifugation	15/15	30.9	24.0	27.6	29.6

Additionally, the volume of the sample, the elution solvents, the size of the sponge, loading and elution time, and the pH were optimized. The extraction was conducted with two different volume samples (1.5 and 3 g) that were spiked with the same amount of the target analytes. The results revealed a decrease in the extraction efficiency by increasing the volume of the sample.

With regards to the elution, methanol (MeOH) and acetonitrile (ACN) were tested both separately and in mixture. It is obvious from the results that the mixed solution increases the efficiency of the elution. In order to succeed better results, 1% acetic acid was added. The addition of acetic acid was successful and the optimum volume ratio for the $CH_3COOH/ACN/MeOH$ solution was 50:40:10.

As for the loading and elution time 10, 15, 20 min were tested. From the results it is observed that 10 min are not enough for the loading and the extraction of the target analytes. However, 15 and 20 min yielded similar results, and the shortest time was selected to reduce the process time.

The effect of the pH in the extraction efficiency was tested, adding 0.5 mL of buffer solution into the sample. Table 2 presents the results obtained from the addition of pH 3, 5, 7, and 9 buffer solution in milk sample. It is obvious from the results that the optimum pH is 5, whereas lower or higher pH values results in decrease in the adsorption for all SAs.

Table 2. Effect of the pH on the adsorption efficiency of the four sulfonamides. (SN = sulfanilamide, SDZ = sulfadiazine, STZ = sulfathiazole, SMT = sulfamethizole). Optimum pH value is given in bold.

	Absolute Recovery Rates (R%)			
pH	**SN**	**SDZ**	**STZ**	**SMT**
3	21.8	22.0	31.7	29.3
5	**22.2**	**27.5**	**36.1**	**31.7**
7	12.3	17.1	21.7	17.2
9	15.0	22.0	27.9	19.0

The proposed sample preparation protocol is very simple and rapid, with low consumption of organic solvents and very clean background signal. Figure 6 illustrates the simple pretreatment procedure. Typical chromatograms of a blank and a spiked milk sample are shown in Figure 7a,b. It is clear that the peaks of the substrate do not interfere with the analysis as they elute at different times.

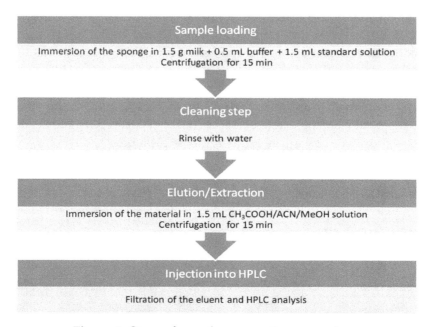

Figure 6. Steps of sample preparation procedure.

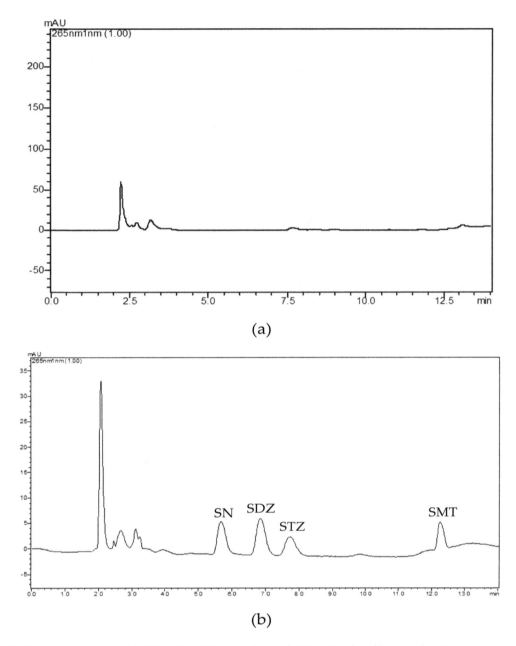

Figure 7. Chromatogram of (**a**) blank milk sample and (**b**) spiked milk sample at a concentration of 300 μg kg^{-1}.

2.5. Method Validation

2.5.1. Selectivity

The good resolution between the chromatographic peaks of analytes and the absence of interferences in the spiked milk samples indicate that a good selectivity was achieved.

2.5.2. Linearity and Sensitivity

Standard solutions showed linearity for all of the target analytes within the range of 0.5 to 10 ng μL^{-1} and showed and good correlation coefficients (0.981–0.999). Moreover, calibration curves were constructed using fortified milk samples after sample preparation, and good coefficients of determination between 0.9969 and 0.999 were achieved over the examined range. (Table 3). Limit of quantification for all analytes in milk was 50 μg kg^{-1}.

Table 3. Linearity data in standard solutions and spiked milk samples. (SN = sulfanilamide, SDZ = sulfadiazine, STZ = sulfathiazole, SMT = sulfamethizole).

Analytes	Calibration Curve	Coefficients of Determination (R^2)
	Standard Solutions	
SN	y = 119367x + 366.66	0.999
SDZ	y = 121371x + 27142	0.995
STZ	y = 64281x + 16245	0.999
SMT	y = 58218x + 13560	0.981
	Milk	
SN	y = 14785x − 1957.4	0.996
SDZ	y = 17012x − 3034.3	0.998
STZ	y = 10182x − 2023.9	0.991
SMT	y = 8489.3x − 2609.2	0.991

2.5.3. Precision and Accuracy

The precision of the method was based on within-day repeatability and between-day precision. The former was assessed by replicate ($n = 4$) measurements from a spiked milk sample at the MRL level for all examined sulfonamides. The recoveries of spiked samples were calculated by comparison of the peak area ratios for extracted compounds toward the values derived from spiked calibration curves. In Between-day reproducibility a triplicate determination was performed for a period of three days (Table 4). Precision and accuracy was determined at three concentration levels according to the 657/2002/EC decision [40].

Table 4. Precision and accuracy parameters of the method for the determination of sulfonamides in milk samples. (SN = sulfanilamide, SDZ = sulfadiazine, STZ = sulfathiazole, SMT = sulfamethizole).

Added Concentration ($\mu g\ kg^{-1}$)	Analyte	Intra-Day $n = 4$		Inter-Day $n = 3 \times 3$	
		R%	RSD	R%	RSD
50	SN	98.2	7.6	97.6	7.1
	SDZ	106.7	6.9	104.3	3.3
	STZ	93.6	8.5	95.6	9.8
	SMT	93.4	10.4	90.8	11.0
100	SN	103.3	4.0	107.7	4.0
	SDZ	112.1	10.8	105.3	0.4
	STZ	96.8	11.0	90.2	10.9
	SMT	92.8	11.8	97.6	12.4
150	SN	100.2	10.4	102.4	7.6
	SDZ	108.7	3.0	100.1	6.0
	STZ	96.6	9.8	92.8	12.0
	SMT	101.7	10.3	95.3	9.5

2.5.4. Decision Limit and Capability of Detection

Decision limit (CCα) is defined as "the limit at and above which it can be concluded with an error probability" and it was calculated after the analysis of 20 spiked milk samples at the MRLs of each compound. The decision limits CCa were 100.2 μg kg^{-1} for SN, 100.3 μg kg^{-1} for SDZ, 100.4 μg kg^{-1}

for STZ, and μg kg^{-1} for 100.3 SMT. Capability of detection (CCb) defined as "the smallest content of the substance that may be detected, identified, and/or quantified in a sample with an error probability of b" and it was calculated after the spiking of 20 blank milk samples at the CCa level of each compound. The capability of detection (CCb) were 110.7 μg kg^{-1} for SN, 109.3 μg kg^{-1} for SDZ, 115.4 μg kg^{-1} for STZ, and 114.3 μg kg^{-1} for SMT.

2.6. Application to Real Samples

The method was applied for the determination of the examined analytes in cow milk samples from local food stores. Five random samples of three different types of milk were collected and analyzed, including full-fat (3.5%), semi-skimmed (1.5%), and skimmed (0%) milk. All analyzed samples were negative in the presence of examined analytes.

2.7. Comparison with Other Methods

The method described in this study was compared with previous analytical approaches for the determination of SAs in milk. The analysis' results are comparable with those attained by other methods, with fairly good recoveries and quite satisfactory sensitivity. Although it provides higher LODs and LOQs than previously reported methods, it is a less costly (no commercial SPE products are needed) and less time-consuming method with easy handling of sponge and does not require highly sophisticated equipment since no MS is used (Table 5).

Table 5. Performance of the presented method in comparison with previously reported analytical methods.

Analytes	Sample Preparation	Analytical Technique	Run Time (min)	LOD-LOQ	Recovery (%)	Ref
4SAs	MSPE	HPLC-AS	N/A	LOD (ng/mL): 2.0–2.5 LOQ (ng/mL): 6.0–7.5	92–105	[41]
38 veterinary drugs (18SAs)	SPE	UHPLC-ESI-MS/MS	13.5	CCα (μg/kg): 109–114 (SAs) CCβ (μg/kg): 116–123 (SAs)	87–119 (all analytes)	[42]
6SAs	SPE	HPLC-DAD	15.3	LOD (μg/kg): 1.9–13.3 LOQ (μg/kg): 5.6–42.2	N/A	[7]
9 SAs	MSPE	HPLC-DAD	35	LOD (μg/L): 7–14	81.8–114.9	[27]
5 SAs	MSPE	HPLC-UV	8	LOD (μg/L): 1.16–1.59 LOQ (μg/L): 3.52–4.81	62.0–104.3	[12]
SMZ, SIX and SDMX	FPSE	HPLC-UV	6.5	CCα (μg/kg): 114.4–116.5 CCβ (μg/kg): 104.1–118.5	93–107	[11]
9 SAs	GMeS microextraction	HPLC-DAD	30	LOQ (μg/kg): 0.31–0.91	90–105	[28]
4 SAs	PU-GO sponge microextraction	HPLC-DAD	14	LOQ: 50 (μg/kg)	90.2–112.1	This study

3. Materials and Methods

3.1. Chemicals and Reagents

Sulfathiazole (STZ), sulfamethizole (SMT), sulfadiazine (SDZ), and sulfanilamide (SN) were purchased from Sigma-Aldrich (Steinheim, Germany). HPLC grade acetonitrile and methanol obtained from Chem-Lab (Zedelgem, Belgium). Formic and acetic acid were of analytical grade and purchased from Chem-Lab (Zedelgem, Belgium) and Merck (Darmstadt, Germany) respectively. Ethanol, reagent grade (Chem-Lab, Zedelgem, Belgium) and ammonia, 25% solution (PANREAC QUIMICA SA,

Barcelona, Spain) were used for the sponge optimization. Polyvinyl alcohol high molecular weight solid, (PVA 98–99 hydrolized) was purchased from A Johnson Company (New Brunswick, NJ, USA).

Graphite was purchased from Sigma Aldrich (St. Louis, MO, USA). Double-deionized water was filtered with 0.45 μm filter membrane before use.

Milk samples were collected from local market (Thessaloniki, Greece). Different fresh milk types were analyzed including skimmed (0% fat), semi-skimmed (1.5% fat), and full-fat milk (3.5% fat). All milk samples were kept refrigerated (at 4 °C) until use.

3.2. Instrumentation

Chromatographic separation and analysis were carried out on a Shimadzu HPLC system coupled to a Diode Array Detector (DAD) (Kyoto, Japan), equipped with Rheodyne 7725i 20 μL loop (Cotati, CA, USA). The system consisted of a Shimadzu LC-10 ADVP pump and a Shimadzu FCV-10ALVP solvent mixer (Kyoto, Japan). The chromatographic separation was achieved using a Merck-Lichrospher RP8e, 5 μm 250 × 4 mm analytical column (Darmstadt, Germany). Degassing of the mobile phase was performed by helium DGU-10B degassing unit by Shimadzu (Kyoto, Japan) directly in the solvent reservoirs. The system was controlled by Shimadzu LabSolutions software (Shimadzu, Kyoto, Japan) which was also used for the data acquisition and analysis.

A glass vacuum filtration apparatus obtained from Alltech Associates (Deerfield, IL, USA), was employed for the filtration of the solvents using cellulose nitrate 0.2 μm membrane filters from Whatman (Maidstone, UK) prior to use. A Glasscol Vortexer (Terre Haute, IN, USA), an ultrasonic bath Transonic 460/H (Elma, Germany), a Reacti-Vap evaporator model from PIERCE (Rockford, IL, USA), and a Hermle centrifugation (Gosheim, Germany) were acquired for the sample preparation. Moreover, a 20–200 μL micropipette ISOLAB Laborgerate GmbH (Wertheim, Germany) was used for the preparation of the standard solutions.

XRD measurements were performed on a Philips PW1820 X-ray diffractometer. The Fourier Transform Infrared Spectra (FTIR) were measured on a Nicolet 560 (Thermo Fisher Scientific Inc., MA, USA) spectrometer.

3.3. Chromatography

The mobile phase consisted of water, containing 0.1% (*v/v*) formic acid (A), acetonitrile (B), and methanol (C). The analytes were separated following a gradient elution program, starting at 80:3:17 (*v/v/v*), turning to 74:6:20 (*v/v/v*) in the next 7.5 min, kept isocratic for 2.5 min, and finally changing to 50:10:40 (*v/v/v*) in the last three minutes. The flow rate was set at 1.0 mL min^{-1}, while monitoring of the analytes was set at 265 nm.

3.4. Functionalization of Sponges

A commercially available polyurethane sponge was cut into cubes, immersed into ethanol/water solution, and placed in an ultrasonic bath for 20 min. The sponge was left at room temperature to dry and then it was dipped in a GO mixture for 24 h to be stirred mechanically. The mixture was prepared by the addition of 0.12 g GO in 60 mL NH_3/EtOH solution (4:1, *v/v*). When mechanical stirring was completed, the sponge was left to dry in room temperature. Subsequently it was rinsed with water and PVA solvent was added as a final step. The PU-GO sponge is shown in Figure 8.

Figure 8. Image of the polyurethane-graphene oxide (PU-GO) sponge.

3.5. Sample Preparation

In the present study, defatted bovine milk was used and the proteins' precipitation was achieved by adding 3 mL of ACN in 1.5 g of milk. The pH was adjusted to 5.0 by using 0.5 mL of buffer solution (70% CH_3COONa 0.2 M/30% CH_3COOH 0.2 M). The sponge was initially placed in a vial containing 1.5 g of milk and the system was centrifuged at low rpm for 15 min. The material was rinsed with deionized water and then squeezed to wash the water away. Subsequently, 1.5 mL of 1% CH_3COOH/ACN/MeOH solution (50:40:10 $v/v/v$) was added to the sponge and the analytes were eluted by centrifugation at low rpm for 15 min. The eluent was filtered and injected in the HPLC column.

In the case of fat containing milk samples, centrifugation was applied for fat removal prior to deproteinization. Moreover, sample preconcentration was applied by evaporation of elution solvent prior to HPLC analysis and reconstitution to 100 μL when necessary and in order to reach the legislation demands.

3.6. Standard Solution Preparation

For the chromatographic analysis, stock standard solutions of each analyte were prepared at a concentration of 100 ng μL^{-1} using a solvent with the same composition as the mobile phase. Stock standard solutions were stable for six months at 4 °C, while working standards were prepared on a daily basis. The calibration curves were constructed by the use of solutions being prepared at concentrations of 0.5–10 ng μL^{-1}.

3.7. Method Validation

The method was validated using spiked samples, under the optimal conditions, in terms of linearity, sensitivity, selectivity, and precision (repeatability and between-day precision), decision limit (CCa), decision capability (CCb), and stability according to the European Decision 657/2002/EC [40].

Linearity was studied by triplicate analysis of working standard solutions at concentration levels between 0.5 ng μL^{-1} to 10 ng μL^{-1}. In milk, linearity was examined by triplicate analysis of spiked samples within the range of 50 μg kg^{-1}–10,000 μg kg^{-1} and calibration curves were calculated. Limits of detection (LOD) and quantification (LOQ) were considered as the concentration giving a signal to noise ratio of 3 and 10, respectively. The selectivity of the method was proved by the absence of interference of endogenous compounds in the analysis of blank milk samples.

Precision and accuracy were calculated by analyzing spiked samples at the concentration levels of 50 μg kg^{-1}, 100 μg kg^{-1} and 150 μg kg^{-1}, which correspond to the $\frac{1}{2}$ MRL, MRL, and 1 $\frac{1}{2}$ MRL of sulfonamides [6]. Within-day repeatability was examined by 4 measurements at the above concentration levels. Between-day precision was assessed by performing triplicate analysis at the same concentration levels in three days. The relative recovery was calculated using the formula of the percentage of the ratio of the analyte mass that was found in the spiked sample, to the spiked mass.

Decision limit (CCa) was calculated using the equation CCa = MRL + 1.64 × SD, where SD is the standard deviation of the duplicate measurements of twenty milk samples spiked at MRL concentrations of each analyte. Decision capability (CCb) was calculated using the equation CCb = CCa + 1.64 × SD, with the SD being the standard deviation of the duplicate measurements of twenty milk samples spiked at CCa concentrations of each sulfonamide.

4. Conclusions

In the present study a new novel material was presented. Particularly, a PU-GO sponge was prepared, taking advantage of the unique properties of GO combined with the characteristics of the common PU sponge. This novel material was applied for the sample preparation of milk samples for the determination of sulfonamides prior to HPLC. The easy preparation of the material and the extremely fast, simple, and green sample preparation procedure make the proposed method suitable for the analysis of a complex matrix such as milk. It is the first time that the PU-GO sponge was applied for the determination of sulfonamides in milk samples. Furthermore, it is a less costly and time-consuming method and requires less equipment than previously reported methods.

Author Contributions: Conceptualization, V.F.S. and E.A.D.; methodology, software, validation, formal analysis, investigation, resources, data curation, V.F.S., E.A.D. and M.M. Writing—original draft preparation, V.F.S., E.A.D. and M.M. writing—review and editing, V.F.S. and E.A.D.; supervision, project administration, V.F.S. and E.A.D.

References

1. Samanidou, V.F.; Tolika, E.P.; Papadoyannis, I.N. Development and validation of an HPLC confirmatory method for the residue analysis of four sulfonamides in cow's milk according to the European Union decision 2002/657/EC. *J. Liq. Chromatogr. Relat. Technol.* **2008**, *31*, 1358–1372. [CrossRef]
2. Arroyo-Manzanares, N.; Gámiz-Gracia, L.; García-Campaña, A.M. Alternative sample treatments for the determination of sulfonamides in milk by HPLC with fluorescence detection. *Food Chem.* **2014**, *143*, 459–464. [CrossRef] [PubMed]
3. Karageorgou, E.; Christoforidou, S.; Ioannidou, M.; Psomas, E.; Samouris, G. Detection of β-lactams and chloramphenicol residues in raw milk—development and application of an HPLC-DAD method in comparison with microbial inhibition assays. *Foods* **2018**, *7*, 82. [CrossRef]
4. Bitas, D.; Kabir, A.; Locatelli, M.; Samanidou, V. Food sample preparation for the determination of sulfonamides by high-performance liquid chromatography: State-of-the-art. *Separations* **2018**, *5*, 31. [CrossRef]
5. Dmitrienko, S.G.; Kochuk, E.V.; Apyari, V.V.; Tolmacheva, V.V.; Zolotov, Y.A. Recent advances in sample preparation techniques and methods of sulfonamides detection—A review. *Anal. Chim. Acta* **2014**, *850*, 6–25. [CrossRef]
6. Commission Regulation (EU) Commission Regulation (EU) No 37/2010. *Off. J. Eur. Union* **2010**, *L 15*, 1–72.
7. Kechagia, M.; Samanidou, V.; Kabir, A.; Furton, K.G. One-pot synthesis of a multi-template molecularly imprinted polymer for the extraction of six sulfonamide residues from milk before high-performance liquid chromatography with diode array detection. *Sep. Sci.* **2018**, *41*, 723–741. [CrossRef]
8. Wang, S.; Zhang, H.Y.; Wang, L.; Duan, Z.J.; Kennedy, I. Analysis of sulphonamide residues in edible animal products: A review. *Food Addit. Contam.* **2006**, *23*, 362–384. [CrossRef]
9. Zotou, A.; Vasiliadou, C. LC of sulfonamide residues in poultry muscle and eggs extracts using fluorescence pre-column derivatization and monolithic silica column. *J. Sep. Sci.* **2010**, *33*, 11–22. [CrossRef]
10. McClure, E.L.; Wong, C.S. Solid phase microextraction of macrolide, trimethoprim, and sulfonamide antibiotics in wastewaters. *J. Chromatogr. A* **2007**, *1169*, 53–62. [CrossRef]

11. Karageorgou, E.; Manousi, N.; Samanidou, V.; Kabir, A.; Furton, K.G. Fabric phase sorptive extraction for the fast isolation of sulfonamides residues from raw milk followed by high performance liquid chromatography with ultraviolet detection. *Food Chem.* **2016**, *196*, 428–436. [CrossRef]

12. Li, Y.; Wu, X.; Li, Z.; Zhong, S.; Wang, W.; Wang, A.; Chen, J. Fabrication of $CoFe_2O_4$-graphene nanocomposite and its application in the magnetic solid phase extraction of sulfonamides from milk samples. *Talanta* **2015**, *144*, 1279–1286. [CrossRef]

13. Garrido Frenich, A.; del Mar Aguilera-Luiz, M.; Martínez Vidal, J.L.; Romero-González, R. Comparison of several extraction techniques for multiclass analysis of veterinary drugs in eggs using ultra-high pressure liquid chromatography-tandem mass spectrometry. *Anal. Chim. Acta* **2010**, *661*, 150–160. [CrossRef]

14. Maggira, M.; Samanidou, V. QuEChERS: The dispersive methodology approach for complex matrices. *J. Chromatogr. Sep. Tech.* **2018**, *9*. [CrossRef]

15. Yan, H.; Sun, N.; Liu, S.; Row, K.H.; Song, Y. Miniaturized graphene-based pipette tip extraction coupled with liquid chromatography for the determination of sulfonamide residues in bovine milk. *Food Chem.* **2014**, *158*, 239–244. [CrossRef]

16. Wu, L.; Yu, L.; Ding, X.; Li, P.; Dai, X.; Chen, X.; Zhou, H.; Bai, Y.; Ding, J. Magnetic solid-phase extraction based on graphene oxide for the determination of lignans in sesame oil. *Food Chem.* **2017**, *217*, 320–325. [CrossRef]

17. Liu, Q.; Shi, J.; Zeng, L.; Wang, T.; Cai, Y.; Jiang, G. Evaluation of graphene as an advantageous adsorbent for solid-phase extraction with chlorophenols as model analytes. *J. Chromatogr. A* **2011**, *1218*, 197–204. [CrossRef]

18. Wang, X.; Liu, B.; Lu, Q.; Qu, Q. Graphene-based materials: Fabrication and application for adsorption in analytical chemistry. *J. Chromatogr. A* **2014**, *1362*, 1–15. [CrossRef]

19. Liu, Q.; Shi, J.; Sun, J.; Wang, T.; Zeng, L.; Jiang, G. Graphene and graphene oxide sheets supported on silica as versatile and high-performance adsorbents for solid-phase extraction. *Angew. Chem. Int. Ed.* **2011**, *50*, 5913–5917. [CrossRef]

20. Fumes, B.H.; Silva, M.R.; Andrade, F.N.; Nazario, C.E.D.; Lanças, F.M. Recent advances and future trends in new materials for sample preparation. *TrAC Trends Anal. Chem.* **2015**, *71*, 9–25. [CrossRef]

21. Maggira, M.; Samanidou, V.F. Graphene based materials in sample preparation prior to HPLC analysis and their applications. In *High Performance Liquid Chromatography, Types, Parameters Applications*; Ivan, L., Ed.; Nova Science Publishers, Inc.: Suite N Hauppauge, NY, USA, 2018.

22. Guan, W.; Li, Z.; Zhang, H.; Hong, H.; Rebeyev, N.; Ye, Y.; Ma, Y. Amine modified graphene as reversed-dispersive solid phase extraction materials combined with liquid chromatography-tandem mass spectrometry for pesticide multi-residue analysis in oil crops. *J. Chromatogr. A* **2013**, *1286*, 1–8. [CrossRef]

23. Huang, K.J.; Yu, S.; Li, J.; Wu, Z.W.; Wei, C.Y. Extraction of neurotransmitters from rat brain using graphene as a solid-phase sorbent, and their fluorescent detection by HPLC. *Microchim. Acta* **2012**, *176*, 327–335. [CrossRef]

24. Zhao, G.; Song, S.; Wang, C.; Wu, Q.; Wang, Z. Determination of triazine herbicides in environmental water samples by high-performance liquid chromatography using graphene-coated magnetic nanoparticles as adsorbent. *Anal. Chim. Acta* **2011**, *708*, 155–159. [CrossRef]

25. Wu, Q.; Liu, M.; Ma, X.; Wang, W.; Wang, C.; Zang, X.; Wang, Z. Extraction of phthalate esters from water and beverages using a graphene-based magnetic nanocomposite prior to their determination by HPLC. *Microchim. Acta* **2012**, *177*, 23–30. [CrossRef]

26. Zhang, X.; Niu, J.; Zhang, X.; Xiao, R.; Lu, M.; Cai, Z. Graphene oxide-SiO_2 nanocomposite as the adsorbent for extraction and preconcentration of plant hormones for HPLC analysis. *J. Chromatogr. B Anal. Technol. Biomed. Life Sci.* **2017**, *1046*, 58–64. [CrossRef]

27. Ibarra, I.S.; Miranda, J.M.; Rodriguez, J.A.; Nebot, C.; Cepeda, A. Magnetic solid phase extraction followed by high-performance liquid chromatography for the determination of sulfonamides in milk samples. *Food Chem.* **2014**, *157*, 511–517. [CrossRef]

28. Chatzimitakos, T.; Samanidou, V.; Stalikas, C.D. Graphene-functionalized melamine sponges for microextraction of sulfonamides from food and environmental samples. *J. Chromatogr. A* **2017**, *1522*, 1–8. [CrossRef]

29. Wu, Q.; Zhao, G.; Feng, C.; Wang, C.; Wang, Z. Preparation of a graphene-based magnetic nanocomposite for the extraction of carbamate pesticides from environmental water samples. *J. Chromatogr. A* **2011**, *1218*, 7936–7942. [CrossRef]

30. Su, C.; Yang, H.; Song, S.; Lu, B.; Chen, R. A magnetic superhydrophilic/oleophobic sponge for continuous oil-water separation. *Chem. Eng. J.* **2017**, *309*, 366–371. [CrossRef]

31. Su, C.; Yang, H.; Zhao, H.; Liu, Y.; Chen, R. Recyclable and biodegradable superhydrophobic and superoleophilic chitosan sponge for the effective removal of oily pollutants from water. *Chem. Eng. J.* **2017**, *330*, 423–432. [CrossRef]

32. Meng, H.; Yan, T.; Yu, J.; Jiao, F. Super-hydrophobic and super-lipophilic functionalized graphene oxide/polyurethane sponge applied for oil/water separation. *Chin. J. Chem. Eng.* **2018**, *26*, 957–963. [CrossRef]

33. Sheng, C.; Wenting, B.; Shijian, T.; Yuechuan, W. Electrochromic behaviors of poly (3-*n*-octyloxythiophene). *Polymer (Guildf).* **2008**, 1–6.

34. Kyzas, G.Z.; Travlou, N.A.; Kyzas, G.Z.; Lazaridis, N.K.; Deliyanni, E.A. Functionalization of graphite oxide with magnetic chitosan for the preparation of a nanocomposite dye. Functionalization of graphite oxide with magnetic chitosan for the preparation of a nanocomposite dye adsorbent. *Langmuir* **2015**, *29*, 1657–1668.

35. Travlou, N.A.; Kyzas, G.Z.; Lazaridis, N.K.; Deliyanni, E.A. Graphite oxide/chitosan composite for reactive dye removal. *Chem. Eng. J.* **2013**, *217*, 256–265. [CrossRef]

36. Istanbullu, H.; Ahmed, S.; Sheraz, M.A.; Rehman, I. ur Development and characterization of novel polyurethane films impregnated with tolfenamic acid for therapeutic applications. *Biomed Res. Int.* **2013**, *2013*, 1–8. [CrossRef]

37. Zhou, S.; Hao, G.; Zhou, X.; Jiang, W.; Wang, T.; Zhang, N.; Yu, L. One-pot synthesis of robust superhydrophobic, functionalized graphene/polyurethane sponge for effective continuous oil-water separation. *Chem. Eng. J.* **2016**, *302*, 155–162. [CrossRef]

38. Xu, Y.; Hong, W.; Bai, H.; Li, C.; Shi, G. Strong and ductile poly(vinyl alcohol)/graphene oxide composite films with a layered structure. *Carbon N. Y.* **2009**, *47*, 3538–3543. [CrossRef]

39. Liang, J.; Huang, Y.; Zhang, L.; Wang, Y.; Ma, Y.; Cuo, T.; Chen, Y. Molecular-level dispersion of graphene into poly(vinyl alcohol) and effective reinforcement of their nanocomposites. *Adv. Funct. Mater.* **2009**, *19*, 2297–2302. [CrossRef]

40. Commission Decision, 2002/657/EC. 2002, pp. 8–36. Available online: https://eur-lex.europa.eu/legal-content/EN/ALL/?uri=CELEX%3A32002D0657 (accessed on 31 May 2019).

41. Tolmacheva, V.V.; Apyari, V.V.; Furletov, A.A.; Dmitrienko, S.G.; Zolotov, Y.A. Facile synthesis of magnetic hypercrosslinked polystyrene and its application in the magnetic solid-phase extraction of sulfonamides from water and milk samples before their HPLC determination. *Talanta* **2016**, *152*, 203–210. [CrossRef]

42. Hou, X.; Chen, G.; Zhu, L.; Yang, T.; Zhao, J.; Wang, L.; Wu, Y. Development and validation of an ultra high performance liquid chromatography tandem mass spectrometry method for simultaneous determination of sulfonamides, quinolones and benzimidazoles in bovine milk. *J. Chromatogr. B* **2014**, *962*, 20–29. [CrossRef]

12

ATR-FTIR Spectroscopy, a New Non-Destructive Approach for the Quantitative Determination of Biogenic Silica in Marine Sediments

Dora Melucci [1],*(ID), **Alessandro Zappi** [1](ID), **Francesca Poggioli** [1], **Pietro Morozzi** [1](ID), **Federico Giglio** [2] **and Laura Tositti** [1](ID)

[1]   Department of Chemistry "G. Ciamician", University of Bologna, 40126 Bologna, Italy;
      alessandro.zappi4@unibo.it (A.Z.); francesca.poggioli3@studio.unibo.it (F.P.);
      pietro.morozzi2@unibo.it (P.M.); laura.tositti@unibo.it (L.T.)
[2]   Polar Science Institute-National Research Council ISP-CNR, Via P. Gobetti 101, 40129 Bologna, Italy;
      federico.giglio@cnr.it
*   Correspondence: dora.melucci@unibo.it

Abstract: Biogenic silica is the major component of the external skeleton of marine micro-organisms, such as diatoms, which, after the organisms death, settle down onto the seabed. These micro-organisms are involved in the CO_2 cycle because they remove it from the atmosphere through photosynthesis. The biogenic silica content in marine sediments, therefore, is an indicator of primary productivity in present and past epochs, which is useful to study the CO_2 trends. Quantification of biosilica in sediments is traditionally carried out by wet chemistry followed by spectrophotometry, a time-consuming analytical method that, besides being destructive, is affected by a strong risk of analytical biases owing to the dissolution of other silicatic components in the mineral matrix. In the present work, the biosilica content was directly evaluated in sediment samples, without chemically altering them, by attenuated total reflection Fourier transform infrared (ATR-FTIR) spectroscopy. Quantification was performed by combining the multivariate standard addition method (MSAM) with the net analyte signal (NAS) procedure to solve the strong matrix effect of sediment samples. Twenty-one sediment samples from a sediment core and one reference standard sample were analyzed, and the results (extrapolated concentrations) were found to be comparable to those obtained by the traditional wet method, thus demonstrating the feasibility of the ATR-FTIR-MSAM-NAS approach as an alternative method for the quantification of biosilica. Future developments will cover in depth investigation on biosilica from other biogenic sources, the extension of the method to sediments of other provenance, and the use higher resolution IR spectrometers.

Keywords: diatoms; biogenic silica; ATR-FTIR; chemometrics; NAS

1. Introduction

In the present work, we introduce an innovative and non-destructive method for the quantification of biogenic silica in marine sediments through the use of infrared spectroscopy combined with chemometrics. The proposed method was applied to sediment samples coming from Terra Nova Bay, Antarctica.

Antarctica is a unique natural laboratory because it is the coldest, driest, highest, windiest, and most isolated continent. Therefore, it is almost unaffected by anthropogenic influence [1]. The Southern Ocean allows the diffusion of atmospheric carbon dioxide into the deep sea, which is partially used by sea plants for growth and for the production of organic matter [2]. Therefore, this region is one of the most important for the study of climate changes and conditions of the ocean [3].

In particular, an important tool to control the chemical composition of seawater and to reconstruct paleo-ocean conditions is represented by marine sediments, which are a reservoir and a sink of chemical species involved and cycled in the marine food chain [1]. Among nutrients, silicon is an essential element in the ocean ecosystems, because it is responsible for the growth of Radiolaria, Sponges, Phaeodaria, and particularly Diatoms, which represent a major portion of planktonic primary producers [4]. Diatoms are planktonic unicellular microalgae, known to form an external skeleton called frustule, constituted by amorphous silica and organic components (usually including long-chain polyamines and silaffins) [5,6]. After their death, the diatom siliceous skeleton settles down through the water column. The extent of diatom deposition in the sediments will be a function of the sea bottom depth and of the degree of solubilization of opal silica in the water column [7]. Siliceous microfossils, therefore, can represent a large part of the mass of biogenic sediments accumulating on the deep-sea floor [8].

In the whole Southern Ocean, the Ross Sea is the region of the most widely extensive algal blooms, usually initiating in the Ross Sea polynya [9], an ice-free area of enhanced bio-productivity that can be considered as a biological "hot spot" compared with the surrounding waters. This area extends to the open sea surface as soon as the austral summer develops and the sea ice melts [10,11]. It plays a key climatic role on a global scale. Indeed, the Ross Sea is one of the main sink areas for the tropospheric CO_2, widely contributing to counterbalancing its budget and the associated role in climate change [12,13]. In the western Ross Sea, the polynya of Terra Nova Bay (TNB) is an area of high accumulation of biogenic silica in the sediments [14,15].

Biogenic silica (BSi) content in marine sediment can be considered as a good proxy to characterize the bio-productivity of the Southern Ocean [16,17]. However, the quantification of BSi is complicated by the presence of lithogenic silica, which is chemically equivalent to BSi (SiO_2), with the only difference being crystalline, while BSi is amorphous. Several methods have been proposed to estimate BSi in marine sediments: (1) X-ray diffraction after the conversion of opal to cristobalite at a high temperature [18]; (2) direct X-ray diffraction of amorphous silica [19]; (3) direct infrared spectroscopy of amorphous opal [20]; (4) elemental partitioning of sediment chemistry [21,22]; (5) microfossil counts [23,24]; and (6) several wet-alkaline extraction methods [7,25,26].

Among the above-mentioned techniques, the wet alkaline methods are the most popular because they are the most sensitive techniques for BSi assessment. According to these methods, BSi is extracted and distinguished from lithogenic silica based on hot alkaline solutions [7]. Wet methods exploit a different rate of dissolution of lithogenic and biogenic silica in alkaline solution, with BSi dissolving more quickly than the mineral component. Solubilized BSi can, therefore, be collected in the supernatant of the solution, and subsequently determined by spectrophotometry. Such separations are extremely demanding and time-consuming, and above all, they do not ensure the quantitative recovery of BSi, owing to inherent systematic problems; that is, dependence on matrix effects, incomplete opal recovery, and contamination by non-biogenic silica [17,23].

The increasing success of chemometric tools applied to basic spectrophotometric techniques such as Fourier transform infrared (FTIR), together with the compelling need for understanding key biogeochemical processes of global importance, have recently inspired the introduction of an alternative approach to solve the problem of BSi assessment. In particular, FTIR spectroscopy has been applied to lacustrine sediments for the analysis of silica and other minerals by Rosén et al. [27,28]. Vogel et al. and Rosén et al. [29,30] showed that FTIR spectroscopy in the mid-infrared region is highly sensitive to chemical components present in minerogenic and organic material, such as sediments; this fact provides an efficient tool for quali- and quantitative characterization of these fundamental, but complex environmental matrices.

Moreover, a method based on attenuated total reflectance (ATR)-FTIR measurements has also been proposed in the literature to quantify inorganic components in marine sediments [31,32]. ATR-FTIR spectroscopy is particularly appealing for the analysis of sediments because no chemical sample pre-treatment is required: it may in principle by-pass all the drawbacks of the wet-chemical method; moreover, it works with small amounts of sample material (0.05–0.1 g, dry weight) and it is rapid, inexpensive, and efficient. Besides, ATR-FTIR is a non-destructive method, allowing to recover the sample for further analyses, and it can be carried out even off the lab.

In the present work, we developed and present an analytical method based on ATR-FTIR for the quantitative determination of biosilica content in marine sediments. The feasibility of the method was evaluated by quantifying BSi in a series of sediment samples collected in the Ross Sea. Optimization of the experimental procedures such as the drying process, homogenization, and deposition of the sample on the ATR crystal are discussed in detail, in order to provide a reliable background useful to solve reproducibility problems, which may constitute a drawback of such a simple instrumental approach. Furthermore, the strong matrix effect intrinsic to environmental samples is faced and solved by applying a multivariate standard addition method (MSAM) [33], improved by net analyte signal (NAS) computation [34,35].

2. Results and Discussion

2.1. ATR Spectra

For each of the 22 analyzed sediment samples (21 coming from Mooring D and one $53\%_{w/w}$ reference standard), four standard-added (add.x) samples were prepared: the zero-added sample (add.0) is the pure sample, add.1 has an added concentration of diatomite at $5\%_{w/w}$, add.2 at $10\%_{w/w}$, and add.3 at $15\%_{w/w}$. All added samples (and a pure diatomite sample) were analyzed by ATR-FTIR.

In the ATR spectra of marine sediments, the contribution of silica, both biogenic and lithogenic, is dominant. Such spectra exhibit four characteristic vibrational bands. The two main bands at 1100 and 471 cm^{-1} are attributed to triply degenerated stretching and bending vibration modes, respectively, of the [SiO$_4$] tetrahedron [36]. The band at 800 cm^{-1} corresponds to an inter-tetrahedral Si–O–Si bending vibration mode, and the band near 945 cm^{-1} to an Si–OH vibration mode [37]. Previous studies have shown that the absorbance centered around 1640 cm^{-1} and between 3000 and 3750 cm^{-1} can be attributed to hydroxyl vibrations because hydroxyl ions are major constituents of clay minerals, opal, and organic compounds present in marine sediments [38]. However, these bands are not specific for silica, their intensity is generally low (about one-tenth of the main band); moreover, they are overlapped with the residual absorption bands of H$_2$O. Therefore, to reduce the noise in the spectral data acquired, we decided to discard the IR region between 4000 and 1300 cm^{-1} and to apply the chemometric procedure only in the region between 1300 and 400 cm^{-1}.

Figure 1 shows the raw spectra of sample D10 (as an example of the spectra obtained for all sediment samples) and the replicates of a pure-diatomite sample. In Figure 1a, spectra obtained by instrumental analysis are shown, while Figure 1b highlights the effect of the spectral pre-treatments: uninformative-band removal and MSC.

On the spectra reported in Figure 1b, the two chemometric procedures described in Section 3.4 were carried out for all sediment samples, and the results are reported in Table 1. The expected values reported in Table 1 are the BSi concentrations obtained by wet analyses that were carried out only on five sediment samples (and on $53\%_{w/w}$-standard): D4, D6, D9, D18, and D21.

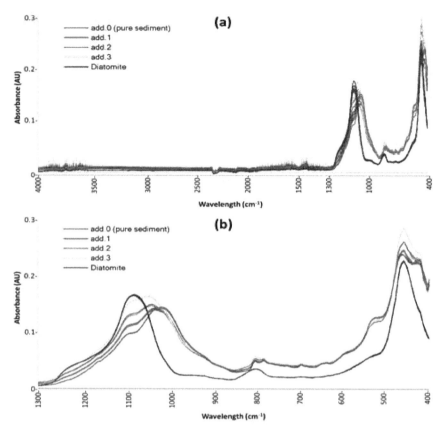

Figure 1. (a) ATR raw spectra of sample D10, as obtained from the spectrophotometer; (b) the same spectra after band removal (discarding the IR range 4000–1300 cm^{-1}) and after multiplicative scatter correction (MSC) pre-treatment. "add." in the legends indicates standard added samples.

Table 1. Net analyte signal (NAS) results for the two pre-processing methods. All the numbers are formatted with three significant digits to allow for a detailed comparison. LoD, limit of detection.

SAMPLE CODE	Expected Value ± Standard Deviation ($\%_{w/w}$)	Procedure 1				Procedure 2			
		NAS Extr. C ($\%_{w/w}$)	Standard Deviation	R^2	LoD	NAS Extr. C ($\%_{w/w}$)	Standard Deviation	R^2	LoD
Std 53%	53 ± 3	53.6	6.02	0.992	1.01	50.2	2.74	0.988	7.76
D0	-	3.23	0.903	0.996	0.0265	8.64	3.01	0.956	2.39
D1	-	3.24	1.92	0.993	0.190	-	-	-	-
D2	-	9.88	1.70	0.994	0.416	9.28	1.23	0.990	6.24
D3	-	7.55	0.315	0.993	0.743	8.40	0.436	0.999	4.08
D4	13 ± 2	12.0	1.16	0.988	0.469	12.8	2.01	0.988	3.58
D5	-	5.41	1.10	0.993	0.0214	5.38	1.75	0.995	2.88
D6	14 ± 2	14.2	1.54	0.989	0.00946	14.2	2.71	0.997	1.41
D7	-	5.87	1.27	0.999	0.646	-	-	-	-
D8	-	9.45	0.671	0.993	0.380	-	-	-	-
D9	8 ± 1	8.26	3.16	0.986	0.654	8.91	0.514	0.982	4.68
D10	-	12.0	1.18	0.997	0.515	12.1	2.12	0.995	0.267
D11	-	2.94	0.646	0.994	0.0666	3.84	0.803	0.993	1.76
D12	-	-	-	-	-	10.6	2.32	0.995	2.32
D13	-	-	-	-	-	3.72	1.81	0.987	3.82
D14	-	9.00	0.146	0.998	0.133	10.9	0.433	0.998	7.62
D15	-	3.25	0.184	0.993	0.0784	5.19	0.122	0.996	2.22
D16	-	-	-	-	-	5.63	1.38	0.989	3.60
D17	-	2.71	0.535	0.994	0.0647	13.7	1.41	0.984	1.93
D18	4.2 ± 0.6	4.80	1.67	0.996	0.0900	5.16	1.70	0.992	4.34
D19	-	4.04	0.332	0.998	0.0459	3.22	0.345	0.997	1.23
D20	-	2.36	0.535	0.999	0.610	4.26	1.19	0.991	8.46
D21	3.4 ± 0.5	3.12	1.86	0.998	0.653	4.12	0.296	0.993	3.93

2.2. Results: Procedure 1

To assess the reliability of the new methodology proposed here, the BSi results obtained here can be compared to those achieved through the traditional wet method, taken as a reference. Table 1 shows that the results obtained with *Procedure 1* (band removal and MSC) are in good agreement with the expected values obtained when samples were analyzed with the wet method. Indeed, the confidence intervals obtained for these five sediment samples by NAS are not significantly different from the ones obtained by the wet method (at 0.05 significance). Also, the result obtained for the standard sample (Std 53%$_{w/w}$) is in agreement with the expected concentration. The coefficients R^2 of the NAS standard addition lines are all higher than 0.98, indicating a good correlation between added concentrations (dependent variable) and the pseudo-univariate NAS values calculated by the chemometric procedure. The LoD values are also very good, being, in general, in the order of magnitude of one-tenth (or even lower) with respect to the corresponding extrapolated BSi concentration.

Moreover, Frignani et al. [15] reported that BSi concentration in surface sediments in this area is usually relatively low, <10%$_{w/w}$; the results obtained by NAS are in agreement with this consideration, as D0, D1, D2, and D3 extrapolated concentrations are lower than the indicated value.

All these considerations confirm that NAS applied to ATR spectra of the standard added samples can be a valuable and reliable alternative to the time-consuming wet method for the quantification of BSi in marine sediments.

The main drawbacks concern the three samples for which no results were obtained: D12, D13, and D16. In these cases, the NAS standard addition lines had, for all PLS-factors, either a negative slope or intercept, giving negative extrapolated values, or not acceptable R^2 (lower than 0.7), that make any possible result unreliable. The reason for such behavior is still under study, but we can hypothesize that there is still some source of noise in ATR analysis that was not taken into account, although several precautions were taken during instrumental analyses, as described in Section 3.3. We, therefore, decided to proceed with further chemometric assessments, also to test the hypothesis of a possible defect in the NAS procedure.

2.3. Results: Procedure 2

As described in Section 3.4, a variable selection was carried out on baseline-corrected spectra. Correlation loadings on PLS-factor 1 were used to select the most important variables to describe the regression model. Although a different variable selection was carried out for each sediment sample, not always giving the same variables, a general description of the selected IR bands can be drawn and is resumed in Figure 2. High correlation loading values in the PLS-factor 1 are computed in the regions of 1260–1060 cm^{-1}, 830–800 cm^{-1}, and 467–436 cm^{-1}. These regions of the IR spectra correspond to the characteristic SiO$_2$ absorbance maxima as reported by Vogel et al. [29]. On these selected variables, NAS computation was carried out and the results are reported in the last vertical section of Table 1.

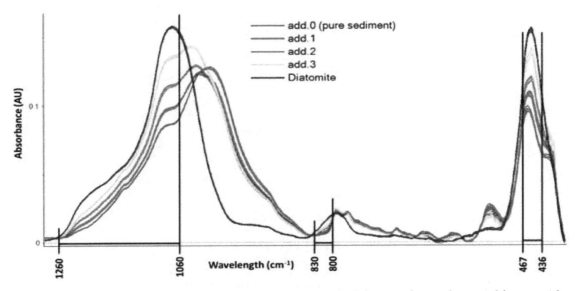

Figure 2. Baseline-corrected ATR spectra of sample D10. Black lines indicate the variables considered most important by partial least square (PLS) correlation loadings.

Again, concentrations extrapolated by NAS are not significantly different from the "wet method-based" values, with high R^2 (>0.95). After the application of this second procedure (baseline correction and variable selection before NAS), significant differences were detected only for D0 and D17, which, in this case, also have a lower R^2 compared with the other samples. In this case, some problems arise from LoDs, which, in most cases, are comparable to the extrapolated value (and also higher than that for D20). Such a drawback might, therefore, be because of the spectral pre-treatment; in order to calculate LoD, a blank spectrum is necessary. However, such a blank spectrum has to be pre-treated as all the other spectra, and in this case, it has to be baseline corrected. In this way, the pre-treatment can likely produce some spikes in the blank spectrum (that is, a noisy signal oscillating around the zero), thus affecting the computation of LoD.

The three samples that did not give results with the computation by *Procedure 1* (D12, D13, and D16) in this case have an acceptable extrapolated concentration. However, there are again three samples (D1, D7, and D8) with no result. This strengthens the hypothesis of the presence of a noise source that was not taken into account. Indeed, variable selection may reduce the noise present in the whole spectrum, but, at the same time, if noise is present in the selected variables, its effect may be enhanced. Therefore, the two chemometric methods presented in this work may be considered to be complementary for this study.

3. Materials and Methods

3.1. Study Area

Sediment samples for the present study were collected in "mooring D" (or "site D"), which is located in Antarctica, in the western sector of the Ross Sea continental shelf within the polynya of Terra Nova Bay at 75°06′ S and 164°28′5′′ E (Figure 3). The box-core, from which the sediments were collected, was sampled at a depth of 972 m during the 2003–2004 Italian PNRA (Programma Nazionale di Ricerca in Antartide) Campaign [39], whose basis was situated in the "Mario Zucchelli" station.

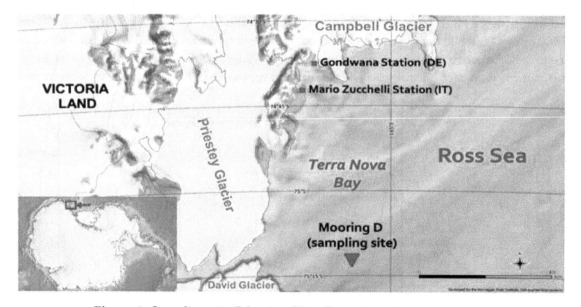

Figure 3. Sampling site (Mooring D) in Terra Nova Bay, Antarctica.

In the Ross Sea, surface sediments are generally composed of unsorted ice-rafted debris, terrigenous silts and clays, and siliceous and calcareous biogenic debris [40]. In site D, in particular, coarse terrigenous deposits are predominant, owing to the proximity of Priestley, David, and Campbell glaciers [15].

3.2. Samples

The sediment collected in site D was sampled using a 1T Oceanic box corer. A sub core 22 cm long was collected by means of a polyvinyl chloride (PVC) liner. The short core was subsampled with a resolution of 1 cm [39]. Twenty-two sediment sections were thus obtained and named with a two-digit code: a letter, "D", indicating the sampling place; and a number, from 0 to 21, indicating the core height, with D0 being the top, corresponding to the sediment surface. Sediments were then stored at −21 °C in a polycarbonate Petri capsule and oven-dried at 50 °C just prior to the analyses. The BSi content of five of these samples was also quantified by a wet method analysis, according to the DeMaster method [7,15], thus providing some comparison values for the ATR analyses. In the absence of a commercial certified reference material for BSi, the "internal reference standard" used in the Polar Science Institute-National Research Council (CNR-ISP) laboratory was adopted for the purpose of this paper. This sample consists of an Antarctic marine sediment analyzed repeatedly both in CNR and other biogeochemical laboratories, resulting in a BSi content of $(53 \pm 3)\%_{w/w}$ and a remaining $47\%_{w/w}$ of alkaline halide.

For the sake of readability, a flowchart concerning the sample preparation is reported in Figure 4. Before sample preparation, all samples were manually ground in an agate mortar for approximately 15 min and heated in a ventilated oven at 105 °C for 1 h to remove atmospheric moisture. Afterward, each sample was split into four aliquots, three of which were added with known amounts ($5\%_{w/w}$, $10\%_{w/w}$, $15\%_{w/w}$) of Diatomite (Celite® 545 AW, Sigma-Aldrich, Darmstadt, Germany), in order to apply the multivariate standard addition method. The total weight of each standard-added sample was 200 mg. Diatomite was chosen as a proxy of standard biogenic silica, because it is composed of frustulae of biogenic silica, similar to what we wanted to quantify in marine sediments. Such a similarity was visually evaluated by analyzing some samples with a scanning electron microscope (SEM) Philips 515B (Philips, Amsterdam, Netherlands), equipped with an EDAX DX4 microanalytical device (EDAX Inc., Mahwah, NJ, USA). Figure 5 shows the pictures obtained by SEM. From Figure 5c,d, it can be seen that samples D1 and D4 contain the same radiolaria present in the Diatomite (Figure 5a) used as a proxy of BSi.

Figure 4. Sample preparation flowchart. ATR-FTIR, attenuated total reflection Fourier transform infrared.

To ensure better homogenization of the powders, a Mixer Mill "MM20" (Retsch Inc., Düsseldorf, Germany) was used. Each added sample was placed in stainless steel cylinders of 1.5 mL volume and left in the ball mill for 60 min at 20 Hz. Before the instrumental analysis, samples were kept in a desiccator filled with silica gel to prevent the absorption of atmospheric moisture. Standard added samples were then analyzed by ATR-FTIR spectroscopy.

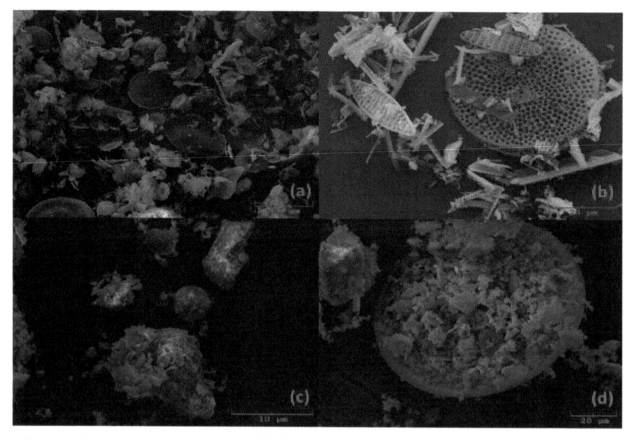

Figure 5. Scanning electron microscope (SEM) images of samples (**a**) pure diatomite; (**b**) 53% standard; (**c**) D1 sample, which is characterized by the presence of both radiolaria and bulks of sedimentary material; and (**d**) D4 sample.

3.3. ATR-FTIR Analysis

Attenuated total reflection spectra were collected using a Bruker ALPHA FT-IR spectrometer (Brucker Optics GmbH, Billerica, MA, USA) equipped with a single-reflection diamond ATR accessory (Bruker Platinum ATR, Billerica, MA, USA) with an approximately 0.6 mm × 0.6 mm active area and a mercury–cadmium–telluride detector. Spectra were collected in the mid-IR range, 400–4000 cm^{-1}, with an optical resolution of 4 cm^{-1}; the registered spectrum is the mean of 64 scans, executed in 3 min. For each sample aliquot, five replicate spectra were recorded to assess precision and ensure the reproducibility of each sample. All measurements were performed at ambient conditions. Before spectra acquisition, a background spectrum (air) was collected with the same operational parameters. Such a background was automatically subtracted to each sample spectrum.

To optimize the analytical reproducibility, some precautions were taken for ATR analysis. Indeed, it is widely reported in the literature how an imprecise sample preparation (especially drying process), sample deposition, and instrumental calibration may cause poor instrumental repeatability and accuracy, fundamental characteristics for quantification purposes [31,41]. For these reasons, a suitable experimental protocol was developed and evaluated (Figure 4).

Before the instrumental analysis, samples were manually ground again for 5 min in an agate mortar, in order to homogenize powder granulometry. Moreover, for each added sample, an aliquot (53 mg) was carefully weighted and lodged in a steel ring of 1 cm in diameter, which was then placed over the spectrophotometer probe. The same amount of material was taken for all the analyzed samples, to maximize reproducibility and reduce scattering and other problems resulting from not optimal (or not constant) contact between the sample and crystal. These problems become relevant in ATR analysis when used for quantification purposes owing to the geometry of ATR irradiation and reflectance, which need an accurate evaluation and the adoption of a suitable experimental protocol [42].

3.4. Chemometrics

Prior to chemometric analysis, the five replicated spectra of each added sample were pre-treated by multiplicative scatter correction (MSC) [43]. MSC allows the reduction of the effects of scattering noise on IR spectra, increasing the reproducibility of sample replicates.

Subsequently, in order to calculate the BSi content in each sediment sample by MSAM, the NAS procedure was applied [35,44]. NAS is a mathematical procedure that allows extracting, from a multivariate signal (in this case, an ATR spectrum), that part of the signal that is only due to the analyte, removing the other signals due to the other interfering species present in the matrix [35]. In this way, the multivariate problem can be reduced to a pseudo-univariate problem, whose results can be obtained by a univariate treatment. NAS computations were performed as follows.

The NAS procedure starts from a partial least square (PLS) regression [45], using the ATR spectra as independent variables (X) and the added concentrations vector as dependent one (y). The best PLS-factor (A) has to be selected and the corresponding PLS-regression coefficient vector (b_A) is used to compute a projection matrix (H) as follows:

$$H = b_A\left(b_A b_A^t\right)^{-1} b_A^t,$$ (1)

where t indicates transpose and superscript "-1" indicates matrix inversion. H matrix is then used to compute NAS vectors (x_i^*):

$$x_i^* = Hx_i,$$ (2)

where x_i are the rows of matrix X, which means samples of ATR spectra. Each calculated x_i^* corresponds to the net signal (devoid of interfering signals) of each replicate of the added samples. The Euclidean norms of such net signals can be then used as pseudo-univariate signals to compute a univariate standard addition linear regression line, from which BSi concentration can be obtained by extrapolation.

The selection of the optimal PLS-factor (A) is a crucial point of the procedure, because, in most of the cases, the final extrapolated concentration varies (also dramatically) while varying A. Therefore, A was chosen (sample by sample) as the PLS-factor that optimizes both the PLS root mean squared error (RMSE), by minimizing it, and the determination coefficient (R^2) of the final pseudo-univariate line, by maximizing it. When these two conditions were not simultaneously achievable for one PLS-factor, A was chosen as the factor giving the best compromise between these two parameters, based on the highest R^2.

The standard deviations of the extrapolated concentration values were computed by the *jackknife* method [46]. Once the optimal PLS-factor is selected, the *jackknife* procedure replicates the NAS computation as many times as the number of objects (x_i), each time keeping out one object. In this way, i different extrapolated values are obtained for each NAS computation and the overall standard deviation is estimated as the standard deviation of the *jackknife*-extrapolated values.

Limits of detection (LoDs) were computed collecting five replicates (the same number of the other samples) of a blank spectrum (empty sample holder) and projecting them onto the NAS space by Equation (2) as if they were real samples [47]. The so obtained NAS-blank signals were mediated to obtain the vector ε, and the LoD was computed as follows [47]:

$$\text{LoD} = 3\frac{\|\varepsilon\|}{\|b_A\|},$$ (3)

where $\|\cdot\|$ indicates the Euclidean norm.

The so far described procedure (*Procedure 1*) was applied to raw spectra, as they were obtained from the spectrophotometer. This procedure gave reasonable results for the majority of the samples, while it failed for three of them; in those cases, for all PLS-factors, the final NAS standard addition line had either a negative slope or intercept, producing a negative extrapolated concentration. The reason behind such behavior is still under evaluation. In order to obtain a result for each sediment sample, another chemometric procedure (*Procedure 2*) has thus been developed. Instead of using raw spectra, a baseline correction was applied directly by the software controlling the instrument, OPUS v.7.2 (Bruker). MSC was applied to baseline-corrected spectra and, before NAS computation, a variable selection was applied. For variable selection, another PLS regression was computed (previously to the one used for NAS). Only factor 1, always retaining more than 95% of the explained variance, was considered, and the variables giving correlation loadings [48] higher than 0.7 (in absolute value) were retained as important. NAS computation was then applied only with these selected variables. The standard deviations on the extrapolated values and LoDs were calculated as before.

MSC and variable selection pre-processing were performed by the software The Unscrambler v.10.3 (CAMO, Olso, Norway), while NAS and *jackknife* procedures were computed by a homemade code in R environment (R Core Team, Vienna, Austria).

4. Conclusions

In this study, we demonstrated the feasibility of a new approach for the quantification of biogenic silica based on IR spectroscopy coupled with chemometrics. Biogenic silica content in marine sediments from Terra Nova Bay in West Antarctica was evaluated with Fourier-transform infrared spectroscopy in attenuated total reflection mode (ATR-FTIR). For quantification, the multivariate standard addition method (MSAM) was applied, and the net analyte signal (NAS) procedure was used to solve the problems deriving from the strong matrix effect affecting such analyses.

Twenty-one subsequent core samples and one reference standard were analyzed. Reliable results were obtained, as observed from the comparison with homologous data from the traditional wet method.

Some drawbacks remain. The chemometric procedure did not give acceptable results for some samples, even if a variable selection was carried out. Moreover, the limits of detection, and perhaps also standard deviations, are in some cases still too high.

However, it has to be taken into account that the quantification of biosilica, in this work, has been carried out with an analytical technique (ATR) that has several intrinsic drawbacks when performing quantitative analysis. In particular, owing to the optical behavior of photons at such low angles as in ATR, extremely careful handling of samples and highly reproducible sample geometry are required when analyzing powdered samples. Moreover, the analyzed samples are powders, which, despite all the precautions taken before and during the analysis, can still have some problems concerning homogeneity and granulometry. The BSi content was also evaluated in natural samples without any chemical pre-treatment, thus its analytical signal may be strongly affected by the presence of lithogenic silica, besides all the other species composing the sediments.

Considering all these aspects, the analytical and chemometric procedure presented in this work, although requiring some more refinements, can be considered a promising alternative to the traditional time-consuming wet method for the quantification of biosilica in marine sediments. In paleolimnological research, the ATR-FTIR technique is seldom used. The results presented here, as well as the fact that this method is fast and cost-effective, requiring only small quantities of sediment sample, should encourage more researchers to use it. Moreover, marine sediments are precious samples, which are difficult to collect; thus, a not-destructive method would be preferable to analyze them, although ATR-FTIR cannot yet entirely replace conventional analytical tools in paleolimnology.

Author Contributions: Conceptualization, D.M., F.G., and L.T.; Methodology, D.M., A.Z., F.G., and F.P.; Software, A.Z. and F.P.; Validation, A.Z. and P.M.; Formal Analysis, F.P., A.Z., and P.M.; Investigation, F.P.; Resources, F.G. and L.T.; Data Curation, D.M., A.Z., and F.P.; Writing—Original Draft Preparation, A.Z. and F.P.; Writing—Review & Editing, D.M., A.Z., P.M., and L.T.; Visualization, F.P. and F.G.; Supervision, D.M.; Project Administration, L.T.

Acknowledgments: We are indebted to the Italian participants to the XIX Expedition to Antarctica, in the year 2003/04, for carrying out the samplings. We are deeply grateful to Prof. Giorgio Gasparotto, of the Department of

Biological, Geological, and Environmental Sciences, University of Bologna for kindly providing the access to his SEM facility and the precious support in producing the diatom images employed herein.

References

1. Casalino, C.E.; Malandrino, M.; Giacomino, A.; Abollino, O. Total and fractionation metal contents obtained with sequential extraction procedures in a sediment core from Terra Nova Bay, West Antarctica. *Antarct. Sci.* **2013**, *25*, 83–98. [CrossRef]
2. Toggweiler, J.R.; Russell, J.L.; Carson, S.R. Midlatitude westerlies, atmospheric CO_2, and climate change during the ice ages. *Paleoceanography* **2006**, *21*. [CrossRef]
3. Sprenk, D.; Weber, M.E.; Kuhn, G.; Rosén, P.; Frank, M.; Molina-Kescher, M.; Liebetrau, V.; Röhling, H.-G. Southern Ocean bioproductivity during the last glacial cycle—New detection method and decadal-scale insight from the Scotia Sea. *Geol. Soc. Lond. Spec. Publ.* **2013**, *381*, 245–261. [CrossRef]
4. Kamatani, A.; Oku, O. Measuring biogenic silica in marine sediments. *Mar. Chem.* **2000**, *68*, 219–229. [CrossRef]
5. Kröger, N.; Poulsen, N. Diatoms—From Cell Wall Biogenesis to Nanotechnology. *Annu. Rev. Genet.* **2008**, *42*, 83–107. [CrossRef]
6. Sumper, M. A phase separation model for the nanopatterning of diatom biosilica. *Science* **2002**, *295*, 2430–2433. [CrossRef]
7. DeMaster, D.J. The supply and accumulation of silica in the marine environment. *Geochim. Cosmochim. Acta* **1981**, *45*, 1715–1732. [CrossRef]
8. Broecker, W.S. Tracers in the sea. *Geochim. Cosmochim. Acta* **1983**, *47*, 1336.
9. Sullivan, C.W.; Arrigo, K.R.; McClain, C.R.; Comiso, J.C.; Firestone, J. Distributions of phytoplankton blooms in the Southern Ocean. *Science* **1993**, *262*, 1832–1837. [CrossRef] [PubMed]
10. Smith, W.O.; Gordon, L.I. Hyperproductivity of the Ross Sea (Antarctica) polynya during austral spring. *Geophys. Res. Lett.* **1997**, *24*, 233–236. [CrossRef]
11. Arrigo, K.R.; DiTullio, G.R.; Dunbar, R.B.; Robinson, D.H.; VanWoert, M.; Worthen, D.L.; Lizotte, M.P. Phytoplankton taxonomic variability in nutrient utilization and primary production in the Ross Sea. *J. Geophys. Res. Ocean.* **2000**, *105*, 8827–8846. [CrossRef]
12. Sandrini, S.; Ait-Ameur, N.; Rivaro, P.; Massolo, S.; Touratier, F.; Tositti, L.; Goyet, C. Anthropogenic carbon distribution in the Ross Sea, Antarctica. *Antarct. Sci.* **2007**, *19*, 395–407. [CrossRef]
13. Manno, C.; Sandrini, S.; Tositti, L.; Accornero, A. First stages of degradation of Limacina helicina shells observed above the aragonite chemical lysocline in Terra Nova Bay (Antarctica). *J. Mar. Syst.* **2007**, *68*, 91–102. [CrossRef]
14. Ledford-Hoffman, P.A.; Demaster, D.J.; Nittrouer, C.A. Biogenic-silica accumulation in the Ross Sea and the importance of Antarctic continental-shelf deposits in the marine silica budget. *Geochim. Cosmochim. Acta* **1986**, *50*, 2099–2110. [CrossRef]
15. Frignani, M.; Giglio, F.; Accornero, A.; Langone, L.; Ravaioli, M. Sediment characteristics at selected sites of the Ross Sea continental shelf: Does the sedimentary record reflect water column fluxes? *Antarct. Sci.* **2003**, *15*, 133–139. [CrossRef]
16. Petrovskii, S.K.; Stepanova, O.G.; Vorobyeva, S.S.; Pogodaeva, T.V.; Fedotov, A.P. The use of FTIR methods for rapid determination of contents of mineral and biogenic components in lake bottom sediments, based on studying of East Siberian lakes. *Environ. Earth Sci.* **2016**, *75*, 226. [CrossRef]
17. Maldonado, M.; López-Acosta, M.; Sitjà, C.; García-Puig, M.; Galobart, C.; Ercilla, G.; Leynaert, A. Sponge skeletons as an important sink of silicon in the global oceans. *Nat. Geosci.* **2019**, *12*, 815–822. [CrossRef]
18. Calvert, S.E. Accumulation of diatomaceous silica in the sediment of the Gulf of California. *Geol. Soc. Am. Bull.* **1966**, *77*, 569–572. [CrossRef]
19. Eisma, D.; Van Der Gaast, S.J. Determination of opal in marine sediments by X-ray diffraction. *Netherlands J. Sea Res.* **1971**, *5*, 382–389. [CrossRef]
20. Chester, R.; Elderfield, H. The infrared determination of opal in siliceous deep-sea sediments. *Geochim. Cosmochim. Acta* **1968**, *32*, 1128–1140. [CrossRef]
21. Nancy Ann, B. Chapter 18 The Determination of Biogenic Opal in High Latitude Deep Sea Sediments. *Dev. Sedimentol.* **2008**, *36*, 317–331.
22. Leinen, M. A normative calculation technique for determining opal in deep-sea sediments. *Geochim. Cosmochim. Acta* **1977** *41*, 671–676. [CrossRef]

23. Leinen, M. Techniques for determining opal in deep-sea sediments: A comparison of radiolarian counts and x-ray diffraction data. *Mar. Micropaleontol.* **1985**, *9*, 375–383. [CrossRef]

24. Pokras, E.M. Preservation of fossil diatoms in Atlantic sediment cores: Control by supply rate. *Deep Sea Res. Part A Oceanogr. Res. Pap.* **1986**, *33*, 893–902. [CrossRef]

25. Donald, W.; Eggimann, F.T.M. Dissolution and Analysis of Amorphous Silica in Marine Sediments. *SEPM J. Sediment. Res.* **2003**, *50*, 215–225.

26. Mortlock, R.A.; Froelich, P.N. A simple method for the rapid determination of biogenic opal in pelagic marine sediments. *Deep Sea Res. Part A Oceanogr. Res. Pap.* **1989**, *36*, 1415–1426. [CrossRef]

27. Rosén, P.; Dåbakk, E.; Renberg, I.; Nilsson, M.; Hall, R. Near-infrared spectrometry (NIRS): A new tool for inferring past climatic changes from lake sediments. *Holocene* **2000**, *10*, 161–166. [CrossRef]

28. Rosén, P. Total organic carbon (TOC) of lake water during the Holocene inferred from lake sediments and near-infrared spectroscopy (NIRS) in eight lakes from northern Sweden. *Biogeochemistry* **2005**, *76*, 503–516. [CrossRef]

29. Vogel, H.; Rosén, P.; Wagner, B.; Melles, M.; Persson, P. Fourier transform infrared spectroscopy, a new cost-effective tool for quantitative analysis of biogeochemical properties in long sediment records. *J. Paleolimnol.* **2008**, *40*, 689–702. [CrossRef]

30. Rosén, P.; Vogel, H.; Cunningham, L.; Hahn, A.; Hausmann, S.; Pienitz, R.; Zolitschka, B.; Wagner, B.; Persson, P. Universally applicable model for the quantitative determination of lake sediment composition using fourier transform infrared spectroscopy. *Environ. Sci. Technol.* **2011**, *45*, 8858–8865. [CrossRef]

31. Mecozzi, M.; Pietrantonio, E.; Amici, M.; Romanelli, G. Determination of carbonate in marine solid samples by FTIR-ATR spectroscopy. *Analyst* **2001**, *126*, 144–146. [CrossRef] [PubMed]

32. Khoshmanesh, A.; Cook, P.L.M.; Wood, B.R. Quantitative determination of polyphosphate in sediments using attenuated total reflectance-fourier transform infrared (ATR-FTIR) spectroscopy and partial least squares regression. *Analyst* **2012**, *137*, 3704–3709. [CrossRef] [PubMed]

33. Melucci, D.; Locatelli, C. Multivariate calibration in differential pulse stripping voltammetry using a home-made carbon-nanotubes paste electrode. *J. Electroanal. Chem.* **2012**, *675*, 25–31. [CrossRef]

34. Lorber, A.; Faber, K.; Kowalski, B.R. Net Analyte Signal Calculation in Multivariate Calibration. *Anal. Chem.* **1997**, *69*, 1620–1626. [CrossRef]

35. Bro, R.; Andersen, C.M. Theory of net analyte signal vectors in inverse regression. *J. Chemom.* **2003**, *17*, 646–652. [CrossRef]

36. Malandrino, M.; Mentasti, E.; Giacomino, A.; Abollino, O.; Dinelli, E.; Sandrini, S.; Tositti, L. Temporal variability and environmental availability of inorganic constituents in an antarctic marine sediment core from a polynya area in the Ross Sea. *Toxicol. Environ. Chem.* **2010**, *92*, 453–475. [CrossRef]

37. Dunbar, R.B.; Anderson, J.B.; Domack, E.W.; Jacobs, S.S. Oceanographic influences on sedimentation along the Antarctic continental shelf. In *Antarctic Research Series*; American Geophysical Union: Washington, DC, USA, 1985; pp. 291–312.

38. Liu, W.; Sun, Z.; Ranheimer, M.; Forsling, W.; Tang, H. A flexible method of carbonate determination using an automatic gas analyzer equipped with an FTIR photoacoustic measurement chamber. *Analyst* **1999**, *124*, 361–365. [CrossRef]

39. Ramer, G.; Lendl, B. Attenuated Total Reflection Fourier Transform Infrared Spectroscopy. In *Encyclopedia of Analytical Chemistry*; John Wiley & Sons: Hoboken, NJ, USA, 2013; pp. 1–27.

40. Rinnan, Å.; van den Berg, F.; Engelsen, S.B. Review of the most common pre-processing techniques for near-infrared spectra. *TrAC Trends Anal. Chem.* **2009**, *28*, 1201–1222. [CrossRef]

41. Lorber, A. Error Propagation and Figures of Merit for Quantification by Solving Matrix Equations. *Anal. Chem.* **1986**, *58*, 1167–1172. [CrossRef]

42. Wold, S.; Sjöström, M.; Eriksson, L. PLS-regression: A basic tool of chemometrics. *Chemom. Intell. Lab. Syst.* **2001**, *58*, 109–130. [CrossRef]

43. Stute, W. The jackknife estimate of variance of a Kaplan-Meier integral. *Ann. Stat.* **1996**, *24*, 2679–2704. [CrossRef]

44. Hemmateenejad, B.; Yousefinejad, S. Multivariate standard addition method solved by net analyte signal calculation and rank annihilation factor analysis. *Anal. Bioanal. Chem.* **2009**, *394*, 1965–1975. [CrossRef] [PubMed]

45. Lorho, G.; Westad, F.; Bro, R. Generalized correlation loadings. Extending correlation loadings to congruence and to multi-way models. *Chemom. Intell. Lab. Syst.* **2006**, *84*, 119–125. [CrossRef]

46. Gendron-Badou, A.; Coradin, T.; Maquet, J.; Fröhlich, F.; Livage, J. Spectroscopic characterization of biogenic silica. *J. Non Cryst Solids* **2003**, *316*, 331–337. [CrossRef]

47. Meyer-Jacob, C.; Vogel, H.; Boxberg, F.; Rosén, P.; Weber, M.E.; Bindler, R. Independent measurement of biogenic silica in sediments by FTIR spectroscopy and PLS regression. *J. Paleolimnol.* **2014**, *52*, 245–255. [CrossRef]

48. Colthup, N.B.; Daly, L.H.; Wiberley, S.E. *Introduction to Infrared and Raman Spectroscopy*, 3rd ed.; Academic Press: San Diego, CA, USA, 1990; ISBN 978-0-12-182554-6.

13

Effects of Harvest Time on Phytochemical Constituents and Biological Activities of *Panax ginseng* Berry Extracts

Seung-Yeap Song [1,†], Dae-Hun Park [2,†], Seong-Wook Seo [3], Kyung-Mok Park [4], Chun-Sik Bae [5], Hong-Seok Son [6], Hyung-Gyun Kim [7], Jung-Hee Lee [7], Goo Yoon [1], Jung-Hyun Shim [1], Eunok Im [3], Sang Hoon Rhee [8], In-Soo Yoon [3,*] and Seung-Sik Cho [1,*]

[1] Department of Pharmacy, College of Pharmacy, Mokpo National University, Jeonnam 58554, Korea; tgb1007@naver.com (S.-Y.S.); gyoon@mokpo.ac.kr (G.Y.); s10004jh@gmail.com (J.-H.S.)
[2] Department of Nursing, Dongshin University, Jeonnam 58245, Korea; dhj1221@hanmail.net
[3] Department of Pharmacy, College of Pharmacy, Pusan National University, Busan 46241, Korea; sswook@pusan.ac.kr (S.-W.S.); eoim@pusan.ac.kr (E.I.)
[4] Department of Pharmaceutical Engineering, Dongshin University, Jeonnam 58245, Korea; parkkm@dsu.ac.kr
[5] College of Veterinary Medicine, Chonnam National University, Gwangju 61186, Korea; csbae210@chonnam.ac.kr
[6] School of Korean Medicine, Dongshin University, Jeonnam 58245, Korea; hsson@dsu.ac.kr
[7] Department of Research Planning, Mokpo Marine Food-industry Research Center, Jeonnam 58621, Korea; khg8279@naver.com (H.-G.K.); bluebabyi@nate.com (J.-H.L.)
[8] Department of Biological Sciences, Oakland University, Rochester, MI 48309, USA; srhee@oakland.edu
* Correspondence: insoo.yoon@pusan.ac.kr (I.-S.Y.); sscho@mokpo.ac.kr (S.-S.C.)

† These authors contributed equally to this work.

Academic Editors: Marcello Locatelli, Angela Tartaglia, Dora Melucci, Abuzar Kabir, Halil Ibrahim Ulusoy and Victoria Samanidou

Abstract: Ginseng (*Panax ginseng*) has long been used as a traditional medicine for the prevention and treatment of various diseases. Generally, the harvest time and age of ginseng have been regarded as important factors determining the efficacy of ginseng. However, most studies have mainly focused on the root of ginseng, while studies on other parts of ginseng such as its berry have been relatively limited. Thus, the aim of this study iss to determine effects of harvest time on yields, phenolics/ginsenosides contents, and the antioxidant/anti-elastase activities of ethanol extracts of three- and four-year-old ginseng berry. In both three- and fourfour-year-old ginseng berry extracts, antioxidant and anti-elastase activities tended to increase as berries ripen from the first week to the last week of July. Liquid chromatography-tandem mass spectrometry analysis has revealed that contents of ginsenosides except Rg1 tend to be the highest in fourfour-year-old ginseng berries harvested in early July. These results indicate that biological activities and ginsenoside profiles of ginseng berry extracts depend on their age and harvest time in July, suggesting the importance of harvest time in the development of functional foods and medicinal products containing ginseng berry extracts. To the best of our knowledge, this is the first report on the influence of harvest time on the biological activity and ginsenoside contents of ginseng berry extracts.

Keywords: ginseng berry; harvest time; ginsenoside; antioxidant activity; anti-elastase activity

1. Introduction

Ginseng (*Panax ginseng*) has long been used as a traditional medicine for the prevention and treatment of various diseases, including cancer, diabetes, inflammation, allergy, and cardiovascular diseases, in the East Asia, particularly in Korea and China [1,2]. Ginseng is one of the best known and most recognized medicinal herbs. Its pharmacological effects have been successfully demonstrated by numerous studies worldwide [3]. However, most studies have focused mainly on the root of ginseng, while studies on other parts of ginseng such as its berry and leaf are relatively limited [4].

More than sixty different ginsenosides have been identified from various parts of ginseng [2]. In particular, ginseng berry is known to have a distinct phytochemical profile. It contains significantly higher ginsenoside content than ginseng root [5,6]. Oral bioavailability of ginsenosides is generally very low. It is only 0.64% for Rb1 and 3.29% for Rg1 in rats [7,8]. However, oral absorption of ginsenoside Re is significantly higher (by 1.18–3.95 fold) after oral ingestion of a ginseng berry extract than pure ginsenoside Re [9]. To date, a few in vitro and in vivo studies have reported a variety of biological activities of ginseng berry on cancer, diabetes, sexual dysfunction, skin whitening, immunity, and liver injury. These studies are summarized in Table 1 [3,4,10–16]. A randomized and placebo-controlled clinical trial has also been performed to evaluate the safety and efficacy of ginseng berry extract on glycemic control [17].

Table 1. Chemical constituents and pharmacological activities of ginseng berry extracts reported in previous literature.

Ext. Solvent	Constituent	Activity	Region	Effective Dose (mg/kg) (route/animal/day)	Estimated Human Dose (mg/60 kg/day)	Ref.
Ethanol Water	Rb1, Rb2, Rd,Re,Rf, Rg1, Rg2, 20SRg3, Rg6, Rh1, Rh4,Rk1,Rk3, F1,F4	Hepatoprotective	South Korea	100–500 (PO/rat)	972.4–4862	[3]
Ethanol	Polysaccharide K	Anti-immunosenescent		30 (PO/mouse)	146	[4]
Butanol	Re	Antidiabetic	China	150 as ext.5–20 as Re(PO/mouse)	729 as ext.24.3–97.3 as Re	[10]
ND	Polysaccharides	Antidiabetic	USA	150 (IP/mouse)		[11]
Ethylacetate	Re	Antidiabetic	South Korea	20–50 (PO/mouse)	97.3–243.3	[12]
70% ethanol	Rb1, Rb2, Rc, Rd, Re, Rg1, Rg2	Penile erection	South Korea	20–150 (PO/rat)	194.5–1458.7	[13]
70% ethanol		Antipigmentation		In vitro		[14]
Butanol	Rg1, Re, Rh1, Rg2, Rb1, Rc, Rb2, Rb3, Rd, Rg3, 20R-Rg3, Rh2	Anticancer	USA	50 (PO/mouse)	243.3	[15]
Water	Rb1, Rb2, Rc, Rd, Re, Rf	Blood circulation		50–150 (PO/rat)	486.2–1458.7	[16]

A recent study reported the alterations of metabolomes during five different ginseng berry maturation stages and their effects on the functional bioactive compounds in ginseng [18]. Thus, information regarding the optimal harvest time of ginseng berry is needed to standardize the collection and pretreatment process of the plant material for its further development as functional foods or medicinal preparations. However, only a few studies have reported the influence of harvest time on the chemical and biological properties of ginseng berry. In a previous study, five different flower and berry development stages (flower bud, flowering, early berry, green berry, and red berry) were tested with respect to ginsenoside biosynthetic gene expression and ginsenoside contents in biochemical and molecular aspects [19]. However, we focused on the effect of harvest time on biological activities and chemical profiles of green-to-red ginseng berries, which is more relevant to the agricultural and industrial aspects. Here, the objective of the present study is to determine the effects of harvest time on yields, phenolic contents, ginsenoside contents, antioxidant activity, and the elastase inhibitory activity of ethanol extracts of three and four years old ginseng berry.

2. Results and Discussion

2.1. Drying and Extraction Yields of Ginseng Berry Extracts

To date, there have been no reports on yields related to the preparation of ginseng berry extracts. As shown in Table 2, after drying the harvested ginseng berry with hot air, the yield of this process ranged from 29.9% to 34.8% (31.8% on average). After extracting the dried ginseng berry with 70% ethanol, the yield of this process ranged from 8.8% to 12.6% (mean: 10.8%). The overall production yield of the extract was calculated to be 3.4%.

Table 2. Drying and extraction yields of ginseng berry extracts.

Sample	Drying (%, w/w)	Extraction (%, w/w)
3Y1W	29.7	11.2
3Y2W	31.2	11.0
3Y3W	34.8	8.8
3Y4W	32.2	11.9
3Y5W	31.3	11.4
4Y1W	32.0	10.4
4Y2W	33.1	10.8
4Y3W	32.8	11.2
4Y4W	30.7	9.2
4Y5W	29.9	12.6

2.2. Antioxidant Properties of Ginseng Berry Extracts

Antioxidant properties of ginseng berry extracts were assessed by measuring DPPH radical scavenging activity, reducing power, and total phenolic contents. DPPH antioxidant assay is a fast and easy method to evaluate free radical scavenging capacity of a given sample [20]. As shown in Figure 1, DPPH radical scavenging activities of three-year-old ginseng berry extracts tended to increase from 26.8% to 62.5% when the harvest time was delayed. Those of four-year-old ginseng berry extracts also showed similar tendency of increase from 11.0% to 72.7%. Extracts of four-year old ginseng berry harvested in the 3rd and 4th weeks of July exhibited DPPH radical scavenging activity comparable to the positive control (vitamin C), which tended to be higher than other groups (Figure 1). As shown in Figure 2, the reducing power tended to increase as the harvest time was delayed from the 3rd year 1st week (3Y1W) to 4th year 5th week (4Y5W). Extracts of four-year-old ginseng berry harvested in the 3rd week of July exhibited significantly higher reducing power than other groups (Figure 2). As shown in Figure 3, total phenol contents of three-year-old ginseng berry extracts tended to increase from 13.6% to 29.7% as the harvest time was delayed from 1st week to 5th week of July. Those of four-year-old ginseng berry extracts showed a similar tendency, increasing from 3.2% to 13.6%. Extracts of three-year-old ginseng berry harvested in the 4th week of July exhibited significantly higher reducing power than other groups (Figure 2). Although the temporal changes of mean DPPH activity tended to be roughly similar to those of mean total phenols, the harvest time to exhibit the highest DPPH activity (4Y3W and 4Y4W) was different from that for total phenols (3Y4W). This discrepancy could be attributed to other antioxidant phytochemicals besides phenols in the ginseng berry extract, which warrants further investigation.

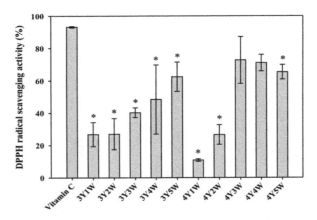

Figure 1. DPPH radical scavenging activity of ginseng berry extracts harvested at various time points. Rectangular bars and their error bars represent means and standard deviations, respectively ($n = 3$). The 'mYnW' on the x-axis means m-year-old ginseng berry harvested in the nth week of July. *, significantly lower than the 'Vitamin C' group (positive control).

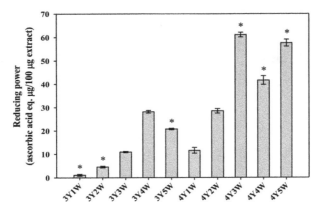

Figure 2. Reducing power of ginseng berry extracts harvested at various time points. Rectangular bars and their error bars represent means and standard deviations, respectively ($n = 3$). The 'mYnW' on the x-axis means m-year-old ginseng berry harvested in the nth week of July. *, significantly different from other groups.

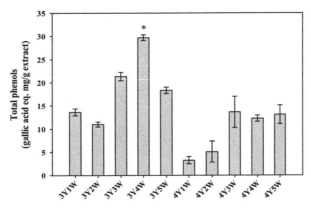

Figure 3. Total phenolic contents of ginseng berry extracts harvested at various time points. Rectangular bars and their error bars represent means and standard deviations, respectively ($n = 3$). The 'mYnW' on the x-axis means m-year-old ginseng berry harvested in the nth week of July. *, significantly different from other groups.

2.3. Elastase Inhibitory Activity of Ginseng Berry Extracts

Figure 4 shows inhibitory effects of ginseng berry extracts on elastase activity. Elastase inhibitory activities of three-year-old ginseng berry extracts tended to increase from 32.5% to 70.0% as the harvest time was delayed from 1st week to 5th week of July. Those of four-year-old ginseng berry extracts showed a similar tendency, increasing from 43.2% to 84.6%. Extracts of three-year-old and four-year-old ginseng berry harvested in the 3rd, 4th, and 5th weeks of July exhibited significantly higher inhibitory activities than the phosphoramidon group (as positive control).

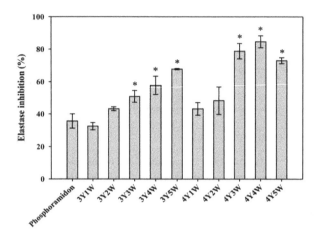

Figure 4. Elastase inhibitory activities of ginseng berry extracts harvested at various time points. Rectangular bars and their error bars represent means and standard deviations, respectively ($n = 3$). The 'mYnW' on the x-axis means m-year-old ginseng berry harvested in the nth week of July. *, significantly higher than the 'phosphoramidon' group as positive control.

2.4. Contents of Ginsenosides in Ginseng Berry Extracts

Contents of ginsenosides Rb3, Rc, Rd, Re, and Rg1 in ginseng berry extracts were determined by LC-MS/MS analysis. Typical mass chromatograms are shown in Figure 5. Contents of five ginsenosides in extracts of ginseng berry harvested at various times are shown in Figure 6. As shown in Figure 6B,C, Rc and Rd contents were significantly higher in extracts of four-year-old ginseng berry harvested in the 1st week of July than those in other groups. Similarly, Rb3 and Re contents tended to be the highest in extracts of four-year-old ginseng berry harvested in early July (Figure 6A,D). However, Rg1 content exhibited a slightly different tendency from other ginsenosides. It tended to be the highest in extracts of three-year-old ginseng berry harvested in the 4th week of July and four-year-old ginseng berry harvested in the 2nd week of July (Figure 6E). Contents of all ginsenosides studied were the lowest in extracts of four-year-old ginseng berry harvested in the last week of July.

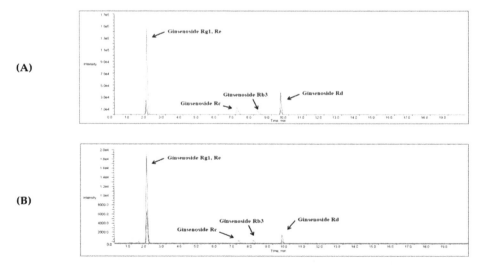

Figure 5. *Cont.*

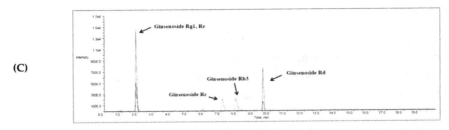

Figure 5. Representative chromatograms of ginsenosides Rb3, Rc, Rd, Re, and Rg1 in calibration standard (**A**), three-year-old ginseng berry extract sample (**B**), and four-year-old ginseng berry extract sample (**C**).

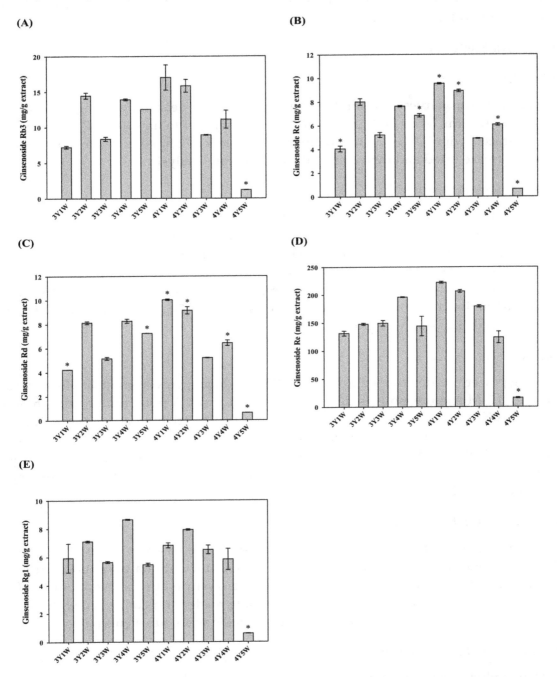

Figure 6. Contents of ginsenosides Rb3 (**A**), Rc (**B**), Rd (**C**), Re (**D**), and Rg1 (**E**) in ginseng berry extracts harvested at various time points. Rectangular bars and their error bars represent means and standard deviations, respectively ($n = 3$). The 'mYnW' on the x-axis means m-year-old ginseng berry harvested in the nth week of July. *, significantly different from other groups.

2.5. Effects of Harvest Time on Chemical Constituents and Biological Activities of Ginseng Berry Extracts

In both 3- and four-year-old ginseng berry extracts, antioxidant (DPPH radical scavenging activity and reducing power) and anti-elastase activities tended to increase as berries ripened from the first week to the last week of July. However, contents of ginsenosides except Rg1 tended to be higher in four-year-old ginseng berries harvested in early July than those in other groups. These results indicate that biological activities and ginsenoside profiles of ginseng berry extracts depend on their age and harvest time in July, suggesting a need to optimize harvest time for the development of functional foods and medicinal products containing ginseng berry extracts. To the best of our knowledge, this is the first study to report the impact of harvest time on antioxidant and anti-elastase activities as well as ginsenoside contents of ginseng berry extracts.

3. Materials and Methods

3.1. Plant Materials

Ginseng berry was harvested from three-year-old and four-year-old ginseng cultivated in a local farm (Healthy Sam-Farm, Jeonbuk, Korea) every week from July 1 to July 30, 2017. Dried ginseng berry of 25 g was extracted with 70% ethanol at room temperature for 72 h. After removing ethanol, residual water part was freeze-dried and then stored at −70 °C before analysis.

3.2. DPPH Free Radical Assay

Antioxidant activity was determined with 2,2-diphenyl-1-picrylhydrazyl (DPPH) radical scavenging assay. Briefly, 1 mL sample solution (final concentration: 1–20 mg/mL; dissolved in DDW) was added to 0.4 mM DPPH sample solution (1 mL; dissolved in methanol) and then vortex-mixed. The resultant mixture was allowed to react at room temperature in the dark for 10 min. Its absorbance at 517 nm was then measured using a microplate reader (Perkin Elmer, Waltham, MA, USA). DPPH free radical scavenging activities of samples in terms of their IC_{50} (μg/mL) values were evaluated. Vitamin C was used as a positive control.

3.3. Reducing Power

Reducing power was determined using a modified reducing power assay. Briefly, sample (0.1 mL) was added to 0.2 M sodium phosphate buffer (0.5 mL) and 1% potassium ferricyanide (0.5 mL), followed by incubation at 50 °C for 20 min. Subsequently, 10% trichloroacetic acid solution (0.5 mL) was added to the reaction mixture followed by centrifugation at 12,000× g for 10 min. The supernatant was mixed with distilled water (0.5 mL) and 0.1% iron (III) chloride solution (0.1 mL). The absorbance of the resulting solution was measured at 700 nm. Reducing powers of samples are expressed as vitamin C equivalents [21].

3.4. Determination of Total Phenolic Content

Total phenolic content was determined by Folin–Ciocalteu assay. Briefly, 1 mL sample (final concentration: 5 mg/mL) was mixed with 1 mL of 2% sodium carbonate solution and 1 mL of 10% Folin–Ciocalteu's phenol reagent. After incubating the mixture at room temperature for 10 min, its absorbance was measured at 750 nm using microplate reader and compared with the calibration curve of gallic acid. Data are expressed as milligrams of gallic acid equivalents per gram of sample [21].

3.5. Determination of Elastase Inhibitory Activity

Elastase inhibitory activity was determined as previously described [22]. Briefly, 10 μL elastase derived from porcine pancreas (10 μg/mL) was mixed with 90 μL of 0.2 M Tris-HCl, 100 μL of STANA (2.5 mM, N-Succinyl-Ala-Ala-Ala-p-nitroanilide), and 50 μL of the sample and incubated at 37 °C for 30 min. The reaction mixture was then centrifuged at 15,000× g for 10 min to obtain supernatant. The absorbance of the supernatant was measured at 405 nm using a microplate reader. Phosphoramidon, an inhibitor of elastase from *Pseudomonas aeruginosa*, was used as a positive control.

3.6. Determination of Ginsenoside Contents

Contents of ginsenosides Rb3, Rc, Rd, Re, and Rg1 were determined by high-performance liquid chromatography-tandem mass spectrometry (LC-MS/MS) analysis. The LC-MS/MS system consisted of a Sciex HPLC system coupled with a triple quadrupole mass spectrometer (Triple Quad 4500, AB Sciex, Framingham, MA, USA). The mobile phase for the HPLC system consisted of water containing 0.1% formic acid (solvent A) and acetonitrile containing 0.1% formic acid (solvent B). It was eluted at 0.4 mL/min. A gradient elution protocol was used: solvent A:solvent B, *v/v* ramped from 72:28 to 65:35 for 6 min; ramped from 65:35 to 0:100 for 4 min; held at 0:100 for 1 min; back to 72:28 for 4 min; and then held at 72:28 form 5 min. Chromatographic separation was performed using a reversed-phase column ZORBAX Eclipse Plus (C18, 3 × 100 mm, particle size 1.8 μm; Agilent, Santa Clara, CA, USA), which was maintained at 40 °C. To avoid contamination by particles, the mobile phase was filtered through a 0.45 μm filter device (PEEK, Supelco, Taufkirchen, Germany) before use. The mass spectrometer was operated in the positive ion mode using multiple reaction monitoring (MRM). The following ion source parameters were used: temperature, 600 °C; collision gas pressure, 9 mTorr; sheath gas pressure, 40 Arb; and auxiliary valve flow rate, 10 Arb. Detailed mass spectrometry parameters are listed in Table 3.

Table 3. Mass spectrometry parameters for the detection of ginsenosides.

Compound	Q1 mass	Q3 mass	Collision Energy (V)
Rb3	969.7	789.7	46
Rc	1101.7	335.0	65
Rd	969.8	789.4	60
Re	1101.7	335.0	65
Rg1	823.5	643.5	50

3.7. Statistical Analysis

A p-value < 0.05 was considered statistically significant using t-test for comparing unpaired two means or analysis of variance (ANOVA) with post-hoc Tukey's HSD test for comparing unpaired three means. All data are rounded to three significant digits and expressed as mean ± standard deviation.

4. Conclusions

The present study demonstrated that antioxidant and anti-elastase activities tended to increase as berries ripened from the first week to the last week of July in both three- and four-year-old ginseng berry extracts, and the contents of ginsenosides except Rg1 tended to be the highest in four-year-old ginseng berries harvested in early July. These findings indicate that biological activities and ginsenoside profiles of ginseng berry extracts are dependent on their age and harvest time in July, suggesting the importance of harvest time in developing functional foods and medicinal products of ginseng berry extracts. To the best of our knowledge, this is the first report on the influence of harvest time on biological activity and ginsenoside contents of ginseng berry extracts.

Author Contributions: Conceptualization, S.-Y.S., D.-H.P., I.-S.Y., and S.-S.C.; Formal analysis, S.-Y.S., D.-H.P., S.-W.S., K.-M.P., C.-S.B., H.-S.S., E.I., S.H.R., I.-S.Y., and S.-S.C.; Funding acquisition, I.-S.Y. and S.-S.C.; Investigation, S.-Y.S., D.-H.P., S.-W.S., H.-G.K., J.-H.L., G.Y., J.-H.S., E.I., S.H.R., I.-S.Y., and S.-S.C.; Methodology, S.-Y.S., D.-H.P., S.-W.S., K.-M.P., C.-S.B., H.-S.S., H.-G.K., J.-H.L., G.Y., J.-H.S., I.-S.Y., and S.-S.C.; Writing—original draft, S.-Y.S., D.-H.P., I.-S.Y., and S.-S.C.; Writing—review and editing, S.-W.S., K.-M.P., C.-S.B., H.-S.S., H.-G.K., J.-H.L., G.Y., J.-H.S., E.I., S.H.R., I.-S.Y., and S.-S.C.

References

1.   Yun, T.K. Panax ginseng—A non-organ-specific cancer preventive? *Lancet Oncol.* **2001**, *2*, 49–55. [CrossRef]
2.   Kim, J.; Cho, S.Y.; Kim, S.H.; Kim, S.; Park, C.-W.; Cho, D.; Seo, D.B.; Shin, S.S. Ginseng berry and its biological effects as a natural phytochemical. *Nat. Prod. Chem. Res.* **2016**, *4*, 209. [CrossRef]
3.   Nam, Y.; Bae, J.; Jeong, J.H.; Ko, S.K.; Sohn, U.D. Protective effect of ultrasonication-processed ginseng berry extract on the D-galactosamine/lipopolysaccharide-induced liver injury model in rats. *J. Ginseng Res.* **2018**, *42*, 540–548. [CrossRef] [PubMed]
4.   Kim, M.; Yi, Y.S.; Kim, J.; Han, S.Y.; Kim, S.H.; Seo, D.B.; Cho, J.Y.; Shin, S.S. Effect of polysaccharides from a Korean ginseng berry on the immunosenescence of aged mice. *J. Ginseng Res.* **2018**, *42*, 447–454. [CrossRef] [PubMed]
5.   Dey, L.; Xie, J.T.; Wang, A.; Wu, J.; Maleckar, S.A.; Yuan, C.S. Anti-hyperglycemic effects of ginseng: Comparison between root and berry. *Phytomedicine* **2003**, *10*, 600–605. [CrossRef] [PubMed]
6.   Kim, Y.K.; Yoo, D.S.; Xu, H.; Park, N.I.; Kim, H.H.; Choi, J.E.; Park, S.U. Ginsenoside content of berries and roots of three typical Korean ginseng (*Panax ginseng*) cultivars. *Nat. Prod. Commun.* **2009**, *4*, 903–906. [CrossRef] [PubMed]
7.   Han, M.; Fang, X.L. Difference in oral absorption of ginsenoside Rg1 between in vitro and in vivo models. *Acta Pharmacol. Sin.* **2006**, *27*, 499–505. [CrossRef]
8.   Han, M.; Sha, X.; Wu, Y.; Fang, X. Oral absorption of ginsenoside Rb1 using in vitro and in vivo models. *Planta Med.* **2006**, *72*, 398–404. [CrossRef]
9.   Joo, K.M.; Lee, J.H.; Jeon, H.Y.; Park, C.W.; Hong, D.K.; Jeong, H.J.; Lee, S.J.; Lee, S.Y.; Lim, K.M. Pharmacokinetic study of ginsenoside Re with pure ginsenoside Re and ginseng berry extracts in mouse using ultra performance liquid chromatography/mass spectrometric method. *J. Pharm. Biomed. Anal.* **2010**, *51*, 278–283. [CrossRef]
10. Attele, A.S.; Zhou, Y.P.; Xie, J.T.; Wu, J.A.; Zhang, L.; Dey, L.; Pugh, W.; Rue, P.A.; Polonsky, K.S.; Yuan, C.S. Antidiabetic effects of *Panax ginseng* berry extract and the identification of an effective component. *Diabetes* **2002**, *51*, 1851–1858. [CrossRef]
11. Xie, J.T.; Wu, J.A.; Mehendale, S.; Aung, H.H.; Yuan, C.S. Anti-hyperglycemic effect of the polysaccharides fraction from American ginseng berry extract in ob/ob mice. *Phytomedicine* **2004**, *11*, 182–187. [CrossRef] [PubMed]
12. Park, C.H.; Park, S.K.; Seung, T.W.; Jin, D.E.; Guo, T.; Heo, H.J. Effect of ginseng (*Panax ginseng*) berry EtOAc fraction on cognitive impairment in C57BL/6 mice under high-fat diet inducement. *Evid. Based Complement. Alternat. Med.* **2015**, *2015*, 316527. [CrossRef] [PubMed]
13. Cho, K.S.; Park, C.W.; Kim, C.K.; Jeon, H.Y.; Kim, W.G.; Lee, S.J.; Kim, Y.M.; Lee, J.Y.; Choi, Y.D. Effects of Korean ginseng berry extract (GB0710) on penile erection: Evidence from in vitro and in vivo studies. *Asian J. Androl.* **2013**, *15*, 503–507. [CrossRef] [PubMed]
14. Kim, J.; Cho, S.Y.; Kim, S.H.; Cho, D.; Kim, S.; Park, C.W.; Shimizu, T.; Cho, J.Y.; Seo, D.B.; Shin, S.S. Effects of Korean ginseng berry on skin antipigmentation and antiaging via FoxO3a activation. *J. Ginseng Res.* **2017**, *41*, 277–283. [CrossRef] [PubMed]
15. Xie, J.T.; Wang, C.Z.; Zhang, B.; Mehendale, S.R.; Li, X.L.; Sun, S.; Han, A.H.; Du, W.; He, T.C.; Yuan, C.S. In vitro and in vivo anticancer effects of American ginseng berry: Exploring representative compounds. *Biol. Pharm. Bull.* **2009**, *32*, 1552–1558. [CrossRef] [PubMed]
16. Kim, M.H.; Lee, J.; Jung, S.; Kim, J.W.; Shin, J.H.; Lee, H.J. The involvement of ginseng berry extract in blood flow via regulation of blood coagulation in rats fed a high-fat diet. *J. Ginseng Res.* **2017**, *41*, 120–126. [CrossRef] [PubMed]
17. Choi, H.S.; Kim, S.; Kim, M.J.; Kim, M.S.; Kim, J.; Park, C.W.; Seo, D.; Shin, S.S.; Oh, S.W. Efficacy and safety of *Panax ginseng* berry extract on glycemic control: A 12-wk randomized, double-blind, and placebo-controlled clinical trial. *J. Ginseng Res.* **2018**, *42*, 90–97. [CrossRef] [PubMed]
18. Lee, M.Y.; Seo, H.S.; Singh, D.; Lee, S.J.; Lee, C.H. Unraveling dynamic metabolomes underlying different maturation stages of berries harvested from Panax ginseng. *J. Ginseng Res.* **2019**. [CrossRef]
19. Kim, Y.K.; Yang, T.J.; Kim, S.-U.; Park, S.U. Biochemical and molecular analysis of ginsenoside biosynthesis in Panax ginseng during flower and berry development. *J. Korean Soc. Appl. Biol. Chem.* **2012**, *55*, 27–34. [CrossRef]

20. Sharma, O.P.; Bhat, T.K. DPPH antioxidant assay revisited. *Food Chem.* **2009**, *113*, 1202–1205. [CrossRef]
21. Song, S.H.; Ki, S.H.; Park, D.H.; Moon, H.S.; Lee, C.D.; Yoon, I.S.; Cho, S.S. Quantitative analysis, extraction optimization, and biological evaluation of *Cudrania tricuspidata* leaf and fruit Extracts. *Molecules* **2017**, *22*, 1489. [CrossRef] [PubMed]
22. Chiocchio, I.; Mandrone, M.; Sanna, C.; Maxia, A.; Tacchini, M.; Poli, F.J.I.C. Products Screening of a hundred plant extracts as tyrosinase and elastase inhibitors, two enzymatic targets of cosmetic interest. *Ind. Crop. Prod.* **2018**, *122*, 498–505. [CrossRef]

14

A Quick and Efficient Non-Targeted Screening Test for Saffron Authentication: Application of Chemometrics to Gas-Chromatographic Data

Pietro Morozzi [1]®, **Alessandro Zappi** [1]®, **Fernando Gottardi** [2], **Marcello Locatelli** [3]® and **Dora Melucci** [1,*]®

[1] Department of Chemistry "G. Ciamician", University of Bologna, 40126 Bologna, Italy
[2] COOP ITALIA Soc. Cooperativa, Casalecchio di Reno, 40033 Bologna, Italy
[3] Department of Pharmacy, University "G. D'Annunzio" of Chieti-Pescara, 66100 Chieti, Italy
*   Correspondence: dora.melucci@unibo.it

Academic Editor: Thomas Letzel

Abstract: Saffron is one of the most adulterated food products all over the world because of its high market prize. Therefore, a non-targeted approach based on the combination of headspace flash gas-chromatography with flame ionization detection (HS-GC-FID) and chemometrics was tested and evaluated to check adulteration of this spice with two of the principal plant-derived adulterants: turmeric (*Curcuma longa* L.) and marigold (*Calendula officinalis* L.). Chemometric models were carried out through both linear discriminant analysis (LDA) and partial least squares discriminant analysis (PLS-DA) from the gas-chromatographic data. These models were also validated by cross validation (CV) and external validation, which were performed by testing both models on pure spices and artificial mixtures capable of simulating adulterations of saffron with the two adulterants examined. These models gave back satisfactory results. Indeed, both models showed functional internal and external prediction ability. The achieved results point out that the method based on a combination of chemometrics with gas-chromatography may provide a rapid and low-cost screening method for the authentication of saffron.

Keywords: saffron; adulteration; food authenticity; gas-chromatography; chemometrics

1. Introduction

The commercial product named "Saffron Powder" is a powdered spice obtained by crushing the filaments of the *Crocus sativus* L. flower [1]. Unfortunately, because of its high market price, this spice is one of the most often adulterated food products worldwide [2]. There are different kinds of possible frauds, the most frequent being the addition of foreign matter, such as derivatives from flowers of other plants, to increase the mass of the final product without adding costly pure saffron. In some cases, even total substitution of saffron powder with adulterants may be found [3].

The high market price of saffron is due to the laborious process required to obtain the spice and the limited areas of production [4]. The flower of *Crocus sativus* L. is indeed cultivated only in some regions of Asia (Kashmir, northern Iran) and Europe (Castilla la Mancha, Spain; Kozani, Greece; Abruzzo and Sardinia, Italy) [5]. Several Protected Designations of Origin (PDOs) have been created to protect the authenticity of saffron (as it has, for example, in the Italian "Zafferano dell'Aquila", one of the major areas in terms of production and global exports) [5]. Galvin-King et al. [6] report that the business volume concerning all herbs and spices is around four billion US dollars; economists soon expect growth up to 50%. As a consequence, the business volume of frauds is estimated to cause economic damage to the global food industry in the order of several tens of billions of US dollars [7].

In order to ensure the authenticity and the quality of saffron, a standard method is proposed by the International Organization for Standardization (ISO). In particular, the last international standard regulation regarding saffron quality (ISO 3632-1:2011) [1] mainly provides a UV-Vis spectrophotometric analysis to conventionally quantify the flavor strength (expressed as concentration of picrocrocin), the aroma strength (concentration of safranal), and the coloring strength (concentration of crocin) of saffron samples. However, this method has sometimes proved incapable of evaluating saffron adulteration [8] related to spectral interferences and to the impossibility to resolve chemicals present in the adulterants that show a similar UV-Vis absorbance.

Consequently, many different analytical methods have been developed to overcome this limitation; a complete and exhaustive description of all the relevant analytical techniques is given by Kiani et al. [9]. In particular, many other spectroscopic techniques [10–13], chromatographic techniques [14–16], and molecular-biological techniques [17–19] have been exploited. Among the molecular-biological techniques, the genome-based approach, usually based on DNA extraction [20], amplification, and sequencing, represents the principal strategy to ensure the food authenticity.

However, many of these procedures are time consuming and expensive, as they require highly specialized personnel and are based on destructive methodologies.

With the aim of by-passing the above-listed drawbacks, a preliminary study for a rapid, simple, and cheap screening test for the assessment of adulterated saffron is herein developed. In particular, a non-targeted approach is used.

The non-targeted approaches are increasingly used in the field of food authenticity because they allow the examining of food fingerprints, which were previously acquired by the use of spectroscopic, spectrometric, or chromatographic techniques. This check is performed holistically and without long, complicated, and problematic identification and quantification of specific and characteristic metabolites [21].

In this work, gas-chromatographic profiles are used as chemical fingerprints, because the patterns of the most volatile compounds are characteristic for odorous spices (such as saffron and their plant adulterants) and, consequently, they may represent important variables for the assessment of saffron authenticity [22–24].

In particular, this study presents a combined application of Heracles II (AlphaMos, Toulouse, France) instrumentation, a headspace flash gas-chromatography with flame ionization detection (HS-GC-FID), and chemometric techniques [25]. Heracles II provides gas-chromatographic profiles of the analyzed samples rapidly and without any chemical sample pre-treatment [25–28]. Thus, the gas-chromatographic fingerprints are subsequently submitted to chemometric modeling through a multivariate approach [29,30], allowing detection of the eventual adulteration of saffron.

The focus of this work is the evaluation of saffron adulteration by two of the most frequently used plant-derived adulterants: turmeric (*Curcuma longa* L.) and marigold (*Calendula officinalis* L.).

2. Results and Discussion

In this work, 61 samples of commercial spices were analyzed by Heracles II flash HS-GC-FID, which meant there were 244 objects or rows of the dataset matrices. Although several peaks were present in the obtained chromatograms, for the non-targeted approach used in this work it was not necessary to associate the identified chromatographic peaks with the corresponding volatile compounds.

Examples of the chromatograms of some analyzed samples are reported in Figure 1. It was evident that the discrimination of pure spices could be directly achieved by simply superimposing the GC chromatograms in Figure 1 without any need of chemometrics. Of course, pure samples are even distinguishable with eyes without any chemical analysis. What is interesting, however, is to discriminate *mixture* samples, which simulate adulterated saffron powders. This can be done only by chemometrics.

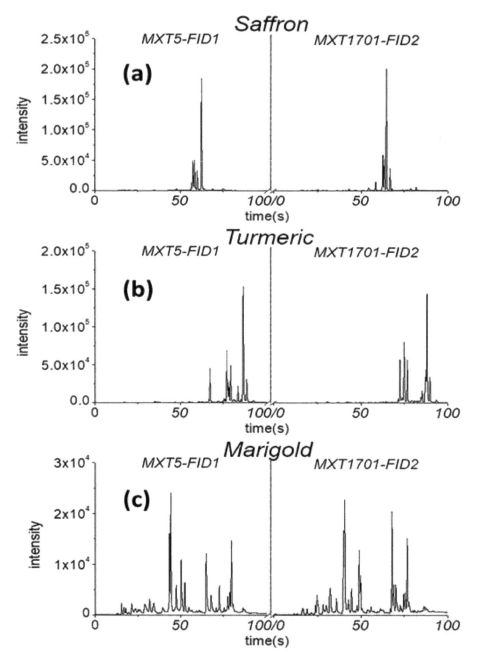

Figure 1. Representative gas-chromatographic (GC) fingerprints of saffron (**a**), turmeric (**b**), and marigold (**c**) obtained by Heracles II instrument. The chromatograms from column MXT5 are reported in the left part of the figure, while the chromatograms from column MXT1701 are reported on the right. These chromatograms were recorded simultaneously by the headspace flash gas-chromatography with flame ionization detection (HS-GC-FID).

Even if distinguishing pure samples is trivial, it is useful to create classification models based on pure standards. In fact, the models allow quantification of the dissimilarity of mixtures with respect to pure classes through parameters that are specific for each multivariate classification method.

From the obtained experimental data, two matrices were constructed: the area dataset (AD, 244 rows × 56 columns) and the intensity dataset (ID, 244 rows × 20,002 columns). More details will be given in the section Materials and Methods, paragraph 3.4 ("Working dataset").

Both matrices, as described previously, were subjected to the following chemometric elaborations (LDA and PLS-DA).

2.1. LDA Model and Results for AD

A preliminary PCA computed on the area dataset led us to find 42 outliers—20 outliers for the "Saffron" class, eight for the "Marigold" class, and 14 for the "Turmeric" class. This brought us to a dataset with dimensions 202 (objects) × 56 (variables). On this dataset, LDA was carried out. Leave-one-out cross validation (LOO-CV) was performed to internally validate the LDA model. The results of LOO-CV, in this case, could be expressed as the percentage of well-classified samples (NER), which for this LDA model was 100%. This result was obvious, since pure samples were considered.

The application of LDA produced the discriminant plot in Figure 2. Three clusters were evidenced, corresponding, as expected (100% NER), to the three a-priori classes (pure spices). In particular, the "Saffron" class was mostly discriminated from "Turmeric" along LD1 and from "Marigold" along LD2. Besides the three clusters, test samples were projected (asterisks). Table 1 summarizes all the test samples.

All the pure samples of the test set (pure_MR, pure_TR, and pure_SF) were assigned to the correct classes. They were correctly put inside the class spaces to which they were referred. What was particularly interesting was the behavior of the mixture samples; their distance from the pure spices clusters was significant. The mixture samples in Figure 2, although close to the "Saffron" class, moved away from it with an increasing percentage of adulterant. Moreover, the turmeric-adulterated samples (SFTR) got closer to the "Turmeric" class, moving along LD1, while the marigold-adulterated samples (SFMR) got closer to the "Marigold" class, moving along LD2. To quantify such behavior, the Euclidean distances between each point and each class centroid were computed, and the results are reported in Table 2. The class centroids were the points whose coordinates were the mean values of the coordinates of all the class objects. Thus, these could be considered as the "most representative" points for each class (although fictitious).

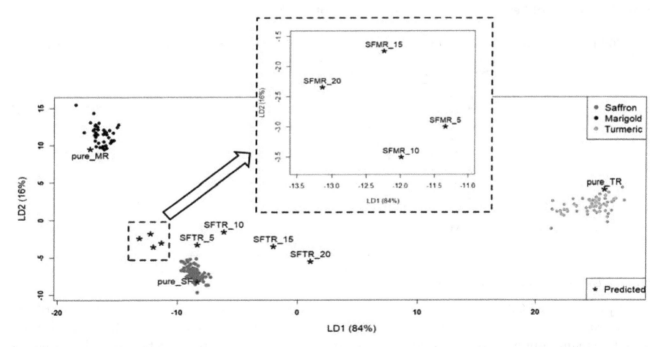

Figure 2. Linear discriminant analysis (LDA) discriminant plot, LD1 vs. LD2. The projected test samples (external validation results) are symbolized by asterisks (*). The graph portion inside the smaller dashed square is magnified into the greater dashed square.

Table 1. The test samples used for external validation: pure spices and artificial mixtures.

Test Samples	%W/W of Saffron Adulteration	Code
Pure Saffron	-	pure_SF
Pure Turmeric	-	pure_TR
Pure Marigold	-	pure_MR
saffron + turmeric	5	SFTR_5
	10	SFTR_10
	15	SFTR_15
	20	SFTR_20
saffron + marigold	5	SFMR_5
	10	SFMR_10
	15	SFMR_15
	20	SFMR_20

From Table 2, it can be seen that the distances of the turmeric-adulterated samples (SFTR) from the "Saffron" class increased, and the distance from the "Turmeric" class decreased with an increasing percentage of adulteration. The situation was a bit more complicated for the SFMR samples, because their distances did not have a "linear" behavior with the adulterant percentage (in particular, SFMR_10 was farther from "Marigold" class than SFMR_5, and SFMR_20 was closer than SFMR_15), as can be seen from Figure 2. However, it is interesting to highlight that the distance of the farthest calibration saffron sample from the "Saffron" class centroid was 2.6. This distance could be considered as a sort of radius of the "Saffron" class, and all the mixture sample distances reported in Table 2 were higher than this value. This meant that, by computing the Euclidean distances of the projected samples from the class centroids, the LDA model could detect (at least qualitatively) a saffron sample adulterated by turmeric or marigold even down to the percentage of adulteration of 5%$_{w/w}$.

Table 2. Euclidean distances of the test samples reported in Table 1 from the three class centroids.

Sample Code	Saffron	Turmeric	Marigold
pure_SF	1.1	34.6	21.1
pure_TR	36.3	2.5	42.8
pure_MR	18.6	42.5	2.2
SFTR_5	3.8	33.4	16.7
SFTR_10	6.2	31.0	16.4
SFTR_15	7.6	27.2	20.7
SFTR_20	9.9	24.7	24.2
SFMR_5	4.8	36.3	15.3
SFMR_10	4.8	37.1	15.6
SFMR_15	6.4	37.1	13.8
SFMR_20	6.4	38.0	14.3

2.2. PLS-DA Model and Results for ID

A preliminary PCA computed on the intensity dataset led to finding four outliers (one sample) for the "Saffron" class and five outliers for the "Turmeric" class. Moreover, to reduce the computational cost while maintaining good data representation, one variable every ten was retained [25]. In this way, the ID dataset on which PLS-DA was carried out had dimensions of 235 × 2001. PLS-DA was chosen instead of LDA for this dataset due to the high number of variables and the high co-linearity between them. LDA requires the computation of the covariance matrix of the dataset, but it is not possible when the variables are co-linear [31]. Figure 3 shows the PLS-DA scores plot. As it can be seen in Figure 3a,

Factor-1 and Factor-2 of PLS-DA together explained 82% of the X-explained variance and 50% of the Y-explained variance, which could be considered satisfactory to describe the dataset. From this scores plot, good discrimination of "Saffron" and "Turmeric" classes could be observed. The "Marigold" class, on the contrary, seemed to be overlapped to the "Saffron" class in the lower left part of the scores plot (third quadrant of the plot). However, when zooming in on this overlap zone, as it can be observed in the scores plot reported in Figure 3b, these two classes were found to be resolved.

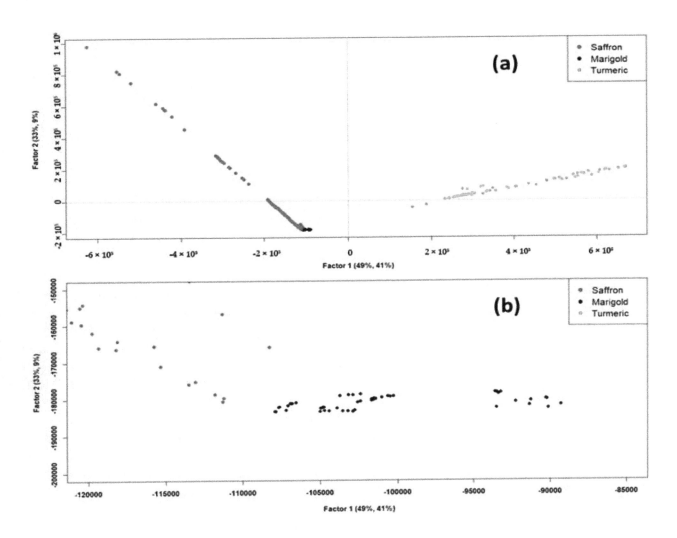

Figure 3. (a) Scores of partial least squares discriminant analysis (PLS-DA) model, Factor-1 vs. Factor-2. **(b)** Zoomed scores plot of the PLS-DA model, Factor-1 vs. Factor-2.

The CV was also performed to internally validate the PLS-DA model. Sensitivity and specificity for each class were computed according to Ballabio and Consonni (2013) [32] using 200 possible threshold values ranging from 0.1 to 1.1. The results are shown in Figure 4. Nine PLS-factors were used for "Saffron" and "Marigold" classes and three factors for "Turmeric" class (from Figure 3, it is easy to see that the discrimination of the "Turmeric" class was easier and required fewer factors than the discrimination of the other two). The vertical dashed lines in Figure 4 represent the chosen thresholds, which were 0.62 for "Saffron", 0.56 for "Turmeric", and 0.58 for "Marigold". Thresholds were chosen as the highest value that maximized both sensitivity and specificity (1.0 or 100%) in order to have a restrictive rule for the class assignment.

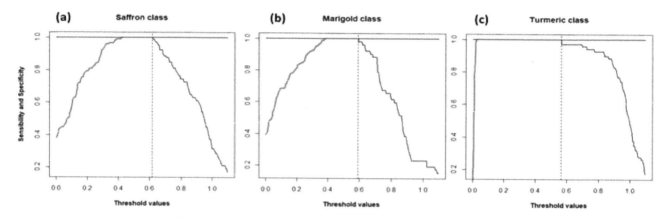

Figure 4. Sensitivity (blue lines) and specificity (red lines) for (**a**) "Saffron"; (**b**) "Marigold"; (**c**) "Turmeric" classes computed for each threshold value. Vertical dashed lines are the chosen thresholds for the corresponding class.

At this point, the test samples reported in Table 1 were projected onto the PLS-DA model to validate it. Table 3 shows the values of the dummy variables (y_marigold, y_turmeric, and y_saffron) and their corresponding standard deviation calculated by the PLS-DA model for the test samples. The pure samples (pure_MR, pure_TR, and pure_SF) could be considered well classified. Indeed, the calculated values of the dummy variables overcame the threshold values (i.e., belonging to the class considered) related to the pertaining class of each sample, while they did not overcome the thresholds (i.e., not belonging to class considered) related to the other classes. In particular, the pure_TR sample was assigned to the "Turmeric" class with a degree of 1.0, while there was still some overlap between "Saffron" and "Marigold" classes, which made the assignment of pure_MR and pure_SF samples to the corresponding class a bit more uncertain, although still satisfactory. The classification results for the adulteration mixtures (SFMR_5, SFMR_10, SFMR_15, SFMR_20, SFTR_5, SFTR_10, SFTR_15, and SFTR_20) instead showed an interesting behavior. The threshold value of 0.62 for the "Saffron" class caused the assignment of almost all the adulterated samples to the "Saffron", except for SFTR_15, SFTR_20, and SFMR_20, and none of the other predicted dummy values overcame the thresholds for the other classes. However, it is interesting to note from Table 3 that the degree of belonging to the "Saffron" class tended to decrease as the percentage of the adulterant increased. At the same time, the degree of belonging to the adulterant class tended to increase. Moreover, the calculated degrees of belonging to the "Saffron" class for all the mixtures were lower than the calculated degree obtained for pure_SF sample (although not significantly different for SFTR_5).

Table 3. External validation results (calculated Ys: degrees of belonging) of the test samples projected on the PLS-DA model. The numbers in brackets are the corresponding standard deviations.

Sample Code	y_saffron	y_turmeric	y_marigold
pure_SF	0.78 (0.03)	0.01 (0.02)	0.21 (0.04)
pure_TR	−0.1 (0.2)	1.0 (0.1)	0.1 (0.2)
pure_MR	0.34 (0.04)	0.03 (0.02)	0.63 (0.04)
SFTR_5	0.71 (0.04)	0.06 (0.02)	0.23 (0.04)
SFTR_10	0.66 (0.07)	0.12 (0.04)	0.22 (0.08)
SFTR_15	0.56 (0.06)	0.26 (0.03)	0.19 (0.06)
SFTR_20	0.51 (0.11)	0.32 (0.06)	0.17 (0.11)
SFMR_5	0.69 (0.04)	0.01 (0.02)	0.30 (0.04)
SFMR_10	0.65 (0.03)	0.02 (0.02)	0.33 (0.04)
SFMR_15	0.63 (0.04)	0.02 (0.02)	0.35 (0.04)
SFMR_20	0.59 (0.04)	0.02 (0.02)	0.39 (0.04)

This meant that the PLS-DA model, except for some uncertainties between "Saffron" and "Marigold", was able to discriminate the three studied spices and to detect both an adulteration with at least $15\%_{w/w}$ of turmeric and at least of $20\%_{w/w}$ of marigold in saffron and, at least qualitatively, some contamination in saffron with the other two spices.

2.3. Comparison between PLS-DA and LDA Models

PLS-DA and LDA models returned good results. Indeed, both models had good performances in LOO-CV, and both were able to determine the adulterations of saffron simulated with the test samples listed in Table 1.

In particular, PLS-DA showed some overlap and some uncertainties of classification between "Saffron" and "Marigold" classes. On the other side, the LDA model did not show any class overlap, and it was better than the PLS-DA model in the identification of the pure test samples. Both methods had good ability in the discrimination of the "Turmeric" class from the other two. However, it is important to underline that, even for pure_MR and pure_SF samples, the PLS-DA model was able to correctly classify them.

Regarding the artificial adulteration mixtures, PLS-DA and LDA had similar performances. In fact, for the mixture samples classified by the PLS-DA model, the calculated values of the dummy variables increased with the percentage of adulteration, although they never reached the thresholds, and some doubts persisted about the assignment to the "Saffron" class of such samples. However, the LDA model, by the calculation of the Euclidean distances between the test samples and the class centroids, showed some uncertainties between "Saffron" and "Marigold" classes, but it showed an excellent visual classification in the discriminant plot.

3. Materials and Methods

3.1. Samples

After an accurate commercial search, it was found that certified standards were not available (with the only exception of saffron pistils). Hence, the training-set samples were purchased in food retails; the reliability of these standards was subsequently verified through chemometric tools (see Paragraph 3.5, principal component analysis (PCA), and Hotelling). The spice samples were taken in the same period (April 2017) from several supermarkets, herbalist's shops, and medicinal herb gardens in Emilia Romagna (Italy). It was verified that these samples arrived at the sales centers within a month before the purchase. Twenty-eight samples of saffron, 19 samples of turmeric, and 14 samples of marigold (61 total samples, "calibration samples") were purchased by the laboratory facilities at Coop Italia. Coop Italia is one of the most important supermarket retail chains in Italy. It also has an internal food quality control laboratory in Casalecchio di Reno (Bologna, Italy), where this work was carried out.

Moreover, three samples of pure saffron, turmeric, and marigold ("test samples") were purchased for validation purposes. The pure saffron sample was taken from a supermarket and was a product certified by the SGS certification authority with the certification "Process Control IT MI. 13.P04 STP 013/24". Additionally, no further analyses by means of the ISO 3632-1:2011 [1] were necessary, because the commercially available samples had been controlled before their packaging and sales. The pure turmeric sample was purchased directly from a producer in the Agricultural fair of Santerno (Imola, Bologna, Italy). The pure marigold sample was taken from the Herb Garden of Casola Valsenio (Ravenna, Italy).

3.2. Sample Preparation

All the spice samples were stored in a dark place at low temperature until instrumental analysis. Analyses were carried out within two weeks after sample acquisition.

Regarding the calibration samples, saffron and turmeric powders did not undergo any pre-treatment, while the petals of marigold samples were powdered with Ultra Turrax Tube Drive control (IKA, Staufen im Breisgau, Germany). An aliquot of the sample was placed inside a 20-mL plastic tube with ten stainless steel spheres (5-mm diameter). The tube was subsequently sealed with the appropriate cap and was subjected to stirring at 6000 rpm for 5 min until a medium-grained powder was obtained.

Moreover, the three test samples of saffron, turmeric, and marigold (pure_SF, pure_TR, and pure_MR) were used to prepare eight artificial mixtures (SFTR_5, SFTR_10, SFTR_15, SFTR_20, SFMR_5, SFMR_10, SFMR_15, and SFMR_20) in order to simulate partial adulterations of saffron with the other spices. These samples were obtained by mixing the pure spices in different proportions to cover a wide range of adulteration degrees. In particular, four different percentages (w/w) of adulteration were examined: 5%, 10%, 15%, and 20%. These pure samples and mixtures did not undergo the chemometric procedure described later but were used to validate the final partial least squares discriminant analysis (PLS-DA) and linear discriminant analysis (LDA) models.

3.3. Flash Gas-Chromatography (Flash-GC)

All samples from both the calibration set (training set) and the test set were analyzed according to the following procedure.

For GC analysis, an aliquot of (30 ± 3) mg of each powdered sample was placed in a 20-mL glass vial sealed with a magnetic cap. Each sample was prepared in quadruplicate to assess the repeatability and the reproducibility of the method as well as to increase the degrees of freedom of statistical problems. The replicate measurements generated four objects (rows of the dataset-matrix) for each sample. Flash HS-GC-FID analysis was performed by Heracles II instrument at Coop Italia Laboratories.

In particular, this instrument was equipped with two capillary chromatographic columns working in parallel, namely a non-polar column (MXT5: 5% diphenyl, 95% methylpolysiloxane, 10 m length, and 180 μm diameter) and a slightly polar column (MXT1701: 14% cyanopropylphenyl, 86% methylpolysiloxane, 10 m length, and 180 μm diameter) and two flame ionization detectors (FIDs) at the end of each column. GC operation, auto sampling, and chromatographic output were managed by Alphasoft V12.4 software (AlphaMos, Toulouse, France).

The parameters of the chromatographic analysis were chosen after an optimization step to avoid significant problems such as low sensitivity, overcoming of full-scale, and low peaks resolution.

The instrument was also equipped with an auto-sampler HS100 (CTC Analytics AG, Zwingen, Switzerland), which managed up to 96 samples in the same program. The sample vials were placed in a shaker oven at 50 °C and 500 rpm for 20 min. Then, the auto-sampler syringe took 5000 μL of the head-space (by piercing the silicone septum of the vial plug). The sample was injected at 100 μL s^{-1} (the injector temperature was 200 °C). The carrier gas was molecular hydrogen (H$_2$) produced by an Alliance High Purity Hydrogen generator (F-dgsi, Évry, France). A solid adsorbing trap Tenax TA 60/80 (Tenax SPA, Verona, Italy) was placed before the chromatographic columns and was maintained at 40 °C and 60 kPa for 65 s while carrier gas was flowing and then heated at 240 °C. This allowed for absorption of the volatile molecules onto the trap and removal of excess air and moisture to concentrate the analytes. Analytes were then introduced into the GC columns by a rotatory valve. The column's initial temperature was 40 °C, which was maintained at such a value for 2 s and then increased by 3 °C s^{-1} until reaching 270 °C, then it was kept at this value for 21 s. The total acquisition time was 100 s, and the signal was digitalized every 0.01 s. While a sample was injected, other samples were shaken; the entire process was automated and managed by the instrument in the absence of personnel. As a result, if 96 samples were analyzed in the same program, the overall time needed was not 20 × 96 min but about 180 min.

3.4. Working Datasets

After flash GC analysis, the gas-chromatographic data obtained were tabled into a source matrix (dataset). The dataset rows represented the replicates of the 61 samples (244 rows or objects, 4 replicates for each sample). The labels of the dataset columns corresponded to GC variables, which were the acquisition times derived from the digitalization of the GC signal. Each dataset cell reported the FID signal registered at the corresponding GC time for the relevant object. A further column was the class variable reporting the *a priori* class to which the relevant object belonged. Objects were grouped into classes based on their labeled identity (saffron, turmeric, and marigold).

In particular, two different datasets were created: the "area dataset" and the "intensity dataset". The "area dataset" (AD) variables corresponded to peak areas (56 columns); these variables corresponded to the chromatographic peaks identified by the automatic integration tool of AlphaSoft. The "intensity dataset" (ID) variables were the full chromatograms recorded by Heracles II (20002 columns); cell values were the electric current intensities of FIDs. The signal was digitalized every 0.01s for 100 s (10,001 signals), and the chromatogram of the second column was appended to the one of the first column.

Both datasets were obtained from the chromatograms elaborated by Alphasoft V12.4 software.

3.5. Chemometrics

Before applying any of the chemometric techniques used in this work, all the data were standardized [33]. In particular, two different scaling methods were applied to the datasets: autoscaling for the "area dataset" and centering for the "intensity dataset".

Two models for the determination of partial or total adulteration of saffron with turmeric and marigold were created and evaluated, LDA [30] and PLS-DA [29,32]. In particular, the LDA model was computed for AD, while the PLS-DA model was computed for ID.

For each dataset, the following chemometric procedure was carried out in parallel. First, for each class, the elimination of the outliers was performed by PCA and Hotelling analysis [34] at a confidence level of 95%, as already described in a previous work [25].

Then, the refined datasets including only statistically significant samples were subsequently subjected to LDA and PLS-DA. Both chemometric models were then validated by internal cross-validation (CV) [29,30] and by projecting the eleven test samples (not used for model creation) [29]. CV is a statistical technique that allowed evaluating the prediction ability of a model (i.e., the ability to determine the values of the response variables from the predictors for the test samples). CV performed the following steps iteratively: exclude some samples (randomly selected) from the training set, build the model without the excluded samples, and classify the excluded samples with this model. During this procedure, each sample of the training set was used as a test sample at least one time. However, the results of CV were different for LDA and PLS-DA.

LDA computed a model characterized by the definition of new variables starting from the original variables (in the case of AD, chromatographic peak areas) as well as in PCA. However, LDA, unlike PCA, defined linear discriminant functions (LDs) rather than principal components (PCs) that were more effective in separating the examined classes [29]. Such a model could classify unknown samples by projecting them in the LDs space. An unknown sample was always assigned to the class for which the calculated posterior probability [35] was higher; however, the distance of objects from the classes needed to be taken into account in order to finely evaluate the degree of membership to a class.

For LDA, the CV output was represented by the confusion matrix. In this matrix, the lines represented the "a priori" classes, and the columns represented the calculated "a posteriori" classes, to which CV reassigned the samples. The ideal situation was a diagonal matrix (i.e., the matrix in which the entries outside the main diagonal were all zero) because it was the situation in which all of the samples were correctly assigned to the corresponding "a priori" classes. Subsequently, starting from the confusion matrix, it was possible to compute the "non-error-rate" (NER) as the ratio between the objects correctly classified and the total number of objects, which represented the ability of the model to correctly recognize its objects.

PLS-DA [32] is instead a regression method in which the predictor variables (X-matrix) were the experimental ones (in the ID case, the full chromatograms), while the responses (Y-matrix) were the so-called "dummy variables". These dummy variables were the degrees of belonging to the examined classes (in this work, saffron, turmeric, and marigold) and assumed the values for calibration objects to be 0 and 1 (where 1 represented the certainty of belonging to the considered class, while 0 represented the certainty of not belonging to the considered class). The projection of an external sample onto a PLS-DA model returned a set of values for the dummy variables that could be considered as "degrees of belonging" to each class.

CV results for a PLS-DA model were represented by the calculated values of dummy variables for each sample, which meant the predicted degree of belonging of each sample to each class. These values could be used to calculate a threshold value for each class that optimized both sensitivity and specificity for the classification. The procedure for computing such threshold values is described by Ballabio and Consonni (2013) [32]. The projected samples of the test set could then be assigned to a class if their corresponding calculated value of the dummy variable overcame the threshold.

Outliers elimination was carried out by the software The Unscrambler V10.4 (Camo, Oslo, Norway), while LDA and PLS-DA were carried out (with relative CV and projections) by the software R V3.4.3 (R Core Team, Vienna, Austria) with the packages "MASS" [31] and "pls" [35].

4. Conclusions

The achieved results illustrate that the herein proposed, non-targeted strategy based on the combined application of chemometrics with Heracles II flash HS-GC-FID may provide a rapid and low-cost screening method for the authentication of saffron.

The samples were analyzed without any preparation or after a rapid grinding operation, allowing us to avoid expensive pre-treatments and any contamination before analysis by gas-chromatography. Furthermore, once the sample is put into the auto-sampler of the instrument, this instrumental analysis is entirely automated and requires a short analysis time (overall, less than 20 min for a single sample and a couple of minutes *per* sample for 96 samples simultaneously put in the auto-sampler).

Finally, with chemometrics, it was possible to use the GC data both as they are produced by the instrument (chromatograms) and by integrating the chromatographic peaks to build classification models (PLS-DA and LDA). These models had good calibration ability, evaluated by cross-validation (CV) and, most of all, good prediction ability, evaluated by projecting external test samples that simulated adulterations of saffron with turmeric and marigold. Moreover, for adulterant additions below $33\%_{w/w}$, the official UV-VIS spectrophotometry method was not able to detect adulteration [8]. On the contrary, Heracles II combined with chemometrics allowed us to go far below this limit; a PLS-DA model able to detect down to $15 \div 20\%_{w/w}$ of adulteration was validated. Moreover, a discriminant plot obtained through LDA showed significant differences between pure samples and adulterated samples down to $5 \div 10\%_{w/w}$.

Another important characteristic of the chemometric approach is that it does not require the identification of the volatile compounds to create a model able to find an adulterated saffron sample. The use of the entire chromatograms ensures that all the possible markers for turmeric or marigold adulteration are taken into account in the model construction.

Author Contributions: Conceptualization, F.G. and D.M.; Methodology, F.G. and P.M.; Software, A.Z. and P.M.; Validation, P.M. and A.Z.; Formal Analysis, P.M. and F.G.; Investigation, P.M.; Resources, F.G. and M.L.; Data Curation, A.Z., P.M. and D.M.; Writing—Original Draft Preparation, P.M.; Writing—Review & Editing, A.Z., M.L. and D.M.; Visualization, F.G.; Supervision, D.M.; Project Administration, F.G.

Acknowledgments: The authors thank Coop Italia for providing the Heracles II flash HS-GC-FID instrument and Paolo De Giorgi for the assistance in the experimental activity.

References

1. International Organization for Standardization. *ISO 3632-1. Spices—Saffron (Crocus sativus L.)*; ISO: Geneva, Switzerland, 2011.
2. Moore, J.C.; Spink, J.; Lipp, M. Development and Application of a Database of Food Ingredient Fraud and Economically Motivated Adulteration from 1980 to 2010. *J. Food Sci.* **2012**, *77*, R118–R126. [CrossRef] [PubMed]
3. Nazari, S.H.; Keifi, N. Saffron and various fraud manners in its production and trades. *Acta Hortic.* **2007**, *739*, 411–416. [CrossRef]
4. Johnson, R. *Food Fraud and "Economically Motivated Adulteration" of Food and Food Ingredients*; Congressional Research Service Report; University of North Texas Libraries: Denton, TX, USA, 2014; pp. 1–40.
5. Bosmali, I.; Ordoudi, S.A.; Tsimidou, M.Z.; Madesis, P. Greek PDO saffron authentication studies using species specific molecular markers. *Food Res. Int.* **2017**, *100*, 899–907. [CrossRef] [PubMed]
6. Galvin-King, P.; Haughey, S.A.; Elliott, C.T. Herb and spice fraud; the drivers, challenges and detection. *Food Control* **2018**, *88*, 85–97. [CrossRef]
7. PwC & SSAFE. Food Fraud Vulnerability Assessment. 2016. Available online: http://www.pwc.com/gx/en/services/food-supply-integrity-services/assets/pwc-food-fraud-vulnerability-assessment-and-mitigation-november.pdf (accessed on 1 July 2019).
8. Sabatino, L.; Scordino, M.; Gargano, M.; Belligno, A.; Traulo, P.; Gagliano, G. HPLC/PDA/ESI-MS evaluation of saffron (*Crocus sativus* L.) adulteration. *Nat. Prod. Commun.* **2011**, *6*, 1873–1876. [CrossRef] [PubMed]
9. Kiani, S.; Minaei, S.; Ghasemi-Varnamkhasti, M. Instrumental approaches and innovative systems for saffron quality assessment. *J. Food Eng.* **2018**, *216*, 1–10. [CrossRef]
10. Petrakis, E.A.; Cagliani, L.R.; Polissiou, M.G.; Consonni, R. Evaluation of saffron (*Crocus sativus* L.) adulteration with plant adulterants by 1H NMR metabolite fingerprinting. *Food Chem.* **2015**, *173*, 890–896. [CrossRef]
11. Petrakis, E.A.; Polissiou, M.G. Assessing saffron (*Crocus sativus* L.) adulteration with plant-derived adulterants by diffuse reflectance infrared Fourier transform spectroscopy coupled with chemometrics. *Talanta* **2017**, *162*, 558–566. [CrossRef]
12. Zalacain, A.; Ordoudi, S.A.; Díaz-Plaza, E.M.; Carmona, M.; Blázquez, I.; Tsimidou, M.Z.; Alonso, G.L. Near-infrared spectroscopy in saffron quality control: Determination of chemical composition and geographical origin. *J. Agric. Food Chem.* **2005**, *53*, 9337–9341. [CrossRef]
13. Ordoudi, S.A.; De Los Mozos Pascual, M.; Tsimidou, M.Z. On the quality control of traded saffron by means of transmission Fourier-transform mid-infrared (FT-MIR) spectroscopy and chemometrics. *Food Chem.* **2014**, *150*, 414–421. [CrossRef]
14. Rubert, J.; Lacina, O.; Zachariasova, M.; Hajslova, J. Saffron authentication based on liquid chromatography high resolution tandem mass spectrometry and multivariate data analysis. *Food Chem.* **2016**, *204*, 201–209. [CrossRef] [PubMed]
15. Nenadis, N.; Heenan, S.; Tsimidou, M.Z.; Van Ruth, S. Applicability of PTR-MS in the quality control of saffron. *Food Chem.* **2016**, *196*, 961–967. [CrossRef] [PubMed]
16. Aliakbarzadeh, G.; Parastar, H.; Sereshti, H. Classification of gas chromatographic fingerprints of saffron using partial least squares discriminant analysis together with different variable selection methods. *Chemom. Intell. Lab. Syst.* **2016**, *158*, 165–173. [CrossRef]
17. Torelli, A.; Marieschi, M.; Bruni, R. Authentication of saffron (*Crocus sativus* L.) in different processed, retail products by means of SCAR markers. *Food Control* **2014**, *36*, 126–131. [CrossRef]
18. Gismondi, A.; Fanali, F.; Martínez Labarga, J.M.; Caiola, M.G.; Canini, A. *Crocus sativus* L. Genomics and different DNA barcode applications. *Plant Syst. Evol.* **2013**, *299*, 1859–1863. [CrossRef]
19. Babaei, S.; Talebi, M.; Bahar, M. Developing an SCAR and ITS reliable multiplex PCR-based assay for safflower adulterant detection in saffron samples. *Food Control* **2014**, *35*, 323–328. [CrossRef]
20. Danezis, G.P.; Tsagkaris, A.S.; Camin, F.; Brusic, V.; Georgiou, C.A. Food authentication: Techniques, trends & emerging approaches. *TrAC Trends Anal. Chem.* **2016**, *85*, 123–132.
21. Esslinger, S.; Riedl, J.; Fauhl-Hassek, C. Potential and limitations of non-targeted fingerprinting for authentication of food in official control. *Food Res. Int.* **2014**, *60*, 189–204. [CrossRef]
22. Matsushita, T.; Zhao, J.J.; Igura, N.; Shimoda, M. Authentication of commercial spices based on the similarities between gas chromatographic fingerprints. *J. Sci. Food Agric.* **2018**, *98*, 2989–3000. [CrossRef]
23. Heidarbeigi, K.; Mohtasebi, S.S.; Foroughirad, A.; Ghasemi-Varnamkhasti, M.; Rafiee, S.; Rezaei, K. Detection of adulteration in saffron samples using electronic nose. *Int. J. Food Prop.* **2015** *18*, 1391–1401. [CrossRef]

24. Carmona, M.; Zalacain, A.; Salinas, M.R.; Alonso, G.L. A new approach to saffron aroma. *Crit. Rev. Food Sci. Nutr.* **2007**, *47*, 145–159. [CrossRef] [PubMed]
25. Melucci, D.; Bendini, A.; Tesini, F.; Barbieri, S.; Zappi, A.; Vichi, S.; Conte, L.; Gallina Toschi, T. Rapid direct analysis to discriminate geographic origin of extra virgin olive oils by flash gas chromatography electronic nose and chemometrics. *Food Chem.* **2016**, *204*, 263–273. [CrossRef] [PubMed]
26. Wiśniewska, P.; Śliwińska, M.; Namieśnik, J.; Wardencki, W.; Dymerski, T. The Verification of the Usefulness of Electronic Nose Based on Ultra-Fast Gas Chromatography and Four Different Chemometric Methods for Rapid Analysis of Spirit Beverages. *J. Anal. Methods Chem.* **2016**, *2016*. [CrossRef] [PubMed]
27. Wojtasik-Kalinowska, I.; Guzek, D.; Górska-Horczyczak, E.; Głabska, D.; Brodowska, M.; Sun, D.W.; Wierzbicka, A. Volatile compounds and fatty acids profile in Longissimus dorsi muscle from pigs fed with feed containing bioactive components. *LWT Food Sci. Technol.* **2016**, *67*, 112–117. [CrossRef]
28. Górska-Horczyczak, E.; Wojtasik-Kalinowska, I.; Guzek, D.; Sun, D.W.; Wierzbicka, A. Differentiation of chill-stored and frozen pork necks using electronic nose with ultra-fast gas chromatography. *J. Food Process Eng.* **2017**, *40*, e12540. [CrossRef]
29. Berrueta, L.A.; Alonso-Salces, R.M.; Héberger, K. Supervised pattern recognition in food analysis. *J. Chromatogr. A* **2007**, *1158*, 196–214. [CrossRef]
30. Bevilacqua, M.; Nescatelli, R.; Bucci, R.; Magrì, A.D.; Magrì, A.L.; Marini, F. Chemometric classification techniques as a tool for solving problems in analytical chemistry. *J. AOAC Int.* **2014**, *97*, 19–28. [CrossRef] [PubMed]
31. Venables, W.N.; Ripley, B.D. *Modern Applied Statistics with S*, 4th ed.; Springer: New York, NY, USA, 2002; Volume 53, ISBN 0387954570.
32. Ballabio, D.; Consonni, V. Classification tools in chemistry. Part 1: Linear models. PLS-DA. *Anal. Methods* **2013**, *5*, 3790–3798. [CrossRef]
33. Van den Berg, R.A.; Hoefsloot, H.C.J.; Westerhuis, J.A.; Smilde, A.K.; van der Werf, M.J. Centering, scaling, and transformations: Improving the biological information content of metabolomics data. *BMC Genom.* **2006**, *7*, 142. [CrossRef]
34. Jolliffe, I.T. *Principal Component Analysis*, 2nd ed.; Springer: Berlin/Heidelberg, Germany, 2002; ISBN 0387954422.
35. Mevik, B.H.; Wehrens, R.; Liland, K.H. pls: Partial Least Squares and Principal Component Regression. R package version 2.5-0. *J. Stat. Softw.* **2015**. Available online: https://www.researchgate.net/deref/http%3A%2F%2Fmevik.net%2Fwork%2Fsoftware%2Fpls.html (accessed on 16 July 2019).

Permissions

All chapters in this book were first published in MDPI; hereby published with permission under the Creative Commons Attribution License or equivalent. Every chapter published in this book has been scrutinized by our experts. Their significance has been extensively debated. The topics covered herein carry significant findings which will fuel the growth of the discipline. They may even be implemented as practical applications or may be referred to as a beginning point for another development.

The contributors of this book come from diverse backgrounds, making this book a truly international effort. This book will bring forth new frontiers with its revolutionizing research information and detailed analysis of the nascent developments around the world.

We would like to thank all the contributing authors for lending their expertise to make the book truly unique. They have played a crucial role in the development of this book. Without their invaluable contributions this book wouldn't have been possible. They have made vital efforts to compile up to date information on the varied aspects of this subject to make this book a valuable addition to the collection of many professionals and students.

This book was conceptualized with the vision of imparting up-to-date information and advanced data in this field. To ensure the same, a matchless editorial board was set up. Every individual on the board went through rigorous rounds of assessment to prove their worth. After which they invested a large part of their time researching and compiling the most relevant data for our readers.

The editorial board has been involved in producing this book since its inception. They have spent rigorous hours researching and exploring the diverse topics which have resulted in the successful publishing of this book. They have passed on their knowledge of decades through this book. To expedite this challenging task, the publisher supported the team at every step. A small team of assistant editors was also appointed to further simplify the editing procedure and attain best results for the readers.

Apart from the editorial board, the designing team has also invested a significant amount of their time in understanding the subject and creating the most relevant covers. They scrutinized every image to scout for the most suitable representation of the subject and create an appropriate cover for the book.

The publishing team has been an ardent support to the editorial, designing and production team. Their endless efforts to recruit the best for this project, has resulted in the accomplishment of this book. They are a veteran in the field of academics and their pool of knowledge is as vast as their experience in printing. Their expertise and guidance has proved useful at every step. Their uncompromising quality standards have made this book an exceptional effort. Their encouragement from time to time has been an inspiration for everyone.

The publisher and the editorial board hope that this book will prove to be a valuable piece of knowledge for researchers, students, practitioners and scholars across the globe.

List of Contributors

Maja Welna, Anna Szymczycha-Madeja and Pawel Pohl
Department of Analytical Chemistry and Chemical Metallurgy, Faculty of Chemistry, Wroclaw University of Science and Technology, Wybrzeze Wyspianskiego 27, 50-370 Wroclaw, Poland

Marta Bystrzanowska and Marek Tobiszewski
Department of Analytical Chemistry, Faculty of Chemistry, Gdańsk University of Technology (GUT), 80-233 Gdańsk, Poland

Eleonora Amante, Alberto Salomone, Eugenio Alladio and Marco Vincenti
Dipartimento di Chimica, Università degli Studi di Torino, Via P. Giuria 7, 10125 Torino, Italy
Centro Regionale Antidoping e di Tossicologia "A. Bertinaria", Regione Gonzole 10/1, 10043 Orbassano, Italy

Francesco Porpiglia
Division of Urology, San Luigi Gonzaga Hospital and University of Torino, 10043 Orbassano, Italy

Rasmus Bro
Department of Food Science, Faculty of Science, University of Copenhagen, Rolighedsvej 30, 1958 Frederiksberg, Denmark

Qin-Qin Wang
Institute of Medicinal Plants, Yunnan Academy of Agricultural Sciences, Kunming 650200, China
College of Traditional Chinese Medicine, Yunnan University of Traditional Chinese Medicine, Kunming 650500, China

Heng-Yu Huang
College of Traditional Chinese Medicine, Yunnan University of Traditional Chinese Medicine, Kunming 650500, China

Yuan-Zhong Wang
Institute of Medicinal Plants, Yunnan Academy of Agricultural Sciences, Kunming 650200, China

Shou-Ying Wang and Qing-Ping Chen
Laboratory of Quality & Safety Risk Assessment for Aquatic products (Shanghai), Ministry of Agriculture and Rural Affairs, East China Sea Fisheries Research Institute, Shanghai 200090, China
College of Food Science & Technology, Shanghai Ocean University, Shanghai 201306, China

Cong Kong and Hui-Juan Yu
Laboratory of Quality & Safety Risk Assessment for Aquatic products (Shanghai), Ministry of Agriculture and Rural Affairs, East China Sea Fisheries Research Institute, Shanghai 200090, China
Key Laboratory of East China Sea Fishery Resources Exploitation, Ministry of Agriculture and Rural Affairs, East China Sea Fisheries Research Institute, Chinese Academy of Fishery Sciences, Shanghai 200090, China

Benedito Roberto de Alvarenga Junior and Renato Lajarim Carneiro
Department of Chemistry, Federal University of São Carlos, São Carlos 13565-905, Brazil

Mara Mandrioli, Matilde Tura and Tullia Gallina Toschi
Department of Agricultural and Food Sciences, Alma Mater Studiorum-University of Bologna, Viale Fanin 40, 40127 Bologna, Italy

Stefano Scotti
Shimadzu Italia, Via G. B. Cassinis 7, 20139 Milano, Italy

Yuanshuai Gan, Yao Xiao, Hongye Guo, Min Liu and Yongsheng Wang
College of Pharmacy, Jilin University, Changchun 130021, China

Shihan Wang
College of Chinese Herbal Medicine, Jilin Agricultural University, Changchun 130118, China

Zhihan Wang
Department of Physical Sciences, Eastern New Mexico University, Portales, NM 88130, USA

Jing Li and Xianqing He
School of Chemistry, Chemical Engineering and Life Science, Wuhan University of Technology, 122 luoshilu, Wuhan 430070, China

Yuanyuan Deng and Chenxi Yang
School of Biological Science & Medical Engineering, Southeast University, No.2 Sipailou, Nanjing 210096, China

Angelo Antonio D'Archivio
Dipartimento di Scienze Fisiche e Chimiche, Università degli Studi dell'Aquila, Via Vetoio, 67100 Coppito, L'Aquila, Italy

Martha Maggira
Laboratory of Analytical Chemistry, Department of Chemistry, Aristotle University of Thessaloniki, GR-541 24 Thessaloniki, Greece

Eleni A. Deliyanni and Victoria F. Samanidou
Laboratory of General and Environmental Technology, Department of Chemistry, Aristotle University of Thessaloniki, GR-541 24 Thessaloniki, Greece

Dora Melucci, Francesca Poggioli, Pietro Morozzi and Laura Tositti
Department of Chemistry "G. Ciamician", University of Bologna, 40126 Bologna, Italy

Federico Giglio
Polar Science Institute-National Research Council ISP-CNR, Via P. Gobetti 101, 40129 Bologna, Italy

Seung-Yeap Song, Goo Yoon, Jung-Hyun Shim and Seung-Sik Cho
Department of Pharmacy, College of Pharmacy, Mokpo National University, Jeonnam 58554, Korea

Dae-Hun Park
Department of Nursing, Dongshin University, Jeonnam 58245, Korea

Seong-Wook Seo, Eunok Im and In-Soo Yoon
Department of Pharmacy, College of Pharmacy, Pusan National University, Busan 46241, Korea

Kyung-Mok Park
Department of Pharmaceutical Engineering, Dongshin University, Jeonnam 58245, Korea

Chun-Sik Bae
College of Veterinary Medicine, Chonnam National University, Gwangju 61186, Korea

Hong-Seok Son
School of Korean Medicine, Dongshin University, Jeonnam 58245, Korea

Hyung-Gyun Kim and Jung-Hee Lee
Department of Research Planning, Mokpo Marine Food-industry Research Center, Jeonnam 58621, Korea

Sang Hoon Rhee
Department of Biological Sciences, Oakland University, Rochester, MI 48309, USA

Pietro Morozzi, Alessandro Zappi and Dora Melucci
Department of Chemistry "G. Ciamician", University of Bologna, 40126 Bologna, Italy

Fernando Gottardi
COOP ITALIA Soc. Cooperativa, Casalecchio di Reno, 40033 Bologna, Italy

Marcello Locatelli
Department of Pharmacy, University "G. D'Annunzio" of Chieti-Pescara, 66100 Chieti, Italy

Index

Printed in the USA
CPSIA information can be obtained
at www.ICGtesting.com
JSHW051405091023
49903JS00006B/289

9 781647 285197